CRC SERIES IN AGRICULTURE

Editor-in-Chief

Angus A. Hanson, Ph.D.
Vice President-Research
W-L Research, Inc.
Highland, Maryland

HANDBOOK OF SOILS AND CLIMATE IN AGRICULTURE

Editor
Victor J. Kilmer (Deceased)
Chief
Soils and Fertilizer Research Branch
National Fertilizer Development Center
Tennessee Valley Authority
Muscle Shoals, Alabama

HANDBOOK OF PLANT SCIENCE IN AGRICULTURE

Editor
B. R. Christie, Ph.D.
Professor
Department of Crop Science
Ontario Agricultural College
University of Guelph
Guelph, Ontario, Canada

HANDBOOK OF PEST MANAGEMENT IN AGRICULTURE

Editor
David Pimentel, Ph.D.
Professor
Department of Entomology
New York College of Agricultural
and Life Sciences
Cornell University
Ithaca, New York

HANDBOOK OF ENGINEERING IN AGRICULTURE

Editor
R. H. Brown, Ph.D.
Chairman
Division of Agricultural Engineering
Agricultural Engineering Center
University of Georgia
Athens, Georgia

HANDBOOK OF TRANSPORTATION AND MARKETING IN AGRICULTURE

Editor
Essex E. Finney, Jr., Ph.D.
Assistant Center Director
Agricultural Research Center
U.S. Department of Agriculture
Beltsville, Maryland

HANDBOOK OF PROCESSING AND UTILIZATION IN AGRICULTURE

Editor
Ivan A. Wolff, Ph.D. (Retired)
Director
Eastern Regional Research Center
Science and Education Administration
U.S. Department of Agriculture
Philadelphia, Pennsylvania

CRC Handbook

of

Engineering
in
Agriculture

Volume III
Environmental Systems
Engineering

Editor
R. H. Brown, P.E.
Chairman Emeritus
Division of Agricultural Engineering
University of Georgia
Athens, Georgia

CRC Series in Agriculture
A. A. Hanson, Editor-in-Chief
Vice President-Research
W-L Research, Inc.
Highland, Maryland

CRC Press, Inc.
Boca Raton, Florida

Library of Congress Cataloging-in-Publication Data

CRC handbook of engineering in agriculture.
  (CRC series in agriculture)
  Includes bibliographies and indexes.
  1. Agricultural engineering—Handbooks, manuals,
etc.   I. Brown, R. H. (Robert H.)   II. Series.
S675.C73 1988   630   87-21870
ISBN 0-8493-3861-1 (v. 1)
ISBN 0-8493-3862-X (v. 2)
ISBN 0-8493-3863-8 (v. 3)

Direct all inquiries to CRC Press, Inc., 2000 Corporate Blvd., N.W., Boca Raton, Florida, 33431.

© 1988 by CRC Press, Inc.

International Standard Book Number 0-8493-3861-1 (v. 1)
International Standard Book Number 0-8493-3862-X (v. 2)
International Standard Book Number 0-8493-3863-8 (v. 3)

Library of Congress Card Number 87-21870
Printed in the United States

EDITOR-IN-CHIEF

Angus A. Hanson, Ph.D., is Vice President-Research, W-L Research, Inc., Highland, Maryland, and has had broad experience in agricultural research and development. He is a graduate of the University of British Columbia, Vancouver, and McGill University, Quebec, and received the Ph.D. degree from the Pennsylvania State University, University Park, in 1951.

An employee of the U.S. Department of Agriculture from 1949 to 1979, Dr. Hanson worked as a Research Geneticist at University Park, Pa., 1949 to 1952, and at Beltsville, Md., serving successively as Research Leader for Grass and Turf Investigations, 1953 to 1965, Chief of the Forage and Range Research Branch, 1965 to 1972, and Director of the Beltsville Agricultural Research Center, 1972 to 1979. He has been appointed to a number of national and regional task forces charged with assessing research needs and priorities, and has participated in reviewing agricultural needs and research programs in various foreign countries. As Director at Beltsville, he was directly responsible for programs that included most dimensions of agricultural research.

In his personal research, he has emphasized the improvement of forage crops, breeding and management of turfgrasses, and the breeding of alfalfa for multiple pest resistance, persistence, quality, and sustained yield. He is the author of over 100 technical and popular articles on forage crops and turfgrasses, and has served as Editor of *Crop Science* and the *Journal of Environmental Quality.*

PREFACE

CRC SERIES IN AGRICULTURE

Agriculture, because of its pivotal role in the development of civilized societies, contributed much to the development of various scientific disciplines. Thus, agricultural pursuits led to the practical application of chemistry, and gave rise to such major disciplines as economics and statistics. The expansion of scientific frontiers, and the concomitant specilization within disciplines, has proceeded to the point where agricultural scientists classify themselves in an array of disciplines and subdisciplines, i.e., nematologist, geneticist, physicist, virologist, and so forth. Nevertheless, within the framework of these various disciplines and mission oriented agricultural research, information of primary interest and concern in the solution of agriculturally oriented problems is generated. Although some of the basic information finds its way into the plethora of reference books available within most disciplines, no attempt has been made to develop a comprehensive handbook series for the agricultural sciences.

It is recognized that there are serious difficulties in developing a meaningful handbook series in agriculture because of the range and complexity of agricultural enterprises. In fact, the single common denominator that applies to all agricultural scientists is their universal concern with at least some aspect of the production and utilization of farm products. The disciplines and resources that are called for in a specific investigation are either the same or similar to those utilized in any area of biological research, or in any one of several fields of scientific endeavor.

The sections in this handbook series reflect the input of different editors and advisory boards, and as a consequence, there is considerable variation in both the depth and coverage offered within a given area. However, an attempt has been made throughout to bring together pertinent information that will serve the needs of nonspecialists, provide a quick reference to material that might otherwise be difficult to locate, and furnish a starting point for further study.

The project was undertaken with the realization that the initial volumes in the series could have some obvious deficiencies that will necessitate subsequent revisions. In the meantime, it is felt that the primary objectives of the Section Editors and their Advisory Boards has been met in this first edition.

A. A. Hanson
Editor-in-Chief

ADVISORY BOARD

CONTRIBUTORS

B. J. Barfield, Ph.D.
Department of Agricultural Engineering
University of Kentucky
Lexington, Kentucky

D. B. Brooker, Ph.D.
Professor Emeritus
Department of Agricultural Engineering
University of Missouri
Columbia, Missouri

R. H. Brown, P.E., Ph.D.
Chairman Emeritus
Division of Agricultural Engineering
University of Georgia
Athens, Georgia

R. R. Bruce, Ph.D.
Soil Scientist
Southern Piedmont Research Center
USDA
Watkinsville, Georgia

J. L. Butler, Ph.D.
Research Leader
Crop Systems Research Unit
USDA
Tifton, Georgia

W. J. Chancellor, Ph.D.
Professor
Department of Agricultural Engineering
University of California
Davis, California

J. L. Chesness, Ph.D.
Professor
Department of Agricultural Engineering
University of Georgia
Athens, Georgia

C. M. Christensen, Ph.D.
Professor Emeritus
Department of Plant Pathology
University of Minnesota
St. Paul, Minnesota

D. S. Chung, Ph.D.
Professor
Department of Agricultural Engineering
Kansas State University
Manhattan, Kansas

C. J. W. Drablos, Ph.D.
Professor
Department of Agricultural Engineering
University of Illinois
Urbana, Illinois

W. C. Fairbank
Extension Agricultural Engineer
Department of Soil and Environmental
Science
University of California
Riverside, California

P. R. Goodrich, Ph.D.
Associate Professor
Department of Agricultural Engineering
University of Minnesota
St. Paul, Minnesota

J. W. Goodrum, Ph.D
Associate Professor
Department of Agricultural Engineering
University of Georgia
Athens, Georgia

W. C. Hammond, Ph.D.
Head
Department of Extension Engineering
Cooperative Extension Services
Athens, Georgia

P. K. Harein, Ph.D.
Department of Entomology, Fisheries, &
Wildlife
University of Minnesota
St. Paul, Minnesota

J. C. Hayes, Ph.D.
Department of Agricultural Engineering
Clemson University
Clemson, South Carolina

J. G. Hendrick
Agricultural Engineer
National Tillage Machinery Lab
USDA
Auburn, Alabama

T. A. Howell, Ph.D.
Agricultural Engineer
Conservation and Production Research Lab
USDA
Bushland, Texas

R. W. Irwin, Ph.D.
Professor
School of Engineering
University of Guelph
Guelph, Ontario, Canada

F. C. Ives, P.E.
Design Engineer
Soil Conservation Service
USDA
Champaign, Illinois

J. M. Laflen, Ph.D.
Research Leader
National Soil Erosion Research Lab
USDA
West Lafayette, Indiana

J. H. Lehr, Ph.D.
Executive Director
National Water Well Association
Dublin, Ohio

W. D. Lembke, Ph.D.
Professor Emeritus
Department of Agricultural Engineering
University of Illinois
Urbana, Illinois

L. Lyles, Ph.D.
Research Leader
Wind Erosion Unit
USDA
Manhattan, Kansas

H. B. Manbeck, Ph.D.
Professor
Department of Agricultural Engineering
Pennsylvania State University
Univerisity Park, Pennsylvania

J. R. Miner, Ph.D.
Associate Director and Professor
Department of International Research and
Development
Oregon State University
Corvallis, Oregon

C. H. Moss, P.E.
Manager Civil Engineer
T. E. Stivers Organization, Inc.
Decatur, Georgia

B. H. Nolte, Ph. D.
Professor
Department of Agricultural Engineering
Ohio State University
Columbus, Ohio

J. C. Nye
Professor
Department of Agricultural Engineering
Louisiana State University
Baton Rouge, Louisiana

C. H. Pair
Engineer
USDA
Boise, Idaho

L. K. Pickett, Ph.D.
Senior Project Engineer
Department of Advanced Engineering
Case International
Hinsdale, Illinois

J. H. Poehlman
Largo, Florida

L. M. Safley, Jr., Ph.D.
Assistant Professor
Department of Agricultural Engineering
University of Tennessee
Knoxville, Tennessee

G. O. Schwab, Ph.D.
Professor Emeritus
Department of Agricultural Engineering
Ohio State University
Columbus, Ohio

Hollis Shull
Engineer
USDA
University of Nebraska
Lincoln, Nebraska

J. W. Simons
Research Associate
Department of Agricultural Engineering
University of Georgia
Athens, Georgia

R. P. Singh, Ph.D.
Professor
Department of Agricultural Engineering
University of California
Davis, California

R. E. Sneed, Ph.D.
Professor
Department of Biological and Agricultural
Engineering
North Carolina State University
Raleigh, North Carolina

J. M. Steichen, Ph.D.
Associate Professor
Department of Agricultural Engineering
Kansas State University
Manhattan, Kansas

C. W. Suggs, Ph.D.
Professor
Department of Biological and Agricultural
Engineering
North Carolina State University
Raleigh, North Carolina

J. M. Sweeten, Ph.D.
Extension Agricultural Engineer
Texas Agricultural Extension Service
Texas A&M University
College Station, Texas

J. R. Talbot
National Soil Engineer
USDA
Washington, D. C.

E. D. Threadgill, Ph.D.
Chairman
Department of Agricultural Engineering
University of Georgia
Athens, Georgia

D. H. Vanderholm
Associate Dean
Institute of Agriculture and Natural Resources
University of Nebraska
Lincoln, Nebraska

G. L. Van Wicklen, Ph.D.
Assistant Professor
Department of Agricultural Engineering
University of Georgia
Athens, Georgia

N. L. West
Project Engineer
John Deere Harvester
East Moline, Illinois

I. L. Winsett
Sales Engineer
Ronk Electrical Indusrties
Nokomis, Illinois

TABLE OF CONTENTS

Volume II

EROSION CONTROL ENGINEERING

DRAINAGE ENGINEERING

Volume III

PHYSIOLOGICAL PARAMETERS AND REQUIREMENTS OF LIVESTOCK POULTRY

STRUCTURAL DESIGNS, REQUIREMENTS, AND SYSTEMS

ELECTRICAL SYSTEMS AND APPLIANCES FOR AGRICULTURAL STRUCTURES

FEED AND CROP STORAGES

APPENDIX

*Physiological Parameters
and Requirements of
Livestock Poultry*

SPACE AND EQUIPMENT REQUIREMENTS OF LIVESTOCK/POULTRY PRODUCTION SYSTEMS

Robert H. Brown and Gary L. Van Wicklen

INTRODUCTION

The purpose of confined housing is to provide animals and poultry with an environment in which a minimum of resources are required for a maximum of production. Feed and water along with desired ventilation, lighting, air temperature, sanitation, and space are recognized considerations. Other factors related to the final result are high/low temperature variations, aerosol concentrations, ammonia, light color and lighted hours, floor surface, and noise.

Daily waste production, slatted-floor designs, and insulation procedures are furnished in the chapters on Insulation and Vapor Barriers and Pit and Slatted Floor Construction.

The industry representatives for agricultural structures and the extension engineers of the State Cooperative Extensive Service are excellent sources of information and assistance when planning structures. Usually, plans and specifications are available upon request. There are new building materials, floor surfaces, and microprocessor-based controls for managing the air handling, temperatures, lighting, feeding, and watering. The technology improves constantly, and contact with local specialists in housing and practices is highly recommended.

After receiving assistance from the specialists in structures and environments, the next procedures involve financial, legal, and utilities (electricity, gas, etc.) aspects of the production structure.

Better decisions related to construction of agricultural structures are reached after reviewing the plans with the contractors, banker, county building inspector, sanitarian or milk inspector in the case of dairy, and electric power supplier representative. Their suggestions and guidance are vital.

POULTRY (FIGURES 1 AND 2)

Poultry production is the most specialized of the livestock enterprises. The entire industry is very competitive, and the production facilities have become very large-volume operations which must be efficient and up-to-date technologically. Three enterprises are involved: commercial egg, broiler meat, and turkey meat.

Efficient housing is expensive, but is long overdue. This is now a major objective of both the grower and the processor. Three housing systems are normally considered: (1) controlled (modified) environment housing, (2) open housing (confinement but with natural ventilation), and (3) range (portable shelters, feeders), for turkey production.

In the case of commercial egg structures, the trend is toward controlled-environment housing and caged-layer systems. Some open housing is in use in the South and Southeast, but in colder climates the "enclosed" housing dominates. Production systems frequently involve flock-units of 12,000 to 30,000 per structure when using cages and 5000 where the layers are placed on the floor. Table 1 contains summary data for initial consideration of poultry structures.

Poultry broilers are grown out on dirt or concrete floors (usually dirt floors). A moisture-absorbing litter is placed on the floor, fresh every brood or every 6 to 12 months, depending upon the management scheme being used. Wood shavings and rice hulls are litter examples.

Table 1 presents the average data for consideration in broiler housing. A space of 0.8 ft²/ per (chick is usual in design but is recognized to be the 8-week-old requirement. Growers

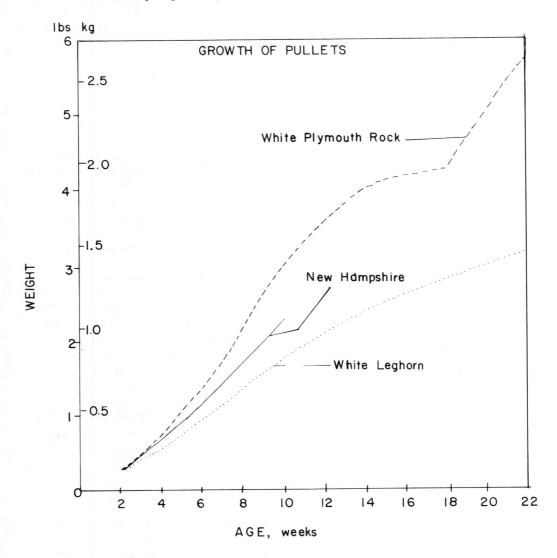

FIGURE 1.   Weight vs. age of pullets.

are presently restricting the 0 to 4-week-old baby chick to a sector of the floor area when heating the structure. At this early age, spaces as small as 0.25 to 0.4 ft² per chick are sufficient.

DAIRY CATTLE (FIGURE 3)

While the type of dairy housing system may vary, certain basic components must be provided. These include a milking area, milk storage, milking equipment cleaning system, feeding system, waste handling, resting area, service areas, and an area for herd replacements. Space and equipment requirements are furnished in Table 2.

Dairy housing systems in the U.S. are almost equally divided between conventional and loose housing. Conventional housing provides a stanchion or tie stall for each animal and is usually found in colder climates. Loose housing includes separate areas for functions such as feeding, milking, and lounging. The lounging area may either be a completely open lot

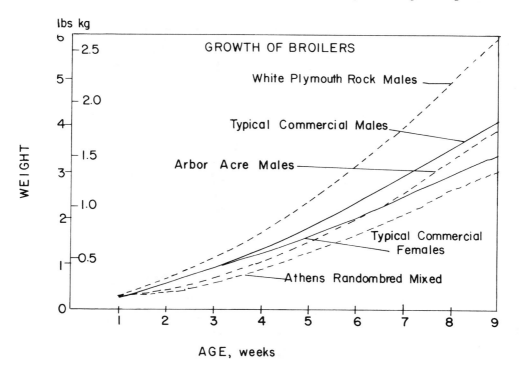

FIGURE 2.   Weight vs. age of broilers

Table 1
REQUIREMENTS FOR POULTRY

Item	Conditions	Requirement
Broilers		
Floor space	Open housing	0.8 — 1.0 ft²/chick
	Closed housing	0.8 ft²/chick
	First 4 weeks	0.25 — 0.4 ft²/chick
Ventilation	Winter	0.1 CFM/chick
	Summer	5.0 CFM/chick
Feeders	Perimeter length	2 in./chick
Waterers	Perimeter length	0.5 in./chick
Layers		
Floor space	Open, loose light breeds	2.5 ft²/bird
	Open, loose heavy breeds	3.5 ft²/bird
	Closed, light	1.5
	Closed, heavy	2.0
Caged system	Light breeds	2 — 8 birds/cage
	Heavy breeds	2 — 4 birds/cage
Ventilation	Winter, min	0.3 CFM/bird
	Summer, min	6 CFM/bird
Feeders		4 in./bird
Waterers		1 in./bird
Water req.		6 gal/day/100 birds

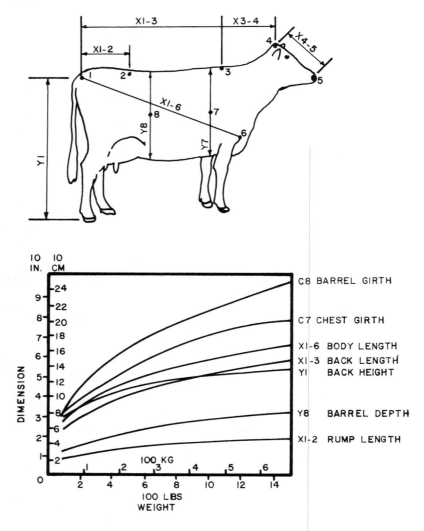

FIGURE 3.   Dimensions of livestock — dairy.

or contain free stalls. Free stall housing is increasingly popular among new construction in warmer climates because of its flexibility. Free stalls may be included in either warm enclosed barns, cold partially enclosed barns, or the open lot with a shading structure.

Conventional housing systems generally contain manure gutters with cleaners for waste removal, feed managers, service alleys, and pipeline milking units. This type of system affords cows individual care and maximum protection, but is costly, inflexible for expansion, and very inefficient for labor. Loose housing uses labor more efficiently, improves cow comfort, and is less costly and more flexible than conventional housing, although contagious disease may spread more easily because cows are in contact with each other.

Milking systems for conventional dairy housing usually consist of pipeline or unit milkers at stanchions or tie stalls. Nearly 70 to 80% of milk cows in the northern U. S. are milked with this type of system. Loose housing requires a milking parlor. The most popular types are the side opening, herringbone, polygon, and rotary parlors. Milking parlors are used to milk 45 to 60% of cows in the warmer climates of the U.S.

Housing for calves depends upon the climate of the location and the purpose for raising the calves. A warm, temperature-controlled environment is important for a veal calf operation

Table 2
SPACE AND EQUIPMENT REQUIREMENTS FOR DAIRY COWS AND CALVES

Item	Conditions	Requirements
Stanchion barn		
Stall	Medium cow	4 ft 2 in. × 5 ft 8 in.
	Large cow	4 ft 6 in. × 6 ft 2 in.
Floor area	Per cow	66 — 76 ft²
Gutters	Width × depth	16 in. × 12 in.
Litter alley	Width	9 ft
Mangers	Width	2 ft
Feed alley	Width	6 ft
Water	Bowl/2 cows	30 gal/cow/day
Stall barn (free stalls)		
Resting area	Per cow, inside	40 ft²
Outside lot	Paved/cow	50 ft²
Free stall	Per cow (large)	4 × 7 ft
	Heifers	3 × 6 ft
Holding area	Milking cow	15 ft²
Litter alley	Width	9 ft
Manger	Width	2 ft
Water	Milking cow	30 gal/day
	Dry cow	20 gal/day
Excrement	1000-lb cow	1.26 ft³/day
Calf spaces		
Individual pens	Per calf	24 ft², 4-ft-high walls
Individual elevated stalls	Per calf	24 in. wide, 4½ ft long
Individual calf hutches	Per calf	4 ft × 8 ft × 3 ft high
Community pens		
Age		
2 — 6 months	Per calf	20 — 25 ft²
6 — 9 months	Per calf	25 — 30 ft²
9 — 15 months	Per calf	30 — 35 ft²
15 — 24 months	Per calf	35 — 40 ft²

		Metal slat spacing (in.)	
Animal wt (lb)	Area (ft²) per calf	Width	Spacing
Slatted floors			
Less than 249	10	4.5	1.25
250 — 374	12	4.5	1.25
375 — 549	14	5.25	1.50
550 — 724	17	6	1.75
Greater than 725	20	6	1.75

to produce an adequate conversion of feed to weight gain. Calves raised as replacements for the dairy herd do not need high feed conversion, but their health and growth is important.

Calves may either be housed in warm, enclosed barns, or in an open, cold type of barn. Several methods of constraining the calves may be suggested whether in warm or cold housing. These include individual pens, individual elevated stalls, individual calf hutches, community pens, and free stalls. Individual pens, elevated stalls, and hutches are suitable for young calves less than 2 months old. Once weaning occurs (usually no later than 10 weeks of age), calves may be grouped together in community pens according to size, with the age difference in any group no more than 3 months. Community pens usually house calves on a bedded pack of manure or on slatted floors made of metal, reinforced concrete, plastic, or hardwood.

BEEF CATTLE (FIGURE 4)

Beef cattle are handled and grown out under many different plans. Accordingly, the production system, utilities, and provision are neither as precise nor as uniform as for other animals. Some general directions are recognized, and these together with the data in Table 3 supply needed guidelines.

Much of the beef cattle activity centers around the cow-calf operations with provisions for ranging and growing out until ready for fattening. The feeding steers (and heifers) are often shipped to areas where the feed is produced, and there they are kept in feed lots until marketed. Feeding animals are fattened to approximately 1000 lb; about 8 lb of feed is required per pound of gain.

Open sheds for animal shelter and to keep the feed dry are normally sufficient. Additional needs are feed lots and maternity pens. These pens should have 60 ft^2 of floor space inside (under cover and partitioned off) if the cow has access to an outside lot. If confined, 100 ft^2 should be allowed for each pen. Extra stalls for calves should also be provided, under cover.

A properly constructed (graded, course gravel-tamped, plus 4 in. of concrete with expansion joints) yard area for feeding and watering is recommended for the feed lot. Dirt floors are used but are most difficult in wet weather. Cleaning and feeding chores are more efficient on a paved surface. When planning the arrangement of open shed and feed yard, tractor removal of manure, adjacent feed storage, feed-handling equipment, portable-feed (hay) bunkers, and access to isolating pens for one or two animals should be considered.

SWINE (FIGURE 5)

In the U.S., the swine industry is concentrated in the cornbelt states of Illinois, Indiana, and Ohio. About 4 lb of grain and supplement is required per pound of gain, with market hogs weighing about 200 lb. Lesser amounts of small grains in other geographical areas and certain processed and transported feeding rations have led to a wider enterprise area for hogs. The South, Southeast, and Midwest now grow out large numbers of hogs. These areas also have certain ambient temperature advantages.

The farrowing houses are well advanced as structures. Many are environmentally controlled, including air conditioning. All have specific provisions for space, water, warmth, and cooling. Baby pigs are well protected. At about 3 weeks of age, the pigs are removed from the farrowing house areas into a nursery barn. Here they are protected environmentally and remain until 8 to 10 weeks of age (about 50 lb). Then they are moved into a growing-finishing barn. (Sometimes a nursery section is added to one end of the farrowing house and the pigs are placed in this area at about 3 weeks and then moved to the growing-finishing structure.) This growing-finishing structure contains the space, waterers, feeder, sanitation, and temperature to enhance feed conversion. Table 4 includes important data and ASAE Standard D270.4 (See next chapter) furnishes design data about waste, ventilation, humidity, and temperature in this finishing area. The pig has short hair and does not sweat. A minimum temperature around 40°F is desirable, but a mature hog can tolerate temperatures below 40°F better than above 80°F. Normally the designed structure will assist in keeping within the range.

A holding or gestation building for sows and another building for boars are usual structures. A range area with small hutch units for sows and pigs is a practice in warm climates.

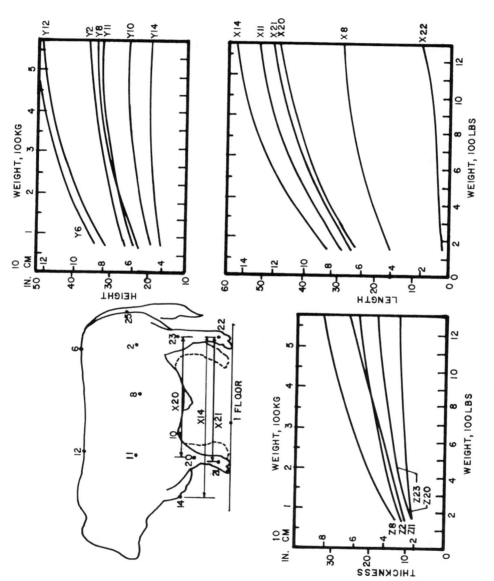

FIGURE 4.   Dimensions of livestock — beef.

Table 3
SPACE AND EQUIPMENT REQUIREMENTS FOR BEEF ANIMALS

Item	Conditions	Requirements
Feeding lot	Paved/head	60 ft²
	Unpaved/head	300 ft²
Shelter	Open front/head	25 ft²
Holding area	1200-lb animal	20 ft²
Calving pen	12 cows	120 ft²
Waterer	1-cup waterer or 18-in. tank perimeter, dry lot area	25 head per waterer
	1-cup waterer or 18-in. tank perimeter, pasture area	15 head per waterer
Water	Feeder cattle	15 gal/day
	Cows	20 gal/day
Feeder perimeter space	Self feed, continuous	1 ft/animal
	Herd eats same time	3 ft/animal

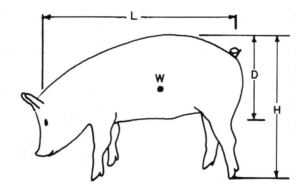

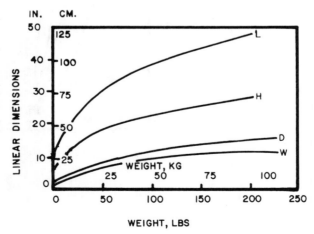

FIGURE 5.   Dimensions of livestock — swine.

Table 4
SPACE AND EQUIPMENT REQUIREMENTS FOR SWINE

Item	Conditions	Requirements
Enclosed housing		
Nursery	Per hog	4 ft^2
Growing finishing	Per hog	8 ft^2
Shed with lot		
Nursery	Inside space/hog	3 ft^2
	Outside space/hog	8 ft^2
Growing finishing	Inside space/hog	6 ft^2
	Outside space/hog	12 ft^2
Pasture area	Sows with litter	6/acre
	Growing finishing hogs	80/acre
Waterer	Unit waterer	10 pigs
Water	Sow and litter	10 gal/day
	Nursery	1 gal/day
	Growing finishing	4 gal/day
	200-lb hogs	5 gal/day

REFERENCES

1. *ASAE Yearbook,* 27th ed., American Society of Agricultural Engineers, St. Joseph, Mich., 1981.
2. The Poultry House Controlled Environment Handbook, ACME Engineering and Manufacturing Corp. Muskogee, Okla., 1967.
3. Farmstead Planning Manuals, Books 1 — 5 Granite City Steel Company, Granite City, Ill.
4. Construction Materials, Series 12 — Insulation-Agriculture, The Dow Chemical Company, Midland, Mich., 1970.
5. **Hellickson, M. A. and Walker, J. N.,** Ventilation of Agricultural Structures, ASAE Monograph No. 6, American Society of Agricultural Engineers, St. Joseph, Mich., 1983.
6. Midwest Plan Service, 11th ed., Iowa State University, Ames, 1983.
7. **Richey, C. B.,** *Agricultural Engineers' Handbook,* McGraw-Hill, New York, 1961.

BASIC DATA AND PROCEDURES FOR DESIGN OF VENTILATION SYSTEMS FOR LIVESTOCK AND POULTRY SHELTERS*

ASAE Data: D270.4

SECTION 1 — PURPOSE AND SCOPE

1.1 These ASAE data were consolidated into one report for the information and convenience of agricultural engineers and other professional persons serving the agricultural industry. They include basic information and technical data, supported by research, for use in designing and/or evaluating ventilation systems for livestock or poultry shelters. Much of the general design information presented in Section 3 will apply to the design of ventilation systems for any structure housing livestock or poultry.

1.2 In addition to the design and configuration of equipment, hazard control and accident prevention are dependent upon the awareness, concern, and prudence of personnel involved in the operation, transport, maintenance, and storage of equipment or in the use and maintenance of facilities.

SECTION 2 — DEFINITION

2.1 Ventilation as used herein is defined as a system of air exchange which accomplishes one or more of the following:

2.1.1 Provides desired amounts of fresh air, without drafts, to all parts of the shelter.

2.1.2 Maintains temperatures in the shelter within desired limits.

2.1.3 Maintains relative humidity in the shelter within desired limits.

SECTION 3 — GENERAL DESIGN INFORMATION

3.1 Weather data — The cold weather outdoor design temperature should be selected for the particular area where the animal shelter will be located. Whereas one temperature at a particular geographical location is normally used for determining heat loss from a building and hence insulation requirements, two temperatures are necessary for designing ventilation systems. The temperature normally used for heat loss computations is also used to determine the minimum continuous air exchange rate. A higher design temperature is used to determine the maximum cold weather ventilation capacity.

3.1.1 Minimum air exchange — When the outdoor temperature is the coldest, it generally contains the least amount of moisture. Thus less of it is required to remove the vaporized moisture produced within a building. Outside design temperatures suitable for calculating winter minimum air exchange rates and heat loss through the building components are shown in Figure 1. Tabular data on outside winter design temperatures can be found in several books and periodicals. Most are derived from the American Society of Heating, Refrigerating and Air Conditioning Engineers Handbood of Fundamentals or older ASHRAE publications. (The 97 $^1/_2$% values from ASHRAE are recommended for animal shelter use.)

3.1.2 Maximum ventilation capacity — Because the ventilation system must also operate

---

* ASAE D270.4. Reproduced with permission. Prepared by the Joint Structures and Environment-Electric Power and Processing Division Committee on Animal Shelter Ventilation; approved by the Electric Power and Processing and Structures and Environment Division Steering Committees; adopted by ASAE February 1963; revised June 1966; December 1968, March 1970; revised April 1975 to incorporate and supersede D249, Effect of Thermal Environment on Production, Heat and Moisture Loss and Feed and Water Requirements of Farm Livestock which was originally adopted by ASAE 1955; reconfirmed for 1 year December 1979, December 1980, December 1981; reconfirmed for 1 year December 1982, December 1984.

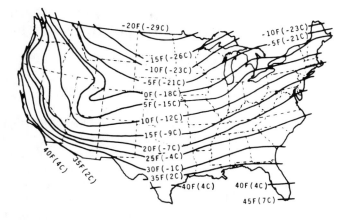

FIGURE 1.   Winter temperature that is exceeded more than 97½% of the time.

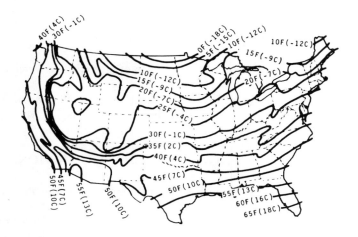

FIGURE 2.   Average daily temperatures for January in the U.S.

satisfactorily during mild winter weather, higher outside design values must be used to establish the maximum winter ventilation capacity. Figure 2 shows the average daily temperatures for January in the U.S. Designing with this temperature will provide enough air exchange to prevent any net accumulation of vaporized moisture in the poultry or livestock shelter. In buildings without openings which depend upon mechanical ventilation during hot weather, air exchange must be sufficient to keep the inside temperature the same or only slightly warmer than the outside temperature. The design maximum ventilation capacity can be computed using Equation 3.3.1-1 and an assumed temperature difference of 2 to 4°F (1 to 2°C) or Equation 3.3.1-2, whichever gives the greater exchange rate.

3.2 Building requirements

3.2.1 Heat transmission through building materials — The rate of heat transmission through building materials depends on the characteristics of the material. Heat transfer coefficients for building components consisting of a limited number of structural combinations are given in Table 1. A more comprehensive list of heat transfer coefficients of building materials is given in the ASHRAE Handbook of Fundamentals.

3.2.2 Need for vapor barriers — Animal shelters of the totally enclosed and ventilated type may be subjected to high moisture conditions. For example, a shelter with inside air

Table 1
U VALUES OF SELECTED STRUCTURAL COMPONENTS

Construction	U value (Btu/hr-ft²-°F)	W/m²°C	R value (R = 1/U English)	R value (metric)
A. Walls				
1. Wood siding (average), building paper, studs	0.06	3.46	1.66	0.29
a. No. 1 plus 3/8-in. plywood inside liner	0.32	1.84	3.10	0.54
b. No. 1a plus 2- to 2 3/4-in. blanket insulation (one airspace)	0.099	0.57	10.10	1.15
c. No. 1a plus 3- to 3 1/2-in. blanket insulation (no airspace)	0.076	0.44	13.13	2.28
d. No. 1 plus 1-in. extruded polystyrene inside liner	0.13	0.75	7.89	1.33
2. 3/8-in. plywood inside and outside, studs, 3- to 3 1/2-in. blanket insulation	0.078	0.45	12.79	2.22
3. Sheet metal siding on wood girts	1.2	6.92	0.85	0.14
a. No. 3 plus 1-in. extruded polystyrene between siding and girts	0.16	0.92	6.11	1.08
b. No. 3 plus 1-in. extruded polystyrene as inside liner (one airspace)	0.14	0.81	7.08	1.24
c. No. 3 plus 2- to 2 3/4-in. blanket insulation, 3/8 plywood inside liner (one airspace)	0.11	0.63	9.29	1.58
4. 8-in., 3-core, sand and gravel concrete block	0.51	2.94	1.96	0.34
5. 8-in., 2-core, lightweight aggregate block	0.34	1.96	3.03	0.51
a. No. 5 plus filled cores	0.17	0.98	5.88	1.02
b. No. 5 plus 1-in. extruded polystyrene bonded to inside	0.12	0.69	8.29	1.45
B. Ceilings (on truss bottom chord)				
1. 1/2-in. plywood	0.55	3.17	1.84	0.32
2. Sheet metal	0.82	4.73	1.22	0.21
a. No. 1 plus approx. 3-in. loose fill mineral fiber	0.092	0.53	10.84	1.89
b. No. 2 plus approx. 3-in. loose fill mineral fiber	0.098	0.56	10.22	1.77
3. 2-in. extruded polystyrene	0.085	0.49	11.74	2.04
C. Doors				
1. 3/4- to 1-in. solid wood	0.64	3.69	1.55	0.27
a. No. 1 plus wood storm door	0.30	1.73	3.33	0.58
2. 3/8-in. plywood liners over 3/4-in. extruded polystyrene	0.18	1.04	5.54	0.96
D. Windows				
1. Single-pane glass, 20% wood sash	1.02	5.88	0.98	0.17
a. No. 1 plus storm windows (1- to 4-in. airspace)	0.50	2.88	2.00	0.35
2. Insulating glass, double, 1/2-in. airspace	0.58	3.34	1.72	0.30

Table 1 (continued)

U VALUES OF SELECTED STRUCTURAL COMPONENTS

Construction	U value (Btu/hr-ft²-°F)	W/m²°C	R value (R = 1/U English)	R value (metric)
E. Roofs (sloped, no separate ceiling)				
1. Sheet metal on wood purlins	1.3	7.49	0.79	0.13
a. No. 1 plus 1-in. extruded polystyrene between roofing and purlins	0.17	0.98	6.05	1.02
b. No. 1 plus 1-in. extruded polystyrene under purlins (one airspace)	0.14	0.81	7.01	1.24
c. No. 1 plus 1/2-in. insulating board between roofing and purlins	0.47	2.71	2.11	0.37
2. Asphalt shingle on 5/8-in. plywood decking	0.48	2.77	2.07	0.36
a. No. 2 plus 1-in. extruded polystyrene under rafters (one airspace)	0.12	0.69	8.29	1.45
b. No. 2 plus 1/2-in. plywood under rafters, 2- to 2 3/4-in. blanket insulation (one airspace)	0.094	0.54	10.65	0.19

Basic data from ASHRAE Handbook of Fundamentals, 1972, chap. 20.

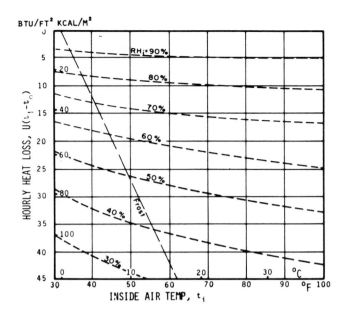

FIGURE 3. Condensation prediction graph.

temperature of 50°F (10°C) and relative humidity of 75% and with outside air temperature of 10°F ($-$12°C) and relative humidity of 85% will be subjected to a winter vapor pressure difference of approximately 3 in. (7.6 cm) of water. A vapor pressure difference of this amount tends to force water vapor out through the components of the enclosure. Water vapor transfer is not a problem as long as the saturated vapor pressure gradient within the enclosure components remains above the actual vapor pressure gradient. Condensation will occur at or near the point in the building component where these gradients intersect. The condensate may migrate to other parts of the enclosure due to gravitational force or capillary attraction. Condensed water vapor is undesirable because it causes decomposition of building components and reduces the insulating capacity of many insulations. This situation increases structural heat loss which in turn hinders moisture control by ventilation. To keep insulation dry and free from condensed vapor, a vapor barrier should be applied on or near the side of the insulation with the highest vapor pressure, which is usually the warm side.

3.2.3 Estimating condensation — Condensation on the inside wall surface may be estimated by using the condensation prediction graph (Figure 3) and the procedure described below. Two of the three variables (heat-loss factor, inside relative humidity, and inside air temperature) must be known or assumed.

3.2.3.1 To estimate at what relative humidities condensation will occur on the inside surface of a particular component (wall, ceiling, floor, window, or door) of the enclosure with given temperature conditions, determine the heat-loss factor $U(t_i - t_o)$, in Btu/hr ft^2, by multiplying the overall heat transfer coefficient for that part of the enclosure (U) by the difference between inside and outside air temperatures. Find the point on the graph (Figure 3) corresponding to this heat-loss factor and inside air temperature. Any relative humidity above the value indicated at this point would be expected to cause condensation.

3.2.3.2 To estimate outside air temperatures that will cause condensation on the inside surface of a particular component with given inside air temperature, find the intersection of the relative humidity curve with the inside air temperature. Determine the heat-loss factor at this intersection. This will be the minimum heat-loss factor that will cause condensation with the given inside air conditions. The minimum outside air temperature may be calculated using the equation for the heat-loss factor.

3.2.3.3 To determine the amount of insulation necessary to prevent condensation with given inside air conditions, determine the intersection of inside relative humidity and air temperature curves. From the heat-loss factor indicated by the point at this intersection, compute the maximum allowable overall heat-transfer coefficient (U) for the given or assumed minimum outside air temperature. The amount of insulation required to limit the coefficient (U) to this value can be determined by calculation or by referring to tables.

3.2.3.4 If any of the above points or intersections fall to the left of the line marked frost, the inside surface temperature of the enclosure will be below 32°F (0°C) and condensation will appear in the form of frost.

3.2.4 *Exposure factor* — Conductive heat loss from an animal shelter can be more readily compared to other shelters when expressed on a per animal basis. Exposure factor is determined from the size of the building, the insulation value of the building, and the number of animals housed and is defined as:

$$EF = \frac{\Sigma(A_i U_i)}{N} \qquad (3.2.4\text{-}1)$$

where EF = exposure factor; U_i = overall coefficient of heat transfer for a particular structural component (i.e., walls, ceiling, doors, windows, etc.), Btu/hr-ft²-°F (W/m²-°C); A_i = area of the specific building component for which the U_i value was determined, ft² (m²); and N = number of animal units housed (an animal unit for cattle is a 1000-lb (454-kg) animal and for swine and poultry it is an individual animal). By definition, the exposure factor is the conductive heat loss from an animal shelter expressed in units of Btu/hr-°F-animal. The total conductive heat loss from the shelter can be obtained by multiplying the exposure factor by the number of animals housed and the difference between inside and outside design temperatures. The normal range of exposure factors is 10 to 50 (5.3 to 26 metric) for cattle, 1 to 5 (0.5 to 2.6 metric) for swine, and 0.2 to 0.5 (0.1 to 0.3 metric) for poultry. The exposure factor can be decreased by increasing the amount of insulation (thus decreasing the U value) or by increasing the livestock density.

3.3 Air handling requirements

3.3.1 Air flow rates — Equations 3.3.1-1 and 3.3.1-2 can be used to determine ventilation rates for temperature control and moisture removal. At low outside temperature, the ventilation rate required for moisture removal is usually greater than that permitted for temperature control. If the moisture removal rate is selected, additional heat may be necessary to maintain the desired indoor temperature. In practice, however, rather than providing supplemental heat the indoor temperature may be allowed to fall below the design temperature for brief periods during the lower portion of the diurnal outdoor temperature curve. As the outside air temperature approaches the inside design air temperature, the ventilation rate for moisture removal becomes less than required for temperature control. In this situation, if the ventilation rate for temperature control is selected, the moisture level will be reduced below the design condition, but this is usually desirable. Air flow rates for constant exposure factor are based on the expression:

$$M = \frac{\bar{v}}{14.4(\Delta t)} [Q_s - EF(\Delta t)] \qquad (3.3.1\text{-}1)$$

where M = ventilation rate,* ft³/min-animal unit; v̄ = specific volume at inside air tem-

---

* Multiply ft³/min by 4.72 × 10⁻⁴ or 28.3 to convert to m³/sec or liters/hr respectively.

perature, ft³/lb dry air; Δt = difference between inside and outside air temperature, °F; Q_s = sensible heat production, Btu/hr-animal unit at a specified inside air temperature, and EF = exposure factor (see paragraph 3.2.4).

Air flow rates for constant inside relative humidity are based on the following expression:

$$M = \frac{\bar{V}}{60}\left[\frac{Q_\ell}{1044(\Delta w)}\right] \qquad (3.3.1\text{-}2)$$

where Q_ℓ = latent heat production, Btu/hr per 1000-pound animal unit, at specified inside air temperature; 1044 = latent heat of vaporization at body temperature, Btu/lb water; and Δw = difference between inside and outside humidity ratios, lb water vapor per lb of dry air.

3.3.2 Design graphs for predicting ventilation performance — Conditions that can be expected to exist for various species and sizes of animals in enclosed structures with a normal range of exposure factors can be predicted using ventilation performance prediction graphs. These graphs for dairy cattle, beef cattle, swine, and laying hens are included in the section for each species as Figures 9 to 11, 21 to 23, 30 to 33, and 45 to 47, respectively. The graphs can be used to select design and operating conditions for cold weather ventilation. Solid lines represent use of the heat balance equation (3.3.1-1) (constant exposure factor). Dotted lines represent use of the moisture balance equation (3.3.1-2) (constant inside relative humidity).

3.3.2.1 Definitions of terms in design graphs — EF = exposure factor, Btu/hr-°F-animal; t_i = indoor dry-bulb air temperature, °F; RH_i = indoor relative humidity, %; RH_o = outdoor relative humidity; Q_s = sensible heat production per animal at t_i, Btu/hr; and Q_ℓ = latent heat production per animal at t_i, Btu/hr.

3.3.2.2 Example application of design graph — Assume that the winter normal cfm ventilation rate is desired for 100-lb pigs, housed at 60°F and a maximum allowable RH_i of 80%. The computed exposure factor of the building is assumed to be 2%. From the applicable graph (Figure 31B) (t = 60°F), the intersection of the EF = 2 curve and the RH_i = 80 curve indicates a ventilation rate of 6 cfm per pig. The outside temperature at which this balance point occurs is 25%F. This is a unique point, the only point at which all the desired conditions can be obtained. If the outside temperature were to increase and the inside temperature remain at 60°F, movement on the graph would be along the EF = 2 curve. At t_o = 40°F, 12 cfm per pig would be required and RH_i = 72%. At t_o = 45°F, 17 cfm per pig would be necessary and RH_i = 70%. This is probably the highest outside temperature which would be considered practical for winter design. Similarly, as the outside temperature decreased below 25°F, movement along the EF = 2 curve shows that RH_i would approach 100%, reaching this condition at t_o = 16°F and a ventilation rate of 4 cfm per pig. If 100 % relative humidity were objectionable or t_o continued to decrease, supplemental heat would have to be added. If t_i were allowed to decrease to 50°F, a ventilation rate of 6 cfm per pig could be maintained at t_o = 7°F and RH_i = 82% (Figure 31A). A workable solution for this example would be installed capacity of 17 cfm per pig with proper controls to limit the minimum rate to 6 cfm per pig.

3.3.3 Air distribution and intake-infiltration data — Ventilation is accomplished in an exhaust system by reducing the pressure within the structure below outside pressure, causing fresh air to enter wherever openings exist. Pressure differences across walls in ventilated shelters should range between 0.02 and 0.107 in. (0.05 to 0.18 cm) of water. The distribution of the fresh air is dependent upon the location, number, and cross-sectional area of the openings. Ideally, the amount of fresh air entering a section of the shelter should be proportional to the amount of heat and moisture produced by the animals in this particular section. In systems where fresh air is supplied through structural leakage, it is difficult to control the distribution of air within the shelter. The larger openings usually will exist as cracks around doors and windows and, therefore, introduce larger quantities of fresh air at these points than in areas of tighter construction. The use of a planned intake system as part of the ventilation system has specific advantages. Proper air distribution can be obtained by location of the inlets and by varying intake cross-sectional area to fit the animal distribution. In this manner, the heat loss from ventilation is kept more uniform throughout the shelter. Automated motorized inlet controls have been developed to adjust the slot intake opening with delivery increments of fan operation. These small units adjust the inlet baffle opening in relation to the exhaust ventilation rate and thus provide better control of fresh air distribution within the shelter. Another method of intake-distribution uses a circulating fan connected to a perforated polyethylene air distribution tube or a rigid duct which allows a combination of heating, circulation, and ventilation to be designed into one system. Equipment manufacturers should be consulted for proper application of this type of equipment.

3.3.4 Inlet design — The rate of air exchange in a building depends upon the ventilation capacity. The uniformity of air distribution throughout the building, however, depends primarily upon the location and size of the air inlet.

3.3.4.1 Location of inlet — For most exhaust ventilation systems slot inlets are used around the perimeter of the ceiling (except near the fan). A large slot, suitable for high cfm requirements during the summer, can be partially closed with an adjustable baffle during the lower cfm winter periods. Air coming from the attic space is desirable for winter ventilation but undesirable during the summer.

3.3.4.2 Size of inlet — Design of inlet-area requirements for stanchion-type dairy barns is possible using the pressure-discharge characteristics shown for barns and inlets in Figures 4 and 5. With these curves the inlet area required to limit the drop in static pressure to a given amount for any rate of air discharge desired can be calculated. The formula for calculating inlet area requirements is

$$A = \frac{(M - I)\,(a)\,(N)}{C} \qquad (3.3.3\text{-}1)$$

where A = total inlet area needed, in.^{2}; M = total quantity of air to be discharged per animal unit, cfm; I = total quantity of air entering by infiltration per animal unit at the pressure selected, cfm; a = area of each inlet selected from Figure 4, in.^{2}; C = flow rate of air entering each inlet at the pressure selected, cfm; and N = number of animal units.

A sample solution will help to explain. A dairy barn, housing 100 animal units (1000-lb units), is to be ventilated at the rate of 150 cfm of air per animal unit. Static pressure differences across the wall are to be no greater than 0.03 in. of water. Assume this barn is judged to be loose construction (see footnote, Figure 5). Figure 3 shows that infiltration can supply more than the desired air movement at well under 0.03 in. of static pressure. This indicates that no planned inlets are required. With this type of construction it is necessary to tighten up the structure to maintain good ventilation control. As an example calculation for a new structure, assume it is planned for 100 animal units with a ventilation rate of 150 cfm/1000-lb animal unit at a static pressure difference limited to 0.03 in. of water. This would be classified as very tight construction. At a static pressure of 0.03 in. water, infiltration will provide 45 cfm/1000-lb animal unit. An additional 105 cfm of inlet air per animal unit is needed. If slot inlets are to be provided, the area of slot required for this barn can be found using Equation 3.3.3-1 and noting that each foot of 1-in. slot admits 43 cfm (Figure 4). Thus:

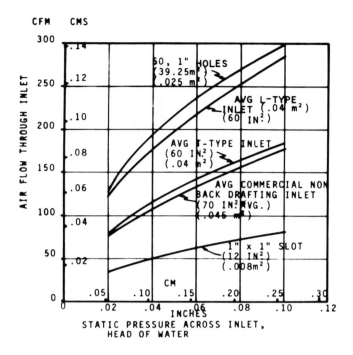

FIGURE 4. Air flow characteristics for typical inlets. (These air flow curves present data in the range of static pressures likely to prevail in animal shelter ventilation installations. These data coupled with information presented in Figure 5 can be used to plan for an appropriate amount of inlet area. It should be noted that the area of opening varies from curve to curve so that direct comparisons cannot be made.)

$$\text{Area of slot required} = \frac{(150 - 45)\,(12)\,(100)}{43}$$

$$= 2930 \text{ in.}^2$$

$$\text{Lineal feet of 1-in. slot} = \frac{2930}{12} = 244 \text{ ft}$$

Hence at least 244 ft of 1-in. slot should be provided for inlet area.

3.3.5 Exhaust fan systems — Most of the variations in exhaust fan systems fall within one of the following classifications.

3.3.5.1 Single fan-intermittent operation — The intermittent system with a single-speed fan requires only a fan and a thermostat or time clock. The capacity of the fan is selected on the basis of the maximum ventilation rate necessary to maintain a given inside temperature for a given outside temperature. Intermittent systems are usually installed with the fan located near the ceiling level. If desired, a floor exhaust duct is used to conserve heat and allow the fan to operate a major part of the time during periods of low outside temperature. A hinged or sliding door, at least the size of the fan intake, should be located at the top of the exhaust duct just back of the fan so that it can be opened in the summertime, allowing the fan to draw air off the ceiling. By closing the door in cold weather the fan will draw its air off the floor.

3.3.5.2 Single fan-continuous operation — The continuously operating, single-speed fan is the simplest system in that no thermostat is used (except, possibly, for low-temperature cut-off). It lacks adjustment in ventilation rate to follow variations in outside temperature.

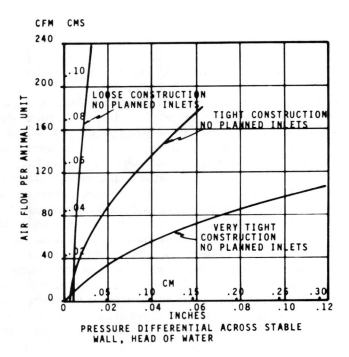

FIGURE 5.   Infiltration rates vs. pressure differential for three Pennsylvania dairy stables. (Loose construction is typified by older barns which have been remodeled to include masonry side walls. Very tight construction is typified by present-day new construction. These data coupled with data from Figure 4 can be used to calculate a planned inlet system for ventilated structures.)

The capacity of the fan is selected on the basis of the minimum rate to maintain a desired inside temperature range for a given exposure factor and given outdoor temperature range. The fan usually is installed in an exhaust duct. Such operation is usually used during winter months only.

3.3.5.3 Single fan-two-volume control by use of a two-speed motor — This system uses a two-stage thermostat to shift from one motor speed to another as the temperature changes. Although it is common for a two-speed motor with two windings to have a low speed equal to two-thirds of the high speed (not true in single winding motors), the air delivery rate at low speed is usually slightly greater than two thirds the rate at high speed. It is good practice to use a low-limit thermostat to shut the fan down in case the low-volume air still lowers the temperature in the building too much.

3.3.5.4 Single fan-two-volume control by throttling — Where the volume of exhausted air is controlled by throttling, the fan has a single-speed motor. The fan is housed in a box-like mounting equipped with shutters on the intake side of the box behind the fan. The shutters are opened and closed by a damper motor controlled by a thermostat. When the shutters are open, the fan exhausts at high volume. When the shutters are closed, air is drawn through a restricted opening in the bottom of the housing. This opening may be adjustable by a second set of shutters in the exhaust duct to provide a third volume which reduces cycling. This type of fan is usually mounted near the ceiling level.

3.3.5.5 Single fan-two-volume control by throttling and changing the exhaust position — This system is the same as that described in paragraph 3.3.5.4 except that it uses a floor exhaust duct for the low-volume exhaust.

3.3.5.6 Single fan-variable-volume by using a variable-speed motor — Such a fan is equipped with a high slip motor with a high resistance rotor. Designed to be controlled by a solid-state variable-speed controller, such a unit smoothly modulates the air flow from minimum to maximum volume. The controller is adjustable to give a mimimum speed or air flow of approximately 10% of the maximum volume of the fan. The dial on the variable-speed controller is set at the desired temperature in the building. The fan will then slow down or speed up to maintain this temperature. Such a fan is usually controlled by the variable-speed controller and a safety shutoff thermostat which will shut the fan off completely should the temperature in the building continue to fall after the fan has reached minimum volume. The fan will then cycle on and off at the minimum speed until the temperature in the building starts to increase above the setting of the speed control.

3.3.5.7 Multiple fan-one or more continuous, remainder intermittent — The multiple-fan system is usually made up of several fans. This system is highly flexible and makes use of standard single-speed fans and standard shutters. One or more fans are selected for continuous operation to exhaust the required minimum amount of air while maintaining the desired inside temperatures for given exposure factor during periods of probable minimum outside temperature. These continuously operating fans may be equipped with an exhaust duct leading down and tapped into the pits on buildings using slats and manure pits. This results in pulling the low-volume air directly from the pit. These continuous-operating fans are equipped with a thermostat to shut them down in case the temperature falls below the minimum desired in the building. The additional single-speed fans are also controlled by a thermostat connected to each fan. The intermittent-operating fans are selected to provide the required capacity to maintain desired inside temperatures during periods of high outdoor temperatures.

3.3.5.8 Multiple fan-intermittent operation — The multiple-fan, intermittent system is similar to the single-fan intermittent system, except that each fan has its own thermostat. This permits setting the thermostats to obtain several rates of ventilation depending upon temperature.

3.3.5.9 Multiple fan-one or more variable speed, remainder intermittent — This is one of the most flexible systems yet devised. The variable-speed fans are controlled by variable-speed controller and a low-limit or safety thermostat. They are selected to give the minimum speed of the fan or fans and are often the only ones operating in colder weather. As the temperature in the building increases from the minimum during cold weather, the variable-speed fans will speed up in an attempt to maintain the desired temperature. After the variable-speed fans have reached their maximum speed, and the temperature in the house increases, the intermittent fans which are controlled by individual thermostats will start to operate.

3.3.5.10 Multiple fan-one or more two-volume fans, remainder intermittent — With this system the two-volume fans are selected to give the required minimum air flow at low volume. The two-volume throttle is controlled by a thermostat, and a low-limit or safety thermostat is used to control the fan. In case the temperature in the building continues to decrease after the two-volume fan has reached minimum volume, the low-limit or safety thermostat will stop the fan completely and will cycle it on and off at low volume until the temperature in the building starts to increase above the setting of the low-limit thermostat. Each of the intermittent fans is controlled by an individual thermostat and they are set to start operating if the temperature in the building continues to increase after the two-volume fans have reached maximum volume.

3.3.5.11 Multiple fans-two-speed fans, remainder intermittent — This system is operated essentially as a multiple-fan, two-volume, remainder intermittent system. The two-speed fans are selected to give the minimum air flow requirements at low speeds. They are usually controlled by a two-stage thermostat to cycle from high to low speeds and a low-limit or safety thermostat to shut the fan off in case the temperature continues to drop after the fans have reached low speed.

3.3.6 Exhaust fan and thermostat locations — Fans should be located so they will not

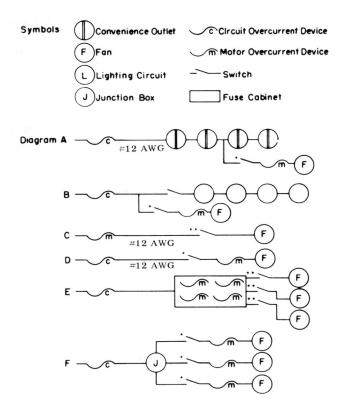

FIGURE 6.   Fan wiring circuits. (Neutral conductors and grounding conductors, as well as the second "hot" conductor in 230-V circuits, are omitted for simplicity.)

exhaust against prevailing winds. If structural or other factors make it necessary to install fans on the windward side, it is important to select fans rated to deliver the required capacity against 1/8 in. (1/3 cm) static pressure and with a motor of sufficient size to withstand a wind velocity of 30 mph (13.4 m/sec), equivalent to a static pressure of 0.435 in. (1.1 cm) of water, without overloading beyond the service factor of the motor. The recommended location for each temperature-control thermostat is at a point in the relative center of the area ventilated by the particular fan served. It should also be located as free as possible from potential physical damage. It should not be placed near an animal, a water pipe, a light, heater exhaust, or any other object which will affect its action. Because the temperature near the ceiling will be somewhat higher than near the floor, the proper setting of the thermostat should be made with reference to temperature indicated by a thermometer at the level occupied by the animals.

3.3.7 Wiring for electric ventilating fans — The National Fire Protection Association Standard No. 70, National Electrical Code, must be adhered to for all wiring and connections. A fan motor of 1/4 hp or less may be connected to a convenience outlet circuit (Figure 6A) or to a lighting circuit if connected ahead of the lighting switch (Figure 6B). A fan motor of 1/3 hp or larger should be on an individual circuit of No. 12 or larger wire (Figure 6C and D). A fan motor of 1/2 hp or larger should be on a 230-V circuit. Two-fan installations with fans located close to each other may be served by a three-wire 115- to 230-V circuit, with one fan connected to each side of the circuit, or may be connected to a 230-V circuit. Each fan must be equipped with a grounding conductor.

3.3.7.1 Multiple-fan installations with fans located near one another may be served by a two-wire or three-wire feeder, subdivided at the fan location into individual circuits (Figure 6E and F).

3.3.7.2 Each fan motor should have individual running overcurrent protection. If a fan motor is equipped with built-in overcurrent protection, it is recommended that this be the manual-reset type.

3.3.7.3 Many thermostats are not rated to carry the starting currents of motors over 1/4 or 1/3 hp. Relays may be required when motors of larger sizes are to be thermostatically controlled. If used, they should be protected from dust.

3.3.7.4 Each fan should be provided with a disconnecting device in sight of and within 50 ft (15 m) from the fan location. Any distance over 50 ft is considered out of sight. The disconnecting device may consist of a safety switch, toggle switch, manual-type motor starter, or circuit breaker and must be capable of being "locked out" while equipment is being serviced.

3.3.7.5 Unless other wiring methods are required by local ordinances or inspection authorities, wiring for fans in livestock shelters should be installed in type UF cable with grounding conductor. Some states do not permit metal-clad wiring systems in farm buildings which house livestock. Whenever possible, fans, controls and wiring connections should be installed in such a way that they can be quickly and easily detached for cleaning and repairing.

3.3.8 Emergency warning systems — The confinement of animals in high-density, windowless shelters in a mechanically controlled environment involves considerable financial risk in the event of power or equipment failure. The failure of ventilating equipment can result in serious impairment of animal health or mortality from suffocation. An adequate alarm system to indicate failure of the ventilation equipment is highly recommended for windowless structures. An automatic standby electric generator should also be considered. There are many types of alarm systems for detecting failure of the ventilation system. These range from inexpensive "power off" alarms to more extensive systems for sensing interruptions of air flow, temperature extremes, and certain gases. Automatic telephone dialing systems have demonstrated effectiveness as alarms and are relatively inexpensive for the protection provided. To insure that the selection and installation of an alarm system will be dependable, it is recommended that the assistance of the manufacturer or other qualified person be sought.

3.3.9 Ventilation for disease control — Filtered-air positive-pressure ventilation systems have been used for preventing airborne diseases in poultry houses and may be appropriate for other livestock confinement units. Microbes capable of causing economically important poultry diseases can be transported via the ventilating air from infected flocks or contaminated premises to susceptible flocks in conventional houses. However, the airborne route of disease transmission can be blocked by filtering the incoming air and maintaining a positive pressure inside the house. More data are needed, but reported evidence suggests the following.

3.3.9.1 Air filters having an efficiency of 95% (based on ASHRAE Standard 52-68, Atmospheric Test) are probably adequate.

3.3.9.2 Positive pressure of 0.25- to 0.30-in. water gauge (0.63 to 0.76 g/cm^2) inside the house (relative to outside static pressure) is probably adequate where the outside wind speed seldom exceeds 25 mph (40 km/hr).

3.3.9.3 To block other routes of disease transmission it is necessary to decontaminate the house before the chickens are put into it, and to prevent microbes from being brought in with the chickens, feed, and other supplies and by the caretaker.

3.4 Supplemental heating and cooling

3.4.1 Supplemental heat — Brooders or heat lamps are usually required for young chicks, ducklings, pigs, poults, and quail and may be beneficial for calves, lambs, and other young animals and birds that are housed apart from their dams in cold weather. Various types of heating equipment can be incorporated into ventilation systems if desired.

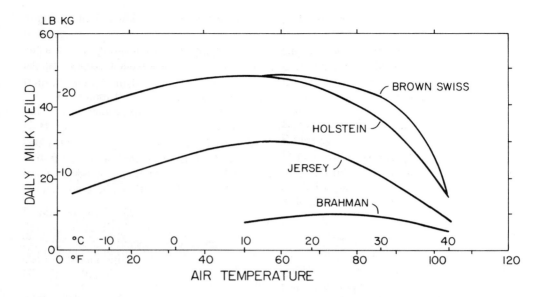

FIGURE 7.   Air temperature effects on milk yield of cows in a constant-temperature laboratory. Relative humidity about 50%.

3.4.2 Supplemental cooling — Cooling, although not as commonly used as supplemental heat, can be of economic benefit in hot climates for dairy cows, fattening hogs, and perhaps for chickens and sows. Under intensive housing conditions supplemental cooling may be necessary during heat waves to prevent heat prostration and mortality. The cooling equipment, like heating equipment, may be incorporated into the ventilation system if desired, and it can consist of supplemental air movement or evaporative cooling.

3.4.2.1 Air movement — Increased air movement during heat stress conditions can increase growth and decrease water consumption of chickens (Figure 38), improve heat tolerance of chickens, improve feed efficiency of beef cattle, and increase growth of swine.

3.4.2.2 Evaporative cooling — Cooling by evaporation can be applied directly to the animals, as with foggers for poultry or sprinklers for swine, or it can be used to cool the air, as with pad-and-fan packaged evaporative coolers for dairy cattle shelters or poultry houses.

SECTION 4 — VENTILATION FOR STALL-TYPE DAIRY BARNS

4.1 Environmental requirements for dairy cattle — Climate requirements for minimal economic loss are broad and include temperatures ranging from 35 to 75°F (2 to 24°C) with coincident relative humidities from 40 to 80%. Figure 7 illustrates air temperature effects on milk yield for various breeds. The temperature-humidity index (THI) provides a reasonable measure of the combined effects of humidity with air temperature above 70°F (21°C) such that:

$$\text{milk production decline} = 2.370 - 1.736\text{NL} + 0.02474(\text{NL})\,(\text{THI})$$

where decline = absolute decline in production, lb/day-cow; NL = normal level of production, lb/day; and THI = daily mean value, obtained from the dry-bulb temperature (t_d, °F) and relative humidity (RH,%) according to the relation THI = t_d − (0.55 − 0.55 RH) (t_d − 58). Below 35°F (2°C), production efficiency declines and management problems increase.

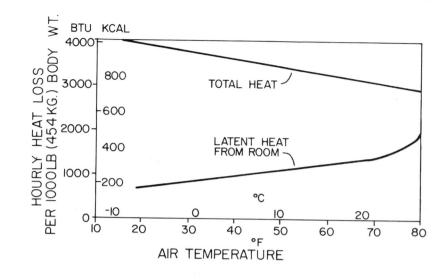

FIGURE 8. Barn heat and moisture dissipation rates with stanchioned dairy cattle on solid concrete where barn relative humidity was 55 to 70%. Total barn heat declined rapidly above 80°F (26.7°C).

4.2 Heat and moisture produced by dairy cattle — Figure 8 provides an estimate of heat and moisture produced by dairy cattle under conditions with varying ambient temperatures.

4.3 Winter ventilation requirements — Ventilation rates for maintaining heat balance or for removal of excess moisture can be determined using procedures described in paragraph 3.3.1.

4.3.1 Graphs for predicting ventilation performance — Figures 9 to 11 were developed to aid in predicting conditions that can be expected to exist in stall barns for dairy cattle in environments with various exposure factors. These graphs present relationships for inside air temperatures of 40, 50, and 60°F (5, 10, and 15°C). Use of the graphs is explained in paragraph 3.3.2.

4.4 Summer ventilation requirements — Ventilation rates for enclosed barns should be adequate to hold inside temperatures within 2 to 4°F (1 to 2°C) of outside temperatures during hot weather. An air change every 1 to 2 min is usually sufficient in insulated structures.

4.5 Aids for summer cooling — At temperatures above 70°F (21°C), cooling can increase productivity and feed efficiency (see paragraph 4.1). Expected milk production losses for June-September, inclusive, with only shades provided are shown in Figure 12 for cows of 50-lb/day (23-kg/day) normal production level.

4.5.1 Evaporative cooling — Adequately designed and maintained systems using wetted pads and fans have potential for economic application in large areas of the U.S. Figure 13 shows expected production benefits from evaporative cooling during June-September, inclusive, for cows with 50-lb/day (23-kg/day) production level. System design information is available in University of Arizona Report P-25.

4.5.2 Other water cooling methods — Specific heat and latent heat of vaporization of water can be utilized directly for increased animal comfort in hot weather by such methods as spraying with sprinklers or foggers and supplying cooled drinking water.

4.5.3 Cooling inhalation air — Outside air cooled by refrigeration to 60°F (15°C), supplied to head enclosures at the rate of 25 to 30 cfm/cow (700 to 850 ℓ/min), has been shown to benefit milk production for cows in hot environments.

4.5.4 Air conditioning of total space — Air conditioning of dairy cow housing can require 0.7 or more tons of refrigeration per cow, depending on local design conditions and the individual situation. Close attention must be given to air filtration, adequate ventilating air,

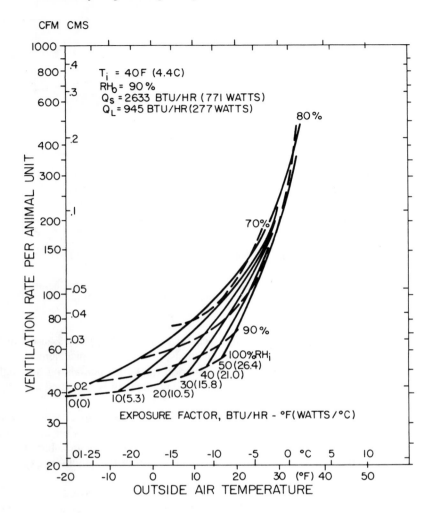

FIGURE 9.   Dairy stable ventilation performance, inside air temperature 40°F.

and to maintenance. Feasibility analyses indicate rather limited potential application to high-producing cows in hot humid areas of the U.S.

SECTION 5 — VENTILATION FOR BEEF CATTLE IN CONFINEMENT

5.1 Definition of the confinement growing of beef cattle — The housing of beef within a structure which is enclosed on all sides (with or without windows or window openings) and is ventilated, insulated, and possibly heated to allow some control of interior temperatures.

5.2 Environmental requirements for beef cattle

5.2.1 Temperature — Temperature influences the feed conversion and weight gain of beef cattle. Figures 14 and 15 show the type of response to temperature found under research conditions where temperature is controlled. Figure 16 illustrates the effects of temperature on feed requirements.

5.2.2 Relative humidity — Relative humidity has not been shown to influence animal performance except when accompanied by thermal stress. Relative humidities consistently below 40% may contribute to excessive dustiness and above 80% may increase building and equipment deterioration.

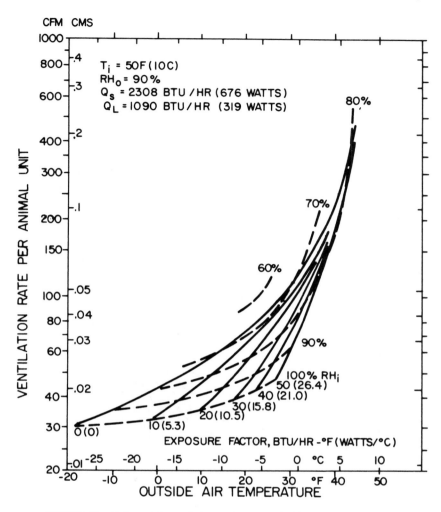

FIGURE 10.   Dairy stable ventilation performance, inside air temperature 50°F.

5.3 Heat and moisture produced by beef cattle — Figures 17 and 18 give total, latent, and sensible heat production values for Ayrshire bull calves ranging in age from 6 to 10 months as affected by temperature and relative humidity. Figure 19 provides total, latent, and sensible heat production from Shorthorn, Brahman, and Santa Gertrudis calves at temperatures of 50 and 80°F (10 to 27°C). These data represent values for the animals on full feed and include heat and moisture transfer effects between the animal environment and a bedded concrete floor. For design purposes the sum of sensible heat from the room and the latent heat from the room is the total heat loss from the beef cattle. However, the ratio of sensible heat to latent heat from the beef cattle may be quite different from the ratio of sensible heat from the room to latent heat from the room because of the utilization of some sensible animal heat to vaporize moisture from the floor. The data in Figure 19 can be used for normal management practices in solid floor buildings. Daily average temperature should be used in estimating heat and moisture production from Figures 17 to 19. Figure 20 illustrates the effect of ration on hourly heat production at environmental temperatures of 68, 86, and 104°F (20, 30, and 40°C).

5.4 Winter ventilation requirements

5.4.1 Design values — Inside air temperature — 45 to 65°F (7 to 18°C). Inside relative humidity — 40 to 80%

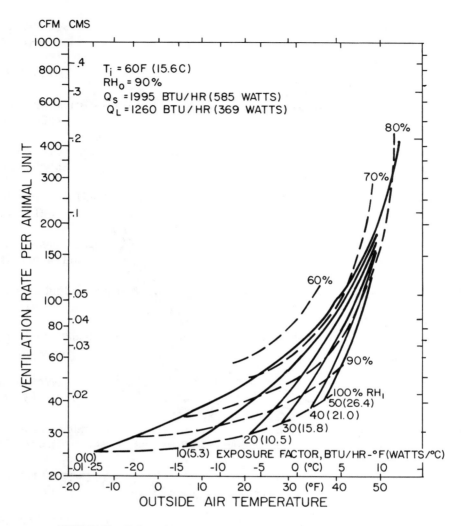

FIGURE 11.   Dairy stable ventilation performance, inside air temperature 60°F.

5.4.2 Procedure — For determining ventilation rates for heat balance and moisture removal, see paragraph 3.3.1.

5.4.3 Design graphs — The accompanying graphs (Figures 21 to 23) aid in predicting conditions that can be expected to exist in totally enclosed beef confinement units with various indoor and outdoor air temperature. Graphs pertain to beef buildings with solid floors. Use of these graphs is explained in paragraph 3.3.2.

5.5 Summer ventilation requirements

5.5.1 Air change rates — Air change rates are adjusted to remove heat produced by the cattle, plus other heat gains, and to minimize inside temperature rise above the outside air temperature. Recommendations are for an air flow rate equivalent to one air change per minute. Consideration must also be given to air distribution and velocity to aid in animal comfort and feed efficiency. Baffles should be employed to direct air flow at animal level. This may be accomplished by baffled center ceiling inlets or wall baffles.

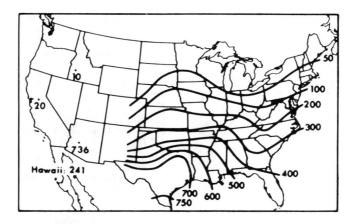

FIGURE 12.   Expected milk production losses (lb/cow) for 122-day summer period for cows of 50-lb/day, (23-kg/day) production level.

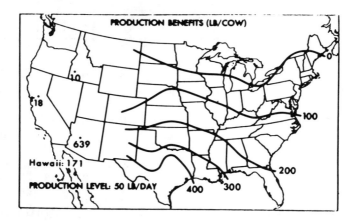

FIGURE 13.   Expected production benefits from evaporative cooling during 122-day summer period for cows of 50-lb/day, (23-kg/day) production level.

SECTION 6 — VENTILATION FOR SWINE IN CONFINEMENT

6.1 Definition of the confinement growing of swine — The housing of swine within a structure which is enclosed on all sides (with or without windows or window openings) and is ventilated, insulated, and possibly heated or cooled to allow some control of interior temperatures.

6.2 Environmental requirements for swine — A comprehensive review of the literature on the effects of environment on swine has been written. Physical environment, social environment, and reported methods for controlling the environment are included. The following physical environmental factors should be considered when designing a ventilation system.

6.2.1 Temperature — Temperature changes within the range of 50 to 75°F (10 to 24°C) do not appear to significantly affect either average daily gain or feed efficiency for "on farm" production practices. At higher and lower temperatures, average daily gain decreases and the amount of feed required to produce a pound of gain increases. Figures 24 and 25 show the response to temperature found under research conditions.

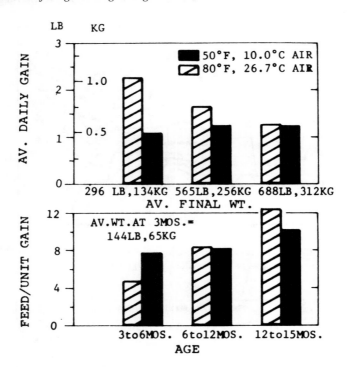

FIGURE 14.   Rate of growth and feed utilization (unit feed per unit gain) of Shorthorns.

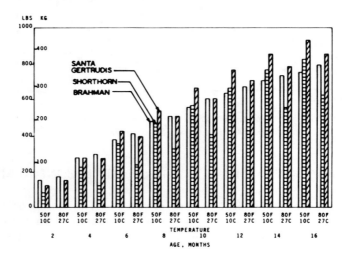

FIGURE 15.   Effect of age and temperature on weight of Brahman, Shorthorn, and Santa Gertrudis calves.

6.2.2 Relative humidity — Relative humidity has not been shown to influence animal performance except when accompanied by thermal stress. Relative humidities consistently below 50% may contribute to excessive dustiness and above 80% may increase building and equipment deterioration.

6.2.3 Light — Studies have shown that light has no effect on feed efficiency, rate of gain, or carcass quality.

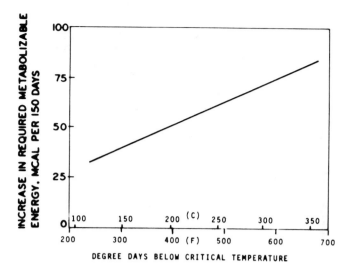

FIGURE 16. Effect of degree-days below critical temperature (6°F) (−14.4°C) on metabolizable energy requirements of a 900-lb (400-kg) beef steer.

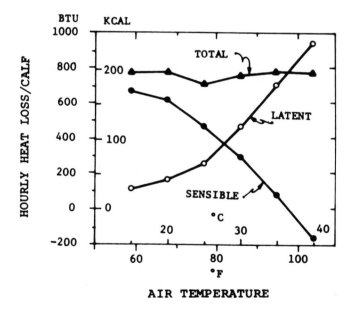

FIGURE 17. Effect of temperature on heat losses of three Ayrshire bull calves 6 to 12 months of age. Vapor pressure 8 mm Hg; dewpoint 46°F (8°C).

6.2.4 Air motion — Most reported values of heat and moisture losses from swine are based on air movements over the animals of 50 fpm (0.254 m/sec) or less. If air motion is greater than this, at least up to 300 fpm (1.53 m/sec), the total animal heat loss will increase, thus increasing the total heat load on the building. The latent heat loss of the animal will increase, causing an increase in moisture production, necessitating a higher ventilation rate than published values. The sensible heat load can either increase or decrease, depending upon the amount of increase of total animal heat. Primarily, one should be aware that varying

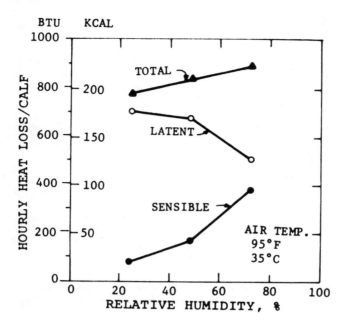

FIGURE 18.   Effect of humidity on heat losses of Ayrshire bull calves 6 to 10 months of age. Air temperature 45°F (7°C).

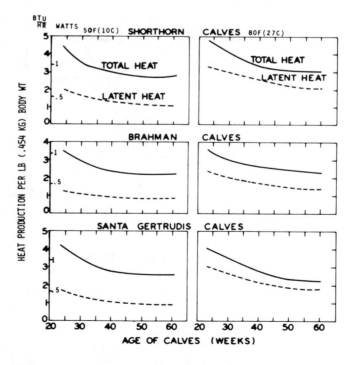

FIGURE 19.   Stable heat and moisture dissipation of beef calves at temperatures of 50°F (10°C) and 80°F (27°C). Latent heat data include vaporization from the animals and the bedding.

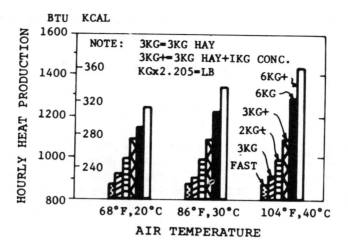

FIGURE 20. Effect of ration on heat production of a grade steer 20 to 24 months of age. Hay was poor quality. Concentrate was cotton cake and barley meal. Fasting was for 72 hr. + indicates addition of 1 kg concentrate to ration.

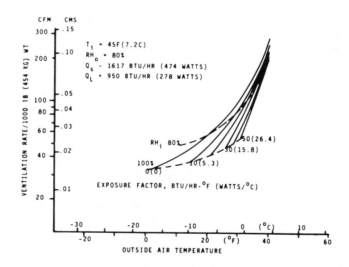

FIGURE 21. Beef building ventilation performance, inside air temperature 45°F (7.2°C).

air velocities within a building can cause errors in the use of published swine heat and moisture loss data or be the cause of unsatisfactory conditions of supposedly well-designed buildings.

6.2.5 Air quality — Some air pollutants (carbon dioxide, methane, sulfur dioxide, ammonia, hydrogen sulfide, amines, amides, mercaptans, sulfides, disulfides, and dust) have been identified in confinement swine buildings. Agitation of the waste storage area below a slotted floor can bring hydrogen sulfide levels to toxic concentrations for both humans and pigs. These hydrogen sulfide concentrations have reached levels in excess of 800 ppm. Studies have shown that animals exposed continuously to levels of about 20 ppm of hydrogen sulfide develop fear of light, loss of appetite, and nervousness. Symptoms at levels between 50 and 200 ppm have included vomiting, nausea, and diarrhea. However, winter ventilation rates designed to remove moisture from the building have been found to maintain the

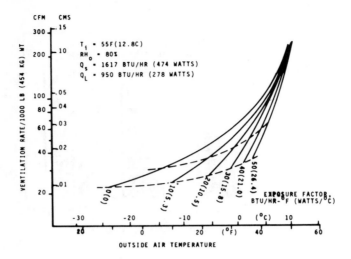

FIGURE 22.   Beef building ventilation performance. Inside air temperature 55°F (12.4°C).

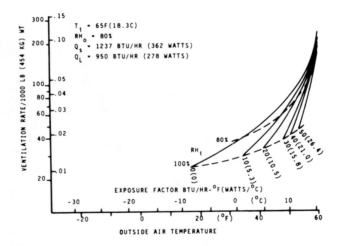

FIGURE 23.   Beef building ventilation performance. Inside air temperature 65°F (−18.3°C).

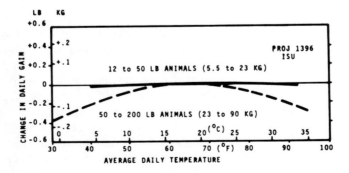

FIGURE 24.   Deviation of average daily gain of swine with temperature.

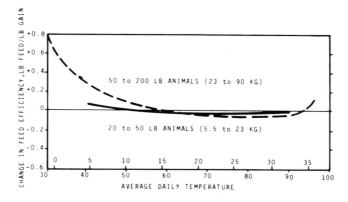

FIGURE 25.   Deviation of pounds feed required per pound gain for swine.

Table 2
TOTAL HEAT
PRODUCTION OF SMALL
PIGS AT 86°F (30°C)

Weight		Heat production	
lb	**kg**	**Btu**	**W**
10	4.5	68	20
15	6.8	120	35
20	9.1	171	50
25	11.4	221	65

hydrogen sulfide level well below 20 ppm and closer to 1 ppm. A tight structure is especially dependent on the reliability of the ventilating system to remove toxic gases. This is particularly true where extensive wall openings are not provided for use in warm weather. Occasional cases of animal deaths, blamed on ventilation system failure for several hours, have been recorded. To protect against such hazards, reliable equipment design, system failure alarms, and emergency ventilation procedures should be considered.

6.3 Heat and moisture produced by swine — Table 2 provides an estimate of total heat loss from small pigs. Figure 26 shows the heat loss per pound, at various air temperatures, for newborn pigs. Figures 27 to 29 give sensible and latent heat production in hog buildings. These data were collected in a chamber having an unbedded solid concrete floor scraped clean twice daily. For design purposes, the sum of sensible heat from the room and the latent heat from the room is the total heat loss from the hogs. However, the ratio of sensible heat to latent heat from the hogs may be quite different from the ratio of sensible heat from the room to latent heat from the room because of the utilization of some sensible animal heat to vaporize moisture from the floor. The data in Figures 27 to 29 can be used for normal management practices in solid floor buildings. The latent heat from buildings may be increased as much as one third (with a corresponding decrease in sensible heat) by such factors as floor flushings, water wastage, temperature, ventilation rate, inside relative humidity, air velocity over the floor surface, and less frequent scraping of floors. Measurements indicate that the total moisture removed in winter by the ventilating system from a solid concrete-floored finishing house correlated closely with the amount predicted by the regression equation developed from calorimeter measurements and shown in Figures 27 and 28. These tests also indicated about three fourths as much moisture was removed from a partially

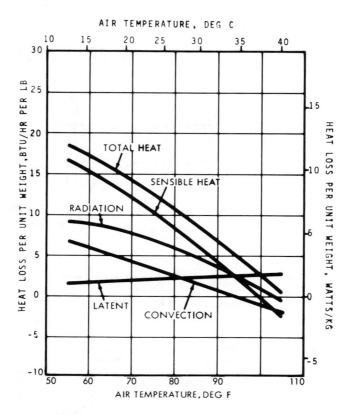

FIGURE 26.   Heat loss per unit weight of newborn pigs vs. air temperature.

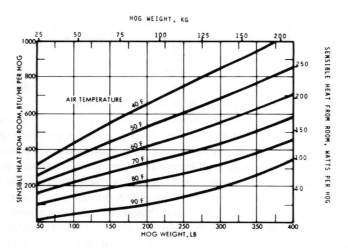

FIGURE 27.   Room sensible heat in a hog house. (Solid concrete floor, scraped daily, no bedding used.)

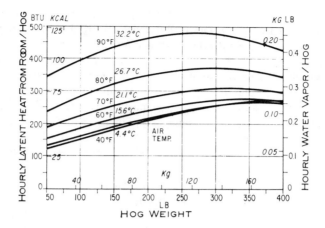

FIGURE 28.   Room latent heat in a hog house. (Solid concrete floor, scraped daily, no bedding used.)

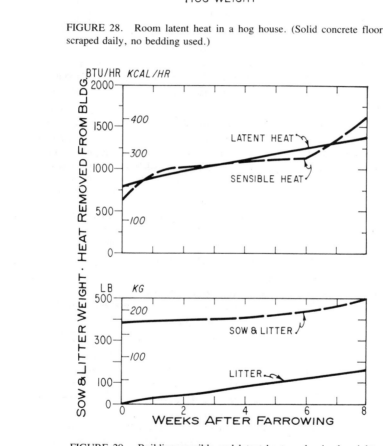

FIGURE 29.   Building sensible and latent heat, and animal weight for sows and litters. (Solid concrete floor, scraped daily, no bedding used; average of observations for three sows with litters.)

slotted floor (approximately 35% slotted) structure, and only one half as much moisture was removed from a totally slotted floor structure under similar conditions. Daily average temperature should be used in estimating heat and moisture production from Figures 27 to 29.

6.4 Winter ventilation requirements

6.4.1 Calculation procedures — For determining ventilation rates for heat balance and moisture removal, see paragraph 3.3.1.

6.4.2 Design graphs — The accompanying graphs (Figures 30 to 33) aid in predicting conditions that can be expected to exist in totally enclosed solid floor swine production units of various exposure factors, varying indoor and outdoor air temperatures, and for pigs ranging in size from 50 to 200 lb (23 to 91 kg). Use of the graphs is explained in paragraph 3.3.2.

6.5 Summer ventilation requirements — Air change rates are adjusted to remove the heat produced by the hogs plus other heat gains to minimize inside air temperature rise above the outside air temperature. Consideration must be given to air distribution and velocity to help in animal comfort, sanitary training of animals, and feed efficiency. When air temperature is higher than animal body temperature, about 103°F (39°C), increased air velocity over the body will increase the heat passing from the environment to the hog.

SECTION 7 — VENTILATION FOR BROILER CHICKENS AND YOUNG TURKEYS

7.1 Definition for broiler housing — Structures used for brooding and/or growing of broiler chickens or young turkeys. These structures are usually fully enclosed (with or without windows) and partially or fully insulated. They permit some control of interior environment by operation of ventilation systems that may or may not include supplemental heating and cooling equipment. Despite this definition, many broilers are reared in curtain-sided houses with no mechanical ventilation and little or no insulation.

7.2 Environmental requirements for broilers — Requirements presented herein cover the brooding period (up to 4 weeks of age) and the growing period (over 4 weeks of age). They reflect the current knowledge of how the immediate environment affects the economic efficiency of broiler meat production, hereafter referred to as performance. Most of the data in this section apply to birds reared on litter floors and may need to be adjusted for birds reared in cages or coops. Where data are given for only one sex, it should be noted that by 8 weeks of age the average weight of male broiler chickens is 1.27 times that of females and that environmental conditions tend to exert somewhat greater influence on growth rates of males than of females.

7.2.1 Temperature — Recommended dry-bulb temperature range in uniformly heated surroundings is 80 to 90°F (29 to 32°C) for the first 3 to 5 days after hatching, then decreases by 1 to 2°F (0.56 to 1.1°C) per day during the brooding period to 65 to 75°F (18 to 24°C) during the growout period. Chicks can be reared in much colder surroundings if supplemental radiant heat is available. The influence of various constant temperatures during growout on final weight of male broilers is indicated in Figures 34 and 35. Feed efficiencies (gain/feed) during growout increase almost linearly (1% per °F) with increasing ambient temperature between 45 and 75°F (7 and 24°C). Growth rates decrease at ambient temperatures above 75°F (24°C) and probably below 50°F (10°C). Dry-bulb temperatures above 96 and 105°F (36 to 41°C) may cause heat prostration and death, especially to unacclimatized chickens. Survival time and exact temperature for lethality depend upon body weight, genetics, nutritional state, and degree of acclimation, as well as upon humidity, air movement, radiant heat, availability of drinking water, and degree of crowding. The influence of diurnally cycled temperatures, as compared to constant temperatures, upon 8-week weight and feed conversion of male broilers in three different experiments is indicated in Figure 36. In moderate to hot environments during growout, broiler growth tends to be stimulated by low-amplitude [≤ 20°F (11.1°C) peak-to-peak] diurnal temperature cycling if the peak of the cycle is ≤80°F (27°C) and the mean is ≤ the constant temperature with which the cycle is compared. The influence of various constant temperatures upon growth and feed conversion of young turkeys is indicated in Figure 37.

7.2.2 Humidity — During the first 2 to 3 weeks of brooding, chicks (especially those from smaller eggs) may require relative humidities of 60% or higher, but during growout

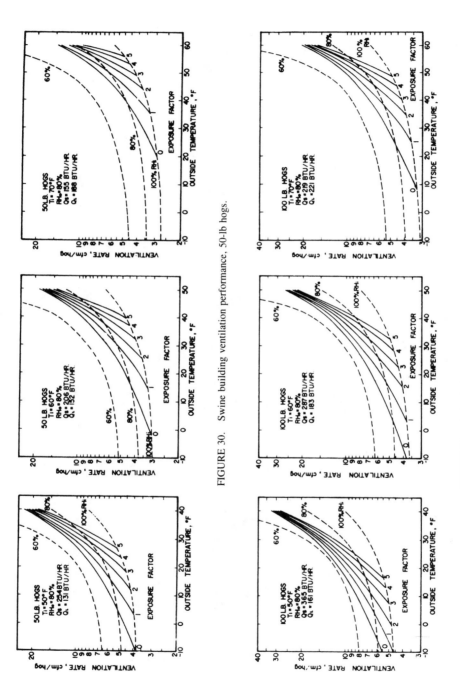

FIGURE 30. Swine building ventilation performance, 50-lb hogs.

FIGURE 31. Swine building ventilation performance, 100-lb hogs.

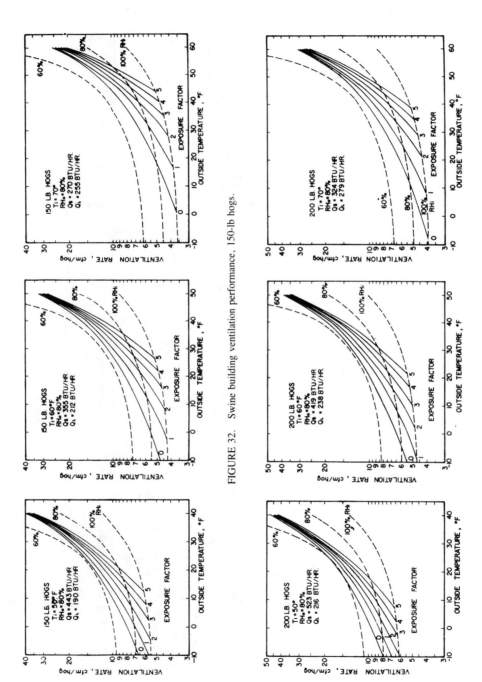

FIGURE 32.   Swine building ventilation performance, 150-lb hogs.

FIGURE 33.   Swine building ventilation performance, 200-lb hogs.

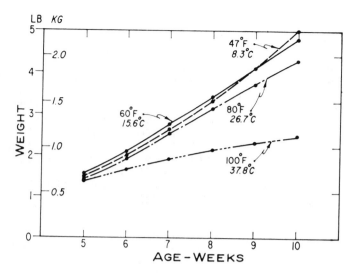

FIGURE 34.   Effect of air temperature on body weight of male broilers. Relative humidity was 60% at all temperatures except for 80% at 47°F (8.3°C).

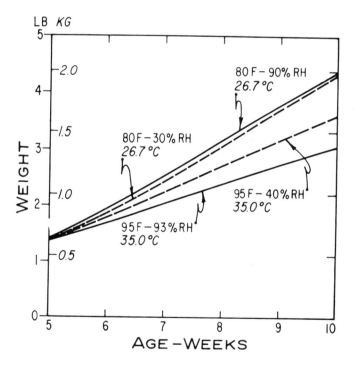

FIGURE 35.   Effect of temperature and humidity on male broiler weight (Winn and Godfrey, 1967). Constant low (30 to 40%) or high (80 to 90%) humidity has little effect on performance if air temperature is below 85°F (29°C).

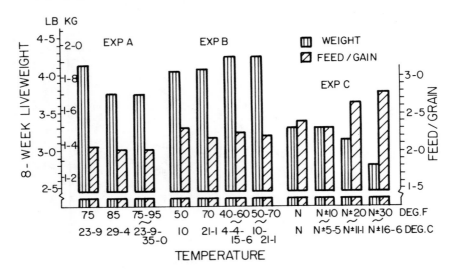

FIGURE 36.   Constant and cyclic temperature effects on growth and feed conversion of male broiler chickens. In Experiment A and Experiment B, commercial chicks were reared on litter floors at normal brooding temperatures for 3 weeks, then held at the indicated constant temperature or 24-hr linear cyclic temperature for 5 weeks. In Experiment C, noncommercial chicks were reared on grid floors for 8 weeks at a normal (N) temperature schedule which started at 90°F (32°C) and decreased 5°F (3.1°C) per week to final temperature of 70°F (21°C) or at one of three 24-hr sinusoidally cyclic temperature schedules, e.g., in the N ± 10 schedule the temperature was varied daily from 10°F above to 10°F below the N temperature for that week.

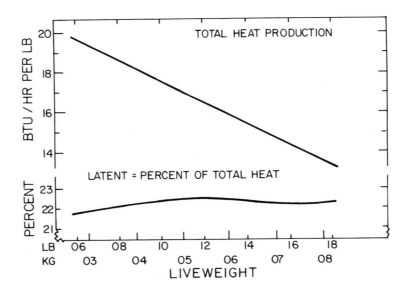

FIGURE 37.   Effect of constant air temperature on weight gain and feed efficiency conversion of Broad-Breasted White and Broad-Breasted Bronze turkeys between 12 and 24 weeks of age. Relative humidity about 50% with 16-hr daylength.

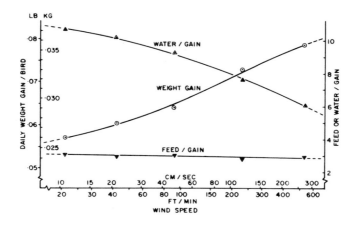

FIGURE 38.   Wind speed effects on growth, feed conversion, and water/
gain of 7-week-old male broilers held for 2 weeks in a diurnally cycling
hot, humid environment. Wind speeds were constant, dry bulb temperature
was sinusoidally cycled 83 ± 13°F (28.3 ± 7.2°C) once every 24 hr, and
dewpoint temperature was constant at 65 ± 4°F (18.3 ± 2.2°C).

the performance of commercial broilers is only slightly affected by relative humidities
between 30 and 80% when the dry-bulb temperature is below 85°F (29°C) (Figure 35). At
8 weeks of age broilers are severely heat stressed by wet-bulb temperatures above 81.5°F
(27.5°C) when dry-bulb temperatures are above 85°F (29°C). At stocking densities of 0.75
to 1.0 ft^2 (695 to 930 cm^2) per broiler on litter floors, the litter becomes too wet (slick,
caked) when the average relative humidity remains above 80% and too dry (dusty) when
the relative humidity remains below 40%. At 0.5 ft^2 (465 cm^2) per broiler it is difficult to
maintain satisfactory litter condition regardless of the humidity.

7.2.3 Light — In windowless broiler houses with continuous "white" incandescent
lighting, an average intensity of 0.1 fc (1.1 m^2) is adequate for maximum broiler growth,
but approximately 0.5 fc (5.4 lm/m^2) is needed by the caretaker. Periodic lighting (e.g.,
lights on for 15 min every 2 hr) and various combinations of natural and artificial lighting
may give as good broiler performance as constant-level lighting, but authorities do not agree
on this. Optimum light for catching broilers for market is approximately 0.0004 fc (0.0043
lm/m^2) in the 450- to 500-nm wavelength range, with the catchers dark-adapted and the
chickens light-adapted.

7.2.4 Air movement — Wind speed greater than 60 fpm (18 m/min) at chicken height is
not recommended for chicks under 2 weeks of age (Payne, 1961), and wind speeds ≤200
fpm (61 m/min) may decrease growth of broilers between 3 to 8 weeks of age at temperatures
in the range of 36 to 37°F (2 to 19°C) (Wilson et al., 1957). Conversely, increased wind
speeds ≤500 fpm (152 m/min) around the chickens have been shown to improve the weight
gains and water-use efficiencies of 3-lb (1.4-kg) broilers when the temperature was cycled
diurnally from 70 to 96°C (21 to 36°C) (Figure 38). Similar increases in wind speed alleviated
heat stress in 8-week-old broilers that were exposed to environmental temperatures ≤105°C
(40.6°C) but tended to exacerbate the stress at higher temperatures.

7.2.5 Radiant heat — As stated in paragraph 7.2.1, supplemental radiant heat is beneficial
to chicks in cool surroundings, but in hot, humid weather the radiant heat from an uninsulated
roof may cause mortality in 8-week-old broilers. Surface temperatures in well-insulated
houses are little higher, if any, than air temperatures, so these surfaces act as radiant heat
sinks, not sources.

7.3 Heat and moisture production

7.3.1 Broiler chickens — Total and latent heat production by birds of various weights

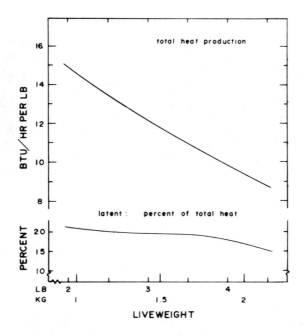

FIGURE 39.   Specific total and specific latent heat production
of growing broilers at 77 ± 2°F (25°C) and near 70% relative
humidity.

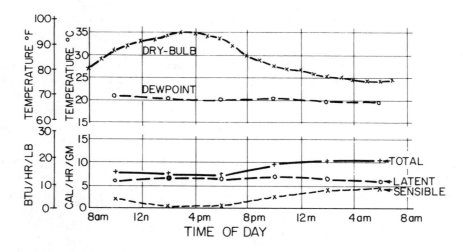

FIGURE 40.   Specific total and specific latent heat production of young broiler chicks at 84
± 2°F (29°C) and near 70% relative humidity.

and at various ambient temperatures are presented in Figures 39 to 43. Heat and moisture
absorption or release by litter are included in Figures 42 and 43, but not in Figures 39 to
41. Therefore, the respired moisture can be calculated from Figures 39 to 41 by dividing
the latent heat production by the heat of vaporization of water at the chicken's body temperature (approximately 1033 Btu/lb of water, 574 cal/g).

   7.3.2 Turkeys — Total and latent heat production by turkeys is presented in Table 3.

   7.4 Cold weather ventilation requirements

   7.4.1 Brooding period — A ventilation rate of 0.1 cfm per chick from 1 to 4 weeks of
age is adequate to remove moisture and supply the necessary fresh air when outside daily

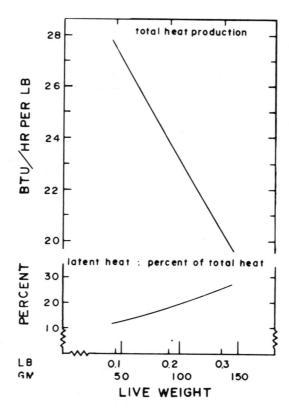

FIGURE 41.   Specific total and specific latent heat production of growing broilers at $67 \pm 2°F$ (19.5°C) and near 70% relative humidity.

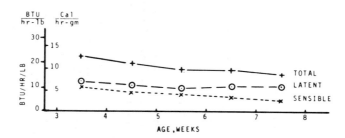

FIGURE 42.   Specific total heat and specific moisture removal from a litter floor, insulated, windowless broiler house operated at an average temperature of 82 and 87°F (27.7 and 30.5°C) at the beginning and end of the test, respectively. Heat loss through ceiling, floor, and walls was negligible.

mean dewpoint temperatures do not exceed 38°F (3.3°C). For higher dewpoints and for body weights greater than 1.35 lb (0.61 kg), use Equation 3.3.1-2 to determine ventilation rate.

   7.4.2 Growing period — Equations in paragraph 3.3.1 are satisfactory for determining the rate of ventilation for heat balance and moisture balance, respectively. Refer to Figures 39 to 42 and other sources that might be available for values of Q_s and Q_ℓ.

7.5 Summer ventilation requirements

   7.5.1 Prior to 1975, a summer design ventilation rate of 1.0 cfm/lb (62 ℓ/min/kg) of

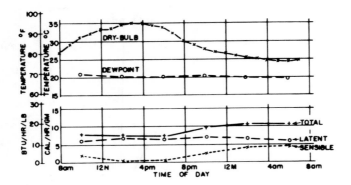

FIGURE 43.   Effects of diurnal temperature variation on heat removal from a litter-floor broiler house. The broilers were 6.5 weeks of age and averaged 2.5 lb per bird. Ventilation rate was 2.5 cfm per bird. The date was July 27, 1968.

Table 3
SPECIFIC TOTAL HEAT AND SPECIFIC MOISTURE PRODUCTION OF TURKEYS

						Heat production						
						Day				Night			
	Temp.			Liveweight		Total heat		Latent		Total		Latent	
Sex	°F	°C	Percent RH	lb	kg	Btu hr-lb	W kg	Btu hr-lb	W kg	Btu hr-lb	W kg	Btu hr-lb	W kg
Wrolstad White Turkey													
M	70	21	42	5.8	2.6	12.3	7.9			8.6	5.6		
F	70	21	42	6.8	3.1	12.0	7.7			9.1	5.9		
M	70	21	42	8.3	3.8	12.1	7.8			8.7	5.6		
Beltsville White Turkey													
M	65	18	76	19.7	8.9	5.9	3.8	1.9	1.2	4.9	3.2	1.5	1.0
F	64	18	80-90	9.6	4.4	6.1	3.9	1.8	1.2	4.0	2.6	1.0	0.6
M	77	25	52	19.7	8.9	5.1	3.3	2.3	1.5	4.3	2.8	1.8	1.2
F	79	26	63	9.6	4.4	6.5	4.2	4.6	3.0	4.3	2.8	3.2	2.1

broiler weight was suggested in Agricultural Engineers Yearbook, without reference to climatic region. The conclusion that 1.0 cfm/lb is adequate even in the mid-South has been reported, but others conclude that a design value of 1.2 cfm/lb (74 ℓ/min/kg) is more appropriate for the hot, humid South Atlantic and the South Central regions of the U.S. This higher ventilation rate is based on Equation 3.3.1-2 (previously recommended for cold weather ventilation calculations) and the following assumptions.

   7.5.1.1 Ventilation rate should be adequate to remove latent heat as fast as it is produced in the house.

   7.5.1.2 With adequate ventilation the inside dry-bulb temperature will not rise more than 0.9°F (0.5°C) above the maximum outside dry bulb temperatures; sensible heat removed from the house decreases to nearly zero during hot afternoons (Figure 43).

   7.5.1.3 Inside wet-bulb temperature should be kept below 81.5°F (27.5°C) to prevent severe heat stress.

   7.5.1.4 Latent heat produced during hot afternoons by 8-week-old broilers on litter floors is 10.6 Btu/hr/lb (5.92 cal/hr/g) body weight.

7.5.1.5 The design maximum concurrent dry-bulb and dewpoint temperatures are, respectively, 99 and 72°F (37.2 and 22.2°C).

7.5.2 The condition stated in paragraph 7.5.1.5 was officially recorded at Meridian, Miss. on July 15, 1971. This dry-bulb temperature is 2°F (1.1°C) higher than the 1% probability design value for Meridian (ASHRAE Guide, 1972), and the dewpoint temperature is equal to the 5% probability design value. Higher values of both dry-bulb and dewpoint temperatures do occur in broiler rearing regions of the U.S., but probabilities that high dry-bulb and high dewpoint temperatures will be concurrent were not found in the literature. Rough approximations to these probabilities can be estimated from U.S. Weather Bureau Series 30 and Series 82 publications.

7.5.3 If a lower outside dry-bulb temperature or a higher outside dewpoint temperature than stated in paragraph 7.5.1.5 and no other assumptions are changed, the ventilation rate calculated by Equation 3.3.1-2 will be even greater than 1.2 cfm/lb (74 ℓ/min/kg) of broiler. Note also (paragraph 7.6) that 1.2 cfm/lb is intermediate between the minimum and maximum air exchange rates recommended for evaporative cooling.

7.5.4 When outside conditions permit faster moisture removal for a given air change rate, rates should be reduced to conserve energy and avoid dusty litter.

7.6 Aids for summer cooling — During heat waves, when outside dewpoint temperatures of about 70°F (21°C) may be accompanied for a few hours by dry-bulb temperatures of 104°F (40°C) or higher, a high ventilation rate alone may not prevent heat prostration of 8-week-old broilers, and some type of supplemental cooling will be needed. Fog nozzles spraying water periodically directly on the chickens provide on economical means of preventing mortality due to heat. The routine use of supplemental cooling (see paragraph 3.4.2) to relieve the milder heat stress that occurs on many summer days should also be considered. Evaporative coolers, if used, should be sized to provide $^3/_4$ to 1 air change per minute in the house or 4.2 to 5.0 cfm (119 to 141 ℓ/min) per 4-lb (1.8-kg) broiler, whichever is larger, with air velocity through the cooling pad 200 fpm (61 m/min) for packaged coolers or 150 fpm (46 m/min) for pad-and-fan coolers. Air from the cooler should be directed toward the chickens.

7.7 Alarm system and/or standby electric generator -- Market-size broilers will become severly heat stressed within 50 min after ventilation failure on a warm summer day and probably within 30 min or less when it is very hot outside (see paragraph 3.3.8 for discussion of alarms).

SECTION 8 — VENTILATION FOR LAYING HENS

8.1 Definition of the confinement housing of laying hens — The housing of laying hens within a structure which is enclosed on all sides (with or without windows) and is ventilated and/or insulated.

8.2 Environmental requirements of laying hens — Maximum egg production usually occurs at or near 55°F (13°C). Temperature influences feed and water consumption as shown in Table 4.

8.3 Heat and moisture produced by laying hens — The total heat produced by laying hens at various liveweights and temperatures is shown in Figure 44. Moisture production by White Leghorn laying hens is presented in Table 5.

8.4 Graphs for predicting the required winter ventilation rate — The accompanying graphs (Figures 45-47) were developed to aid in estimating the ventilation rates required in laying houses having exposure factors from 0.00 to 0.50. These graphs are based on the total sensible heat input (Q_s) and water introduced into the house atmosphere (W_r) on a per bird unit. Inside relative humidity (RH_i) and outside relative humidity (RH_o) are as indicated. The use of these graphs is explained in paragraph 3.3.2.

Table 4

CONSTANTS FOR DETERMINING WATER CONSUMPTION AND FECAL AND WATER ELIMINATION OF LAYING HENS IN RELATION TO FEED CONSUMPTION

	Ambient temperature			
	20—40°F (−6.7—4.4°C)	**50—60°F (10.0—15.6°C)**	**60—80°F (15.6—26.7°C)**	**80—100°F (26.7—37.8°C)**
Water to feed ratio	1.5—1.7	1.7—2.0	2.0—2.5	2.5—5.0
Water-plus-feed to feces ratio	1.7	1.7[a]	1.8[a]	1.9[a]
Percent water content of feces	75	75	77	80
Percent water content of eggs	65	65	65	65
Egg size, oz/doz (g/doz)	24 (680)	24 (680)	24 (680)	24 (680)
Percent free, hygroscopic, and metabolizable water in feed	54	54	54	54
Approx. heat in respired moisture, Btu/lb (cal/g)	1100 (611)	1100 (611)	1100 (611)	1100 (611)
Ratio of respired water to water input				
S.C. White Leghorn hens	0.30—0.33	0.33—0.40	0.40—0.45	0.45—0.55
Rhode Island Red hens	0.22—0.35	0.35	0.35—0.42	0.42—0.55
New Hampshire and Cornish hens	0.25	0.25—0.35	—	—

[a]   For S.C. White Leghorns add 0.03 to these values.

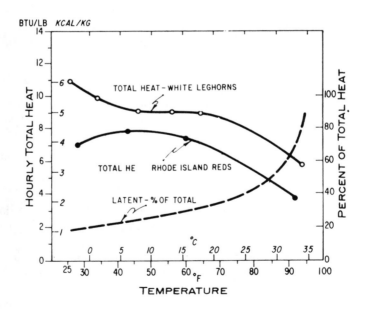

FIGURE 44.   Effect of air temperature on total and latent heat losses of Rhode Island red and white Leghorn laying hens.

Table 5
AVERAGE HOURLY MOISTURE PRODUCTION OF 1000
4-lb (1.8-kg) WHITE LEGHORN LAYING CHICKENS AT
VARIOUS TEMPERATURES

Temperature		Respired		Defecated		Total[a]	
°F	°C	lb	kg	lb	kg	lb	kg
25	−3.9	6.3	2.9	14.5	6.6	22.8	10.4
35	1.7	8.3	3.8	14.5	6.6	24.8	11.3
45	7.2	8.4	3.8	12.9	5.9	23.7	10.8
60	15.6	11.4	5.2	12.7	5.8	26.4	12.0
80	26.7	14.3	6.5	14.4	6.4	31.6	14.3
95	35.0	20.0	9.1	10.3	4.7	33.7	15.3

[a]   Includes drinking water wasted by hens estimated at 10% of water consumed from
25 to 80°F (−3.9 to 36.7°C) and 15% at 95°F (35.0°C).

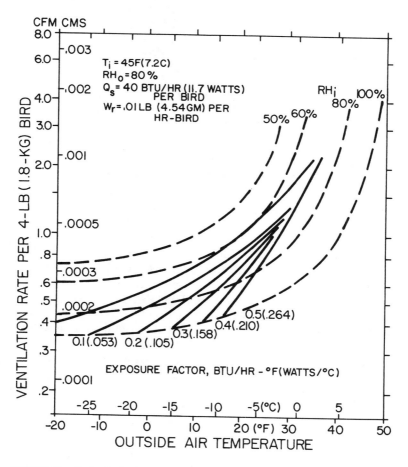

FIGURE 45.   Poultry laying house ventilation performance prediction graph, 45°F (7.2°C).

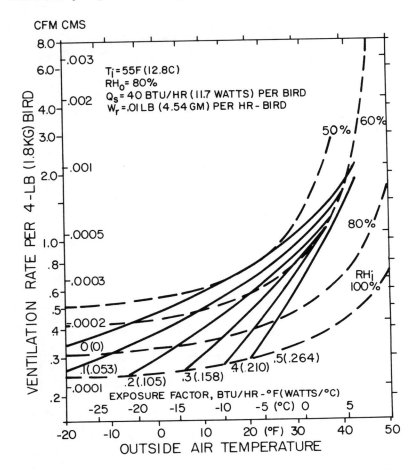

FIGURE 46.   Poultry laying house ventilation performance prediction graph, 55°F (12.8°C).

8.5 Summer ventilation rates — Summer ventilation rates can be calculated using Equations 3.3.1-1 and 3.3.1-2 with values of Q_s and Q_ℓ from Figure 44.

SECTION 9 — EFFECT OF ENVIRONMENT ON SHEEP

9.1 Sheep housing — Sheep are not customarily confined in structures enclosed on all sides, so no data are available on ventilation requirements.

9.2 Environment influences — Effect of environment and fleece length on heat losses from Half-Bred X Dover Cross adult wethers is shown in Figure 48. Data on the effects of wind and fleece length on heat production of adult sheep have been used to generate the equation:

$$Q = (11.47e^{-0.0302F}) V + 463F^{-0.5068}$$

where Q = hourly total heat production, kcal/hr; F = fleece length, mm; V = air velocity, m/hr. Additional work on the effects of fleece on partitioning of heat losses has been done and further information on vapor losses from sheep is available.

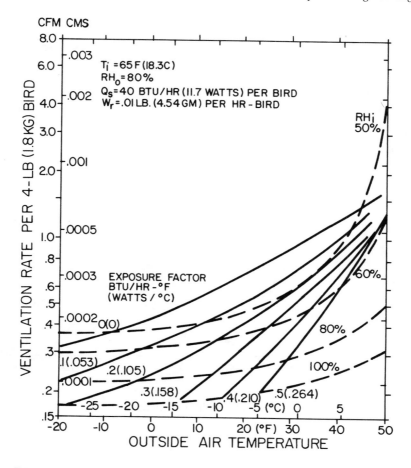

FIGURE 47.   Poultry laying house ventilation performance prediction graph, 65°F (18.3°C).

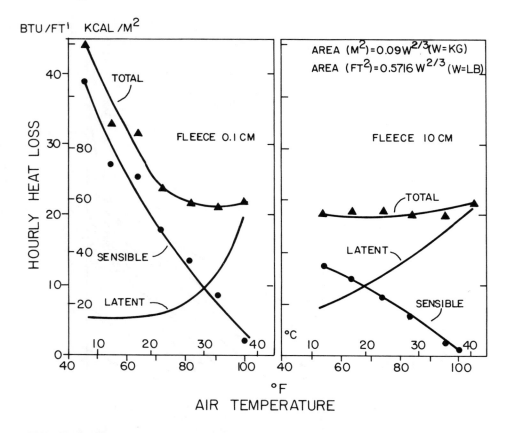

FIGURE 48.   Effect of environment and fleece length on heat losses from sheep (Half Bred × Dover-Cross Adult Wethers). Relative humidity 45 to 54%. Air flow 4 ft³/min (1.9 ℓ/sec).

REFERENCES

Last printed in 1977 Agricultural Engineers Yearbook; list available from ASAE Headquarters.

CITED STANDARDS

ASHRAE 52-68, Atmospheric Test.
NFPA No. 70, National Electrical Code.

Structural Designs, Requirements, and Systems

WIND, SNOW, LIVE, AND DEAD LOADS: DATA AND CALCULATIONS

Joseph W. Simons

INTRODUCTION

Standards governing the design of buildings are essential to provide adequate resistance to weather and safety for occupants under various load conditions. Federal, state, and local codes are upgraded periodically to meet stricter standards, based on the latest research findings.

Snow and wind loads are important factors in the design of buildings. Various engineering societies, associations, and governmental agencies, as well as individuals, have researched and reported on these factors over the years. The American Society of Agricultural Engineers (ASAE), through its Farm Building Construction Standards Committee, has developed a standard for wind and snow loads to be considered in the design of buildings. This is based on authoritative research data and records and is published in the Agricultural Engineers Yearbook of ASAE.* It is one of the most complete and foremost references on the subject. Extensive use of this standard is included in the following discussion of snow and wind loads.

TERMS

Snow pack on the ground is the weight of ice, snow, and water accumulated in an undrifted condition on flat, exposed ground surfaces. Extreme-mile wind is a maximum velocity determined from the least time required for 1 mi of air to pass a fixed point.

LOADS

Dead loads are the weights of all materials used in the building construction such as floor, roof, framing, and covering. Snow loads are vertical loads applied to the horizontal projection of the building roof. Wind loads are those caused by wind blowing from any horizontal direction. Live loads are any loads not constant in their application that a structure carries in addition to its own weight. Examples include people, livestock, vehicles, and products stored in the structure. Also included are wind and snow loads.

In the interpretation of snow and wind data, the mean recurrence interval is the number of years during which on the average an event is expected to occur only once, and it is the reciprocal of the probability of the event occurring during any one year.

Figures 1 and 2 give the snow packs and wind velocities likely to be exceeded only once during the indicated mean recurrence intervals. The maximum snow pack or wind velocity and when it will occur are not predicted.

BUILDING CLASSES AND DESIGN INTERVALS

Buildings are divided by use and occupancy into three classes: (1) Class A — permanent structures with a high degree of wind sensitivity and an unusually high degree of hazard to life and property in case of failure, such as a high-rise office building; (2) Class B — permanent structures with human occupancy, such as public buildings, dwellings, bunk houses; and (3) Class C — structures having no human occupants or where there is negligible

---

* ASAE Standard S288.3, Agricultural Engineers Handbook, American Society of Agricultural Engineers, St. Joseph, Mich., 1979—80, 308.

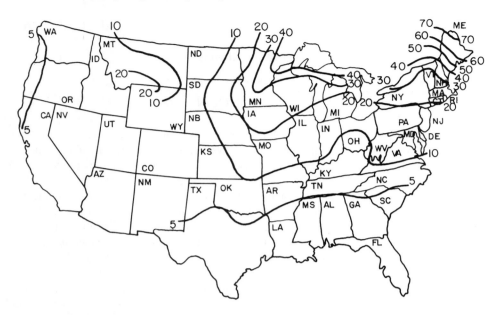

FIGURE 1.   Snow load, S, in lb/ft^2 on the ground, 25-year mean recurrence interval. (From Agricultural Engineers Yearbook, American Society of Agricultural Engineers, St. Joseph, Mich., 1979. With permission.)

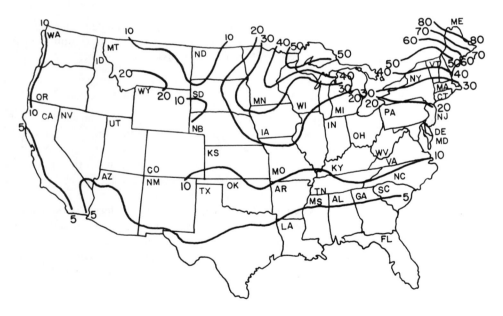

FIGURE 2.   Snow load, S, in lb/ft^2 on the ground, 50-year mean recurrence interval. (From Agricultural Engineers Yearbook, American Society of Agricultural Engineers, St. Joseph, Mich., 1979. With permission.)

risk to human life, such as crop storage buildings, livestock production buildings, farm shops, and machinery storage structures.

Recommended design loads are calculated using criteria and recurrence intervals as follows:

Class of building	Recurrence interval (years)
A	100
B	50
C	25

Maps for recurrence intervals of 25 years are given. For 50- and 100-year recurrence interval maps, refer to American National Standard A58.1-1972, Building Code Requirements for Minimum Design Loads in Buildings and Other Structures.

SNOW LOADS

Estimated snow pack on the ground is shown in Figure 1 for a 25-year mean recurrence interval. Design snow loads are found by the formula

$$L = 0.6(\cos\theta)\,(S)$$

where L = design snow load for the projected area, lb/ft^2 (kg/mi^2);* 0.6 = a dimensionless factor that estimates the proportion of snow pack on the ground that occurs as snow load on a flat roof; θ = angle between the horizontal and the roof slope, degrees; and S = snow pack on the ground as shown by map values in lb/ft^2 (S in kg/mi^2) = (S in lb/ft^2) (4.8824). Exceptions are to be made discretionally for roof shapes subjected to drifting or sliding snow, such as saw-tooth roofs, monitors, sheds at lower elevation than roof above, or valleys; these usually require increased design snow loads. Trusses and arches should be designed to support the design snow load, L, placed as an unbalanced load on each slope independently. Map values should not be used for geographic areas subjected to unusual snow because of elevation or other special geographic features.

WIND LOADS

Estimates of extreme-mile winds shown in Figures 3 and 4 are only for normal conditions of exposure where surface friction is relatively uniform for about 25 mi upwind. Adjustments must be made to map values for special winds such as the Santa Ana; for unusual exposures causing channeling or uplift such as ocean promontories, mountains, or gorges; for tornadoes; and for exposures or elevations where wind records or experience indicates illustrated wind speeds are inadequate.

Wind pressures used as a basis for design are determined by:

$$q = 0.00256V^2k \quad \text{or} \quad [q = 0.00466V^2K]$$

where q = basic velocity pressure in lb/ft^2 (kg/mi^2), V = map value of wind velocity for the extreme-mile wind in miles per hour (V in km/hr) = (F in mi/hr) (1.6093), and k = building height correction factor found by:

$$k = (h/30)^{2/7} \quad \text{or} \quad [k = (h/9.144)^{2/7}]$$

---

* Metric units are shown in parentheses.

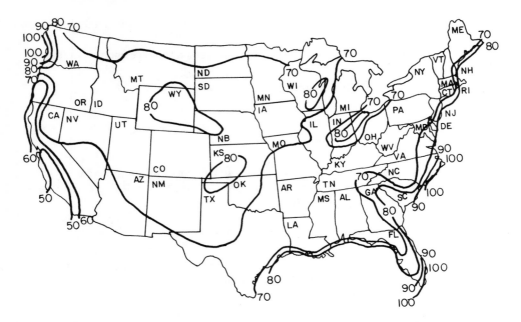

FIGURE 3. Annual extreme-mile wind velocity, V, in mi/hr at 30-ft height, 25-year mean recurrence interval. (From Agricultural Engineers Yearbook, American Society of Agricultural Engineers, St. Joseph, Mich., 1979. With permission.)

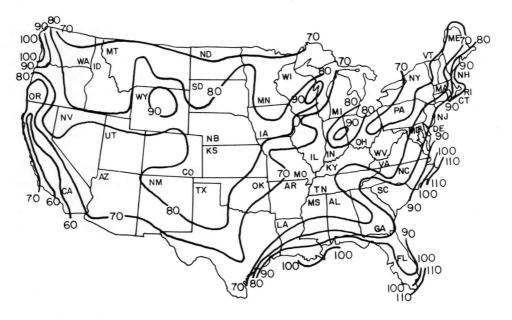

FIGURE 4. Annual extreme-mile wind velocity, V, in mi/hr at 30-ft height, 50-year mean recurrence interval. (From Agricultural Engineers Yearbook, American Society of Agricultural Engineers, St. Joseph, Mich., 1979. With permission.)

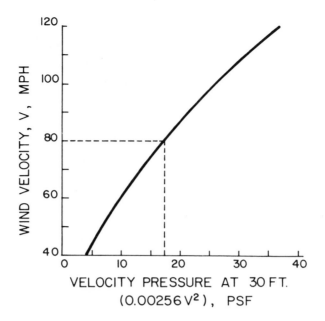

FIGURE 5. Velocity pressure at 30 ft.

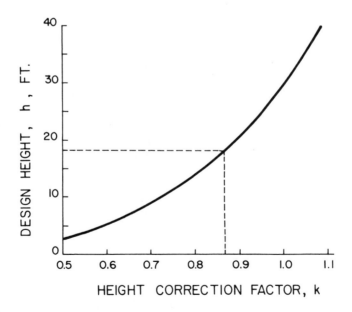

FIGURE 6. Height correction factor.

where h = design height of building, in ft(m) (equal to the eave height, H, except that if the vertical height from eave to the extreme top of the building is equal to or greater than the eave height, then the design height, h, will be the midelevation of the roof; all heights are measured from the average grade line), and 30 = height, in ft, at which wind velocities were reported and at which the isotachs of Figures 3 and 4 were plotted.

Velocity pressure, q, is determined from Figures 5 and 6 using wind velocity, V, and design height, h. Dimensionless shape coefficients, C (Tables 1 to 3), are to be used for design of main frames (trusses, columns, rigid frames, and other main members) and an-

Table 1
SHAPE COEFFICIENTS, C, FOR EXTERNAL WIND LOADS ON SINGLE-SPAN GABLE-TYPE BUILDINGS — TOTALLY ENCLOSED

Ratio eave height to width	Windward wall coef. C_1	Windward roof coef. C_2^a, roof slope						
		1:12	2:12	3:12	4:12	5:12	6:12	7:12
0.10	0.70	0.34	0.24	0.13	0.03	+0.05	+0.12	+0.19
0.15	0.70	0.51	0.35	0.20	0.05	+0.05	+0.12	+0.19
0.20	0.70	0.60	0.47	0.27	0.06	+0.05	+0.12	+0.19
0.25	0.70	0.60	0.59	0.34	0.08	+0.05	+0.12	+0.19
0.30	0.70	0.60	0.60	0.41	0.18	+0.01	+0.08	+0.16
0.35	0.70	0.60	0.60	0.47	0.26	0.07	+0.05	+0.12
0.40	0.70	0.60	0.60	0.53	0.33	0.15	+0.01	+0.09
0.45	0.70	0.60	0.60	0.57	0.39	0.22	0.06	+0.05
0.50	0.70	0.60	0.60	0.60	0.44	0.29	0.14	0.00
0.60	0.72	0.60	0.60	0.60	0.49	0.34	0.20	0.06
0.70	0.74	0.60	0.60	0.60	0.53	0.39	0.25	0.13
0.80	0.76	0.60	0.60	0.60	0.57	0.43	0.30	0.13
0.90	0.78	0.60	0.60	0.60	0.60	0.47	0.35	0.23
1.00[b]	0.80	0.60	0.60	0.60	0.60	0.51	0.39	0.23

Note: For leeward roof $C_3 = -0.50$ for all ratios of H/W. For leeward wall $C_4 = -0.40$ for all ratios of H/W (eave height to width).

[a] C_2 values are negative unless marked positive.
[b]   1.00 or more.

chorage. They are to be multiplied by q to determine design pressures. Vectors thus determined are normal to the surface considered. Wind should be considered from all directions; therefore, the shape coefficients shown for the windward and leeward sidewalls apply to the windward and leeward endwalls.

Shape coefficient (Table 1) for design of main frames of totally enclosed gable-type buildings depends on the ratio of the eave height, H, to the total width of the building, W, and the roof pitch. Tables 2 and 3 give coefficients for main frames of curved-roof buildings with and without vertical sidewalls. Shape coefficients for main frames of buildings with both sides open are as follows (negative values indicate external suction on building surfaces):

For roof slopes of 30° or more:
    Windward slope, C_2 = +0.8
    Leeward slope, C_3 = −0.8
For roof slopes less than 30°:
    Windward slope, C_2 = +0.6
    Leeward slope, C_3 = −0.6

The main frames of class A and B buildings and the anchorages of all classes of buildings, when such buildings are constructed with an open side, should be designed for wind loads as determined for the external surfaces of a totally enclosed building plus the load as determined for the interior surfaces resulting from 0.7 q acting outward at all points.

Purlins, rafters, roof panels, or coverings and their fastenings, whether a structural part of the framing or merely covering, should be designed for pressures equal to 1.25 q acting outward (suction) normal to the surface considered.

Wall girts, studs, and wall paneling or covering should be designed for pressures either inward or outward equal to the design wind pressure, q. Wind forces on shapes other than those specified may be found in "Wind Forces on Structures".*

* Transactions Paper No. 3269, Vol. 126, Part II, American Society of Civil Engineers, St. Joseph, Mich., 1961.

Table 2
SHAPE COEFFICIENT, C, FOR EXTERNAL WIND LOADS ON CURVED ROOF BUILDING WITH VERTICAL SIDEWALLS[a]

f/w	Windward quarters of roof	Center half of roof
0.10	0	−0.80
0.15	0	−0.85
0.20	0	−0.90
0.25	0.10	−0.95
0.30	0.19	−1.00
0.35	0.29	−1.05
0.40	0.39	−1.10
0.45	0.48	−1.15
0.50	0.58	−1.20

Note: (1) Leeward quarter of roof: C = −0.58 for all values of f/w. (2) Windward wall and leeward wall, use Table 1. (3) As previously, negative values of C indicate external suction on building surface.

[a]
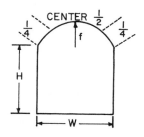

Table 3
SHAPE COEFFICIENT, C, FOR EXTERNAL WIND LOADS ON CURVED ROOF BUILDINGS[a]

f/w	Windward quarter of roof
0.10	0.12
0.15	0.18
0.20	0.25
0.25	0.31
0.30	0.37
0.35	0.43
0.40	0.49
0.45	0.55
0.50	0.60

Note: (1) Center half of roof C = −0.070 for all values of f/w. (2) Leeward quarter of roof: C = −0.58 for all values of f/w. (3) Endwalls: C = +0.70 for all values of f/w.

[a]
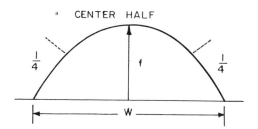

COMBINED LOADS

Combinations of dead load plus snow load, and of dead load plus wind load, are the combinations that usually control the design, but for extreme shapes the designer should consider other combinations. Typical wind loading for agricultural structures in most localities is assumed to be 15 lb/ft². Close observation of Figures 3 and 4 reveals that provision for 80-mi/hr winds provides adequate strength in all but the most severe wind areas of the U.S. Further, consider that most farm structures are either one- or two-story structures with a design height of only 20 ft. Figure 5 indicates that an 80-mi/hr wind would provide a wind pressure of 17 lb/ft². This is reduced for a 20 ft-high structure to 0.9 of the 17-lb/ft² value by consulting Figure 6. Thus the actual expected wind load on a 20-ft-high structure in an 80-mi/hr wind area would be 17 lb/ft² × 0.9 = 15.3 lb/ft². This is very near the 15-lb/ft² wind load mentioned above and is considered adequate. This will not provide for the high winds encountered along the coast where hurricanes are known to occur, nor will it provide protection against tornados. Tornado winds are so destructive that no attempt is made to design for them.

ALLOWABLE STRESSES

The allowable design stresses for materials and their connections should be taken from tha latest editions of publications of associations or societies having responsibility for materials used in construction. Recommended sources are as follows:

1.   For soils engineering in foundation design; American National Standard A56.1, Building Code Requirements for Excavations and Foundations, joint ANSI-ASCE recommendation, also published by American Society of Civil Engineers as ASCE 32.
2.   For design in concrete: American Concrete Institute, ACI 318, Building Code Requirements for Reinforced Concrete.
3.   For design in masonry: ANSI A41.1, Building Code Requirements for Masonry. Structural Clay Products Institute: Brick and Tile Engineering Handbook of Design. Technical Notes on Brick and Tile Construction.
4.   For design in timber: American Institute of Timber Construction, Timber Construction Manual. American Wood Preservers Institute, How to Design Pole-Type Buildings. Canadian Institute of Timber Construction, Timber Construction Manual. American Plywood Association, Plywood Design Specification. National Forest Products Association Standard, National Design Specification for Wood Construction. U.S. Forest Products Laboratory, Wood Handbook, Agricultural Handbook 72, 1955.
5.   For design in metal: American Iron and Steel Institute, Specifications for the Design of Cold-Formed Steel Structural Members. American Institute of Steel Construction, Specifications for the Design, Fabrication, and Erection of Structural Steel for Buildings. American Society for Testing and Materials Standard, ASTM A345, Specifications for Quenched and Tempered Alloy Steel Bolts, Studs, and Other Externally Threaded Fasteners. (Specifically, grade BX.) American Institute of Steel Construction, Specifications for Structural Joints Using ASTM A325 or A490 Bolts. American Society of Civil Engineers, Proceedings Paper No. 3341, Suggested Specifications for Structures of Aluminum Alloys, 6061-T6 and 6062-T6; Proceedings Paper No. 3342, Suggested Specifications for Structures of Aluminum Alloys, 6063-T5 and 6063-T6. American Welding Society, Standard Code for Welding in Building Construction. Steel Deck Institute, Basic Design Specifications. Steel Joist Institute, Standard Specifications for Open Web Steel Joists.

STRUCTURAL CONSIDERATIONS FOR WOOD, STEEL, AND CONCRETE

Harvey B. Manbeck

INTRODUCTION

The most common types of load carrying members in structural framing systems in agriculture are tension members, compression members (columns), flexural members (beams), and beam-columns. Only occasionally will one encounter members with significant torsional loading.

There are two basic philosophies of structural design: elastic and plastic. In elastic design, the design load capacity of a member is evaluated using ideal elastic equations and actual service loads. Discrepancies between real behavior and ideal behavior and uncertainties about the magnitude of service loads are accounted for by reducing the allowable material stresses by appropriate factors of safety. Timber and many steel structural elements are designed using this approach.

In plastic, or ultimate, strength design, the theoretical load capacity of a member is based upon its ultimate strength, i.e., the maximum stress the section can carry without fracturing. The allowable load capacity is found by factoring the theoretical capacity downward to account for uncertainties of material behavior under loading. In ultimate strength design, the service loads are factored upward to account for uncertainties of loading. For example, wind loads are factored upwards more than dead loads since they are more variable and uncertain over the life expectancy of the structure. Most modern reinforced concrete design is based upon ultimate strength concepts.

The primary purposes of this chapter are to:

1. Summarize the elementary strength and deformation equations for each type of load carrying member
2. Note the limitations of the equations
3. Note how real members differ from ideal behavior and how the member strength is affected
4. Direct the reader to resources for in-depth study and procedural information

TENSION MEMBERS

The stress and strain in uniaxially loaded homogeneous tensile members are uniformly distributed over a transverse cross section and are defined as:

$$f_t = P/A \tag{1}$$

$$\epsilon = \delta/L_0 \tag{2}$$

where f_t = tensile stress, P = tensile load, A = original cross-sectional area, ϵ = tensile strain, L_0 = original length of the member, and δ = change in the member length.

If the bar is elastic and loaded below its yield stress, Hooke's Law applies, and stress and strain are related by Young's Modulus, E:

$$f_t = E\epsilon \tag{3}$$

where E = Young's Modulus (modulus of elasticity). In elastic design, the design criterion for strength design is

$$f_t = P/A \leqslant F_y/F.S. = F_t \tag{4}$$

where F_y = yield stress of the material, F_t = allowable tensile stress, and F.S. = factor of safety.

The deformation of an elastic tensile member in terms of the load and the geometric and mechanical properties of the bar is

$$\delta = \frac{PL}{AE} \tag{5}$$

COMPRESSION MEMBERS

There are three classes of axially loaded compression members, namely, short, intermediate, or long. The column classification is dependent upon the mode of failure.

A short column yields and reaches its allowable load before buckling. Thus, the elastic design criterion for strength is

$$f_a = P/A \leqslant \frac{F_y}{F.S.} = F_a \tag{6}$$

where f_a = axial compressive stress, and F_a = allowable axial compressive stress. A long column buckles before the section is loaded to its yield stress, and no part of the section yields during buckling. Euler's equation predicts the stress level at which a long column will buckle:

$$f_a = \frac{P}{A} = \frac{1}{F.S.} [\pi^2 E/(KL/r)^2] = F_a \tag{7}$$

where KL = effective column length, r = radius of gyration of the cross section, and KL/r = largest slenderness ratio of the column.

An intermediate column buckles before the section achieves its full yield load and then yields during buckling. Most governing equations for the allowable load are empirical in nature and are dependent upon the material properties such as the modulus of elasticity, yield stress, and geometric properties of the section. Generally,

$$f_a = P/A \leqslant \frac{1}{F.S.} \text{ (Empirical Code Equation)} \tag{8}$$

An example of an intermediate column equation is the American Institute for Steel Construction equation for hot rolled steel shapes:

$$f_a = P/A \leqslant \frac{1}{F.S.} \left[1 - \frac{(KL/r)^2}{(C_c)^2} \right] F_y = F_a \tag{9}$$

where

$$F.S. = 5/3 + \frac{3KL/r}{8C_c} - \frac{(KL/r)^3}{8C_c^3}$$

$$C_c = \sqrt{2\pi^2 E/F_y}$$

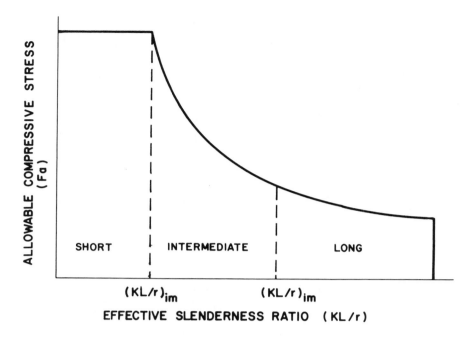

FIGURE 1. Allowable compressive stresses in columns.

The variation of F_a with KL/r is shown in Figure 1. The limiting values of KL/r at which intermediate and long column behavior commence are dependent upon the structural material and the design code being used.

The effective length factor, K, in the column equations depends upon the end fixity of the columns. Typical theoretical and practical values for K are given in Figure 2. The length in all of the column equations is always the unbraced length of the column.

FLEXURAL MEMBERS

There are many criteria that need to be checked in flexural member design, including flexural stresses, shear stresses, bearing stresses at end supports and at concentrated loads, local buckling of thin compression elements of a cross section, lateral torsional buckling, and beam deflections.

The primary factors to consider for each of these criteria are presented herein for beams satisfying the following criteria:

1. The line of action of the flexural loads passes through the shear center of a transverse cross section. If a section is doubly symmetric (has two axes of symmetry), the shear center coincides with the centroidal axis of the section. If the section has one axis of symmetry, the shear center lies somewhere along the axis of symmetry, but does not generally coincide with the centroidal axis. Shear centers for common shapes are available in any good text on advanced mechanics of materials.
2. Elastic materials loaded below their elastic strength.
3. Homogeneous cross sections.
4. Beams of a length at least four to five times as long as the cross-section depth.

Flexural Stresses

The flexural stress criteria control the design of the majority of beams used in agricultural framing systems. The elastic design criterion for flexural stresses is

	a.	b.	c.	d.	e.	f.
BUCKLED SHAPE OF COLUMN IS SHOWN BY DASHED LINE	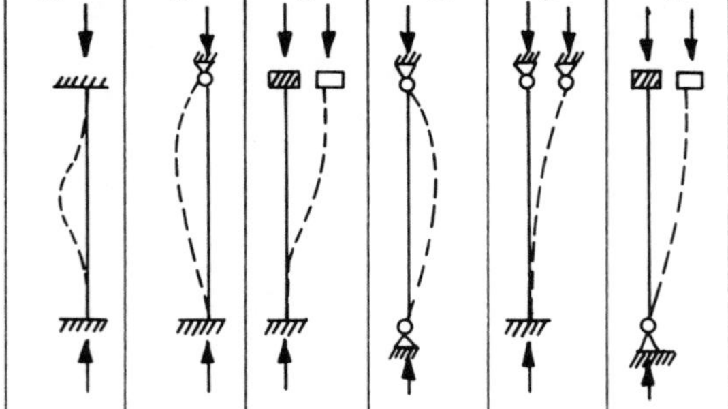					
THEORETICAL K VALUE	0.5	0.7	1.0	1.0	2.0	2.0
DESIGN VALUE OF K WHEN IDEAL CONDITIONS ARE APPROXIMATED	0.65	0.80	1.2	1.0	2.1	2.0
END CONDITION CODE	ROTATION FIXED TRANSLATION FIXED					
	ROTATION FREE TRANSLATION FIXED					
	ROTATION FIXED TRANSLATION FREE					
	ROTATION FREE TRANSLATION FREE					

FIGURE 2.   Effective length factors for columns (From Steel Construction Manual, 8th ed., American Institute of Steel Construction, New York, 1980. With permission.)

$$f_b = \frac{Mc}{I} = M/Z \leq F_b = \frac{F_y}{F.S.} \tag{10}$$

where f_b = maximum flexural stress under service loads, M = maximum bending moment in the flexural member, c = distance from the neutral axis to the outermost fiber of a transverse cross section, I = area moment of inertia of the transverse cross section about the neutral axis (N.A.) of the section (the neutral axis and centroidal axis of the section coincide for pure elastic flexure), Z = I/c = section modulus, and F_b = allowable flexural stress of the beam material.

The factor of safety is dependent upon the beam material, beam cross-section geometry, and the unbraced length of the compression elements of the beam.

Shear Stresses

The shear stress design criteria are

$$(f_v)_{max} = \frac{V_{max} Q_{N.A.}}{It} \leq F_v = \frac{F_y}{F.S.} \tag{11}$$

where f_v = maximum shear stress under service loads, V_{max} = maximum shear force in the beam, t = width of the cross section at the point at which shear stresses are calculated,

Q = first moment of the area of the cross section between the N.A. and the outermost fibers taken about the N.A., and F_v = allowable shear stress of the beam material.

The maximum shear stress occurs at the neutral axis of most common structural shapes. For solid rectangular cross sections, the maximum shear stress is

$$f_v = \frac{3}{2} \frac{V_{max}}{A} \tag{12}$$

where A = area of the entire cross section.

For sections with thin webs and wide flanges, the flange area is usually ignored and the shear stress is nearly constant over the web area. For such sections:

$$f_v = \frac{V_{max}}{A_{web}} \tag{13}$$

where A_{web} = cross-sectional area of the web.

Bearing Stresses

At end supports and at concentrated loads, the bearing stresses may be excessive unless adequate bearing area is provided. The design criterion for bearing stresses is

$$f_{brg} = \frac{R}{A_{brg}} \leq F_{brg} = \frac{F_y}{F.S.} \tag{14}$$

where f_{brg} = maximum bearing stress under service loads, R = end reaction or concentrated load, A_{brg} = bearing area (length times width of the flange area in contact with the supporting structure), and F_{brg} = allowable bearing stress of the material.

Local Buckling

Some structural shapes, such as hot rolled steel shapes and cold-formed steel shapes, are fabricated with relatively thin elements. If the elements are very slender (if the element width to thickness ratio, b/t, is large enough), and if the element is loaded in compression (as is the compression flange of a wide flange steel beam), it can buckle locally before the section reaches its yield moment capacity. In cases where b/t is large, special reductions in allowable stresses or geometric section properties such as I, Z, and A must be made.

Most standard hot rolled steel shapes are proportioned such that local buckling does not occur before the sections develop their yield moment. Cold-formed steel sections tend to have much thinner elements and local buckling does occur before reaching the yield moment capacity. Allowable stresses and design procedures for these sections may be found in the AISI Specification for Cold-Formed Steel. Local buckling is seldom a problem in either wood or concrete flexural members.

Lateral Buckling of Beams

Flexural members, when loaded in a direction parallel and through the weak axis (axis with the larger area moment of inertia) of the cross section, may fail by buckling and twisting in the lateral direction before the allowable yield moment is developed in the beam. Figure 3 illustrates lateral buckling of beams.

The compressive stress level at which lateral buckling commences is primarily a function of the laterally unsupported span of the compression edge of the beam, L_b, and the depth of the beam, d. Figure 4 presents the generalized variation of the compressive stress at which lateral buckling commences.

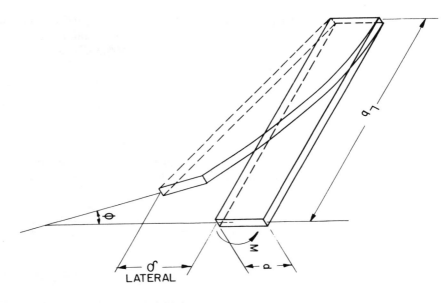

FIGURE 3.   Lateral buckling of a beam.

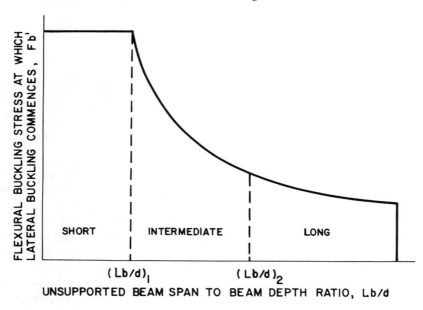

FIGURE 4.   Variation of allowable flexural stress as a function of unsupported beam length and beam depth.

For short unsupported spans, the beam will yield before buckling laterally. Between $(L_b/d)_1$ and $(L_b/d)_2$, the section buckles inelastically, and when $(L_b/d)_1 > (L_b/d)_2$, inelastic buckling occurs.

Lateral buckling is prevented by providing adequate lateral support along the beam length with bridging, decking, or struts. If the bridging is spaced such that $L_b/d < (L_b/d)_1$, lateral buckling will not occur and full yield moment capacities are used in design. If $L_b/d > (L_b/d)_1$, the allowable compressive flexural stresses are reduced to $F_b'/F.S.$ This assures that buckling does not occur at design loads. Thus, for laterally unsupported beams the flexural stress criterion is

$$f_b = \frac{M_{max}\, c}{I} \leq \frac{F'_b}{F.S.} \tag{15}$$

DEFLECTION

Excessive deformation must be avoided in agricultural structures to prevent excessive vibrations, cracking of interior finishes, and loss of structural integrity. Most design codes provide deflection limits on rafters, ceilings, and floors. Common deflection limits are L/240 and L/360 for ceilings and floors, respectively, where L is the beam span. On pitched roofs, the requirement is that no ponding may occur at any point along the slope.

Beam deflections are normally estimated by elastic theory and then adjusted as necessary for long-term creep effects. Typical deflections for some of the more common loadings are given in Table 1.

In steel beams, the design deflection is usually taken as the deflection evaluated with the elastic equations for the full service loads. In timber beams, the total deflection is typically the sum of the live load deflection plus two times the dead load deflection (both the live and dead load deflections are calculated with the elastic equations). In reinforced concrete beams, the total deflection is also taken as the sum of the long- and short-term deflections. The elastic deflection computation, however, is complicated somewhat since effective values of E and I must be approximated before the elastic equations may be used.

COMBINED LOADING

Many structural members are subject to both axial- and flexural-type loadings. An eccentrically loaded column with no lateral loading and a centrically loaded column with lateral loads applied between supports are two common examples of combined loading.

The interaction equation is widely used to evaluate the adequacy of members subject to combined loading. The generalized form of this equation is

$$\frac{\text{Actual Axial Stress}}{\text{Allowable Axial Stress}} + \frac{\text{Actual Flexural Stress}}{\text{Allowable Flexural Stress}} \leq 1.0 \tag{16}$$

For axial tensile loading in combination with flexural loading about the x-axis of the cross-sectional area, the interaction equation is

$$\frac{f_t}{F_t} + \frac{f_{bx}}{F_{bx}} \leq 1.0 \tag{17}$$

For biaxial bending about both the x- and y-axes, the equation expands to:

$$\frac{f_t}{F_t} + \frac{f_{bx}}{F_{bx}} + \frac{f_{by}}{F_{by}} \leq 1.0 \tag{18}$$

In both Equations 17 and 18, F_{bx} and F_{by} are the allowable flexural stresses for pure bending about the x- and y-axes, respectively.

If the member is subject to axial compression and flexural loading, the interaction equation becomes:

$$\frac{f_a}{F_a} + \frac{f_{bx}}{F_{bx}} + \frac{f_{by}}{F_{by}} \leq 1.0 \tag{19}$$

Table 1
TABLE OF BEAM DEFLECTIONS AND SLOPES

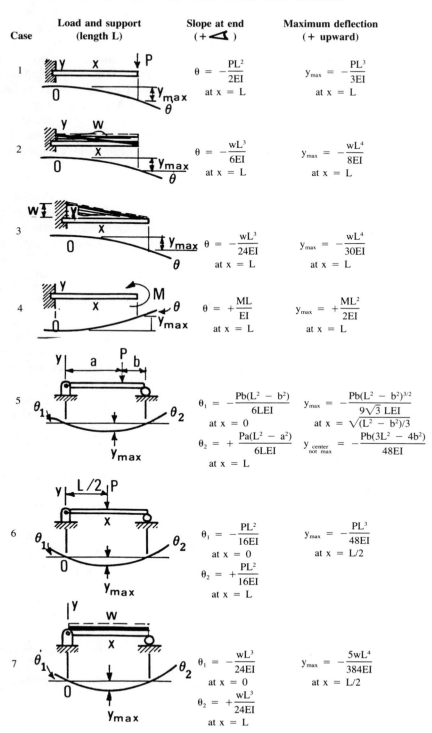

Case	Load and support (length L)	Slope at end (+ ◁)	Maximum deflection (+ upward)
1		$\theta = -\dfrac{PL^2}{2EI}$ at x = L	$y_{max} = -\dfrac{PL^3}{3EI}$ at x = L
2		$\theta = -\dfrac{wL^3}{6EI}$ at x = L	$y_{max} = -\dfrac{wL^4}{8EI}$ at x = L
3		$\theta = -\dfrac{wL^3}{24EI}$ at x = L	$y_{max} = -\dfrac{wL^4}{30EI}$ at x = L
4		$\theta = +\dfrac{ML}{EI}$ at x = L	$y_{max} = +\dfrac{ML^2}{2EI}$ at x = L
5		$\theta_1 = -\dfrac{Pb(L^2-b^2)}{6LEI}$ at x = 0 $\theta_2 = +\dfrac{Pa(L^2-a^2)}{6LEI}$ at x = L	$y_{max} = -\dfrac{Pb(L^2-b^2)^{3/2}}{9\sqrt{3}\,LEI}$ at x = $\sqrt{(L^2-b^2)/3}$ $y_{\substack{center \\ not\ max}} = -\dfrac{Pb(3L^2-4b^2)}{48EI}$
6		$\theta_1 = -\dfrac{PL^2}{16EI}$ at x = 0 $\theta_2 = +\dfrac{PL^2}{16EI}$ at x = L	$y_{max} = -\dfrac{PL^3}{48EI}$ at x = L/2
7		$\theta_1 = -\dfrac{wL^3}{24EI}$ at x = 0 $\theta_2 = +\dfrac{wL^3}{24EI}$ at x = L	$y_{max} = -\dfrac{5wL^4}{384EI}$ at x = L/2

Table 1 (continued)
TABLE OF BEAM DEFLECTIONS AND SLOPES

Case	Load and support (length L)	Slope at end (+ ◢)	Maximum deflection (+ upward)

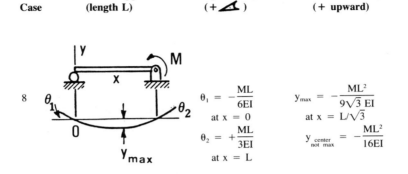

$$\theta_1 = -\frac{ML}{6EI}$$
at $x = 0$

$$\theta_2 = +\frac{ML}{3EI}$$
at $x = L$

$$y_{max} = -\frac{ML^2}{9\sqrt{3}\,EI}$$
at $x = L/\sqrt{3}$

$$y_{\substack{center \\ not\ max}} = -\frac{ML^2}{16EI}$$

Case 8

From Higdon, A. et al., *Mechanics of Materials*, 3rd ed., John Wiley & Sons, New York, 1976, 722. With permission.

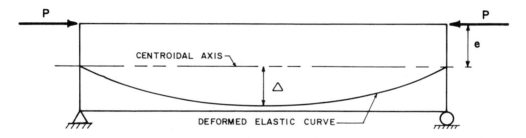

FIGURE 5. The P-Δ effect in beam columns.

In Equation 19, F_a is the allowable compressive stress in pure compression based upon the maximum effective slenderness ratio of the column. The terms f_{bx} and f_{by} are the actual flexural stresses produced by both the flexural loads and the eccentricities of the axial loads. Thus

$$f_{bx} = \frac{M_x\, c}{I_x} + \frac{P\, e_x\, c}{I_x} \tag{20}$$

or

$$f_{bx} = \frac{M_x\, c_x}{A\, r_x^2} + \frac{P\, e_x\, c_x}{A\, r_x^2} \tag{21}$$

Similarly,

$$f_{by} = \frac{M_y\, c_y}{A\, r_y^2} + \frac{P\, e_y\, c_y}{A\, r_y^2} \tag{22}$$

The second term on the right side of Equations 20, 21, and 22 is sometimes increased by a moment magnification term to account for a P-Δ effect. Figure 5 illustrates this effect. In short beam-columns, the deflection Δ is small compared to the load eccentricity. Thus the moment due to eccentricity of loading is $M = P(e + \Delta)$ which is nearly the same as

Pe. In long columns, however, Δ becomes significant compared to e and $M = P(e + \Delta) > Pe$. To account for this, many design codes specify a magnification factor β for longer members. Then

$$f_{bx} = \frac{M_x \, c}{I_x} + \beta_x \frac{P \, e_x \, c}{I_x} \tag{23}$$

The interaction equations are used to design hot rolled steel, cold rolled steel, and timber beam-columns. Reinforced concrete beam-columns are designed by ultimate stress concepts. The approach is beyond the scope of this text, and the reader is directed to the 1983 ACI Code and Commentary and to a standard text on reinforced concrete design for the procedures.

SPECIAL CONSIDERATIONS FOR WOOD

Wood is a nonhomogeneous material and is not linearly elastic. The design philosophy adopted by the National Forest Products Association in their National Design Specification for Wood Structures (NDS) is to assume that the simple elastic equations from elementary mechanics of materials are valid and then to account for discrepancies between ideal and real behavior by adjusting the allowable design stresses.

Timber design stresses are dependent upon a multitude of factors, including lumber species and grade, moisture content, load duration, angle of grain to primary stress, type loading, and member size. Table 2 contains typical design stresses for several grades of southern pine and Douglas fir lumber. Design stresses are given for 15 and 19% moisture content and for several types of stress conditions. Reduced stress values for deep tension members are summarized in Table 3 for normal (10-year) load duration. The section properties for a variety of rectangular solid sawn timber sections are given in Table 4.

Allowable design stresses in solid sawn timber beams depend upon the unsupported length of the compression edge. The NDS defines the allowable stresses for flexural members loaded through their shear center as F_b'.

$$F_b' = F_b \text{ (From Table 2)} \qquad \text{if} \quad c_s < 10 \tag{24}$$

$$F_b' = F_b \left[1 - \frac{1}{3} \left(\frac{c_s}{c_k} \right)^4 \right] \leq F_b \quad \text{if} \quad 10 \leq c_s \leq c_k \tag{25}$$

$$F_b' = \frac{0.4E}{c_s^2} < F_b \qquad \text{if} \quad c_s > c_k \tag{26}$$

where $c_s = \sqrt{L_e \, d/b^2}$, $c_k = \sqrt{3E/5F_b}$, d = beam depth, b = beam width, L_e = unsupported length factor (= $1.61 \, L_u$ for concentrated load at center of a single span, $1.84 \, L_u$ for concentrated end moments at ends of a single span, $1.92 \, L_u$ for uniform load over a single span, $1.69 \, L_u$ for concentrated load at end of a cantilevered beam, $1.06 \, L_u$ for uniform load over a cantilevered beam, $1.92 \, L_u$ for any load for a single span or cantilevered beam — conservative value), and L_u = unsupported length, distance between lateral support.

Equations 27 to 30 define the allowable axial compression stresses, F_c', for short, intermediate, and long solid sawn rectangular timber columns:

$$F_c' = F_c \text{ (Table 2)} \qquad \text{if} \quad L/d < 11 \tag{27}$$

$$F_c' = F_c[1 - (L/d)^4/K] \quad \text{if} \quad 11 \leq L/d \leq K \tag{28}$$

Table 2
ALLOWABLE STRESSES FOR VISUALLY GRADED SOUTHERN PINE AND DOUGLAS FIR STRUCTURAL LUMBER

Species and commercial grade	Size classification	Extreme fiber in bending (F_b) Single-member uses	Extreme fiber in bending (F_b) Repetitive-member uses	Tension parallel to grain (F_t)	Horizontal shear (F_v)	Compression perpendicular to grain (F_{cp})	Compression parallel to grain (F_c)	Modulus of elasticity (E)
Southern pine (surfaced at 15% moisture content, K.D.; used at 15% max. m.c.)								
Select structural	2—4 in. thick	2,150	2,500	1,250	105	405	1,800	1,800,000
Dense select structural	2—4 in. wide	2,500	2,900	1,500	105	475	2,100	1,900,000
No. 1		1,850	2,100	1,050	105	405	1,450	1,800,000
No. 1 dense		2,150	2,450	1,250	105	475	1,700	1,900,000
No. 2		1,550	1,750	900	95	405	1,150	1,600,000
No. 2 dense		1,800	2,050	1,050	95	475	1,350	1,700,000
No. 3		850	975	500	95	405	675	1,500,000
No. 3 dense		1,000	1,150	575	95	475	800	1,500,000
Stud		850	975	500	95	405	675	1,500,000
Construction	2—4 in. thick 4 in. wide	1,100	1,250	650	105	405	1,300	1,500,000
Standard		625	725	375	95	405	1,050	1,500,000
Utility		275	300	175	95	405	675	1,500,000
Select structural	2—4 in. thick 5 in. and wider	1,850	2,150	1,200	95	405	1,600	1,800,000
Dense select structural		2,200	2,500	1,450	95	475	1,850	1,900,000
No. 1		1,600	1,850	1,050	95	405	1,450	1,800,000
No. 1 dense		1,850	2,150	1,250	95	475	1,700	1,900,000
No. 2		1,300	1,500	675	95	405	1,200	1,600,000

Design values (psi)

Table 2 (continued)
ALLOWABLE STRESSES FOR VISUALLY GRADED SOUTHERN PINE AND DOUGLAS FIR STRUCTURAL LUMBER

Design values (psi)

Species and commercial grade	Size classification	Extreme fiber in bending (F_b)		Tension parallel to grain (F_t)	Horizontal shear (F_v)	Compression perpendicular to grain (F_{cp})	Compression parallel to grain (F_c)	Modulus of elasticity (E)
		Single-member uses	Repetitive-member uses					
No. 2 dense		1,550	1,750	800	95	475	1,400	1,700,000
No. 3		750	875	400	95	405	725	1,500,000
No. 3 dense		875	1,000	450	95	475	850	1,500,000
Stud		800	900	400	95	405	725	1,500,000
Southern pine (surfaced dry; used at 19% max. m.c.)								
Select structural	2—4 in. thick 2—4 in. wide	2,000	2,300	1,150	100	405	1,550	1,700,000
Dense select structural		2,350	2,700	1,350	100	475	1,800	1,800,000
No. 1		1,700	1,950	1,000	100	405	1,250	1,700,000
No. 1 Dense		2,000	2,300	1,150	100	475	1,450	1,800,000
No. 2		1,400	1,650	825	90	405	975	1,600,000
No. 2 Dense		1,650	1,900	975	90	475	1,150	1,600,000
No. 3		775	900	450	90	405	575	1,400,000
No. 3 dense		925	1,050	525	90	475	675	1,500,000
Stud		775	900	450	90	405	575	1,400,000
Construction	2—4 in. thick 4 in. wide	1,000	1,150	600	100	405	1,100	1,400,000
Standard		575	675	350	90	405	900	1,400,000
Utility		275	300	150	90	405	575	1,400,000
Select structural	2—4 in. thick 5 in. and wider	1,750	2,000	1,150	90	405	1,350	1,700,000
Dense select structural		2,050	2,350	1,300	90	475	1,600	1,800,000

Grade							
No. 1	1,450	1,700	975	90	405	1,250	1,700,000
No. 1 dense	1,700	2,000	1,150	90	475	1,450	1,800,000
No. 2	1,200	1,400	625	90	405	1,000	1,600,000
No. 2 dense	1,400	1,650	725	90	475	1,200	1,600,000
No. 3	700	800	350	90	405	625	1,400,000
No. 3 Dense	825	925	425	90	475	725	1,500,000
Stud	725	850	350	90	405	625	1,400,000
Southern pine (surfaced green; used any condition)							
Select structural — 2½—4 in. thick	1,600	1,850	925	95	270	1,050	1,500,000
Dense select structural — 2½—4 in. wide	1,850	2,150	1,100	95	315	1,200	1,600,000
No. 1	1,350	1,550	800	95	270	825	1,500,000
No. 1 dense	1,600	1,800	925	95	315	950	1,600,000
No. 2	1,150	1,300	675	85	270	650	1,400,000
No. 2 dense	1,350	1,500	775	85	315	750	1,400,000
No. 3	625	725	375	85	270	400	1,200,000
No. 3 dense	725	850	425	85	315	450	1,300,000
Stud	625	725	375	85	270	400	1,200,000
Construction — 2½—4 in. thick	825	925	475	95	270	725	1,200,000
Standard — 4 in. wide	475	525	275	85	270	600	1,200,000
Utility	200	250	125	85	270	400	1,200,000
Select structural — 2½—4 in. thick	1,400	1,600	900	85	315	900	1,500,000
Dense select structural — 5 in. and wider	1,600	1,850	1,050	85	315	1,050	1,600,000
No. 1	1,200	1,350	775	85	270	825	1,500,000
No. 1 dense	1,400	1,600	925	85	315	950	1,600,000
No. 2	975	1,100	500	85	270	675	1,400,000
No. 2 dense	1,150	1,300	600	85	315	800	1,400,000
No. 3	550	650	300	85	270	425	1,200,000
No. 3 dense	650	750	350	85	315	475	1,300,000
Stud	575	675	300	85	270	425	1,200,000
No. 1 SR — 5 in. and thicker	1,350	—	875	110	270	775	1,500,000
No. 1 dense SR	1,550	—	1,050	110	315	925	1,600,000

Table 2 (continued)

ALLOWABLE STRESSES FOR VISUALLY GRADED SOUTHERN PINE AND DOUGLAS FIR STRUCTURAL LUMBER

Species and commercial grade	Size classification	Design values (psi)						
		Extreme fiber in bending (F_b)		Tension parallel to grain (F_t)	Horizontal shear (F_v)	Compression perpendicular to grain (F_{cp})	Compression parallel to grain (F_c)	Modulus of elasticity (E)
		Single-member uses	Repetitive-member uses					
No. 2 SR	2—4 in. thick	1,100	—	725	95	270	625	1,400,000
No. 2 dense SR		1,250	—	850	95	315	725	1,400,000
Douglas fir-larch (surfaced dry or surfaced green; used at 19% max. m.c.)								
Dense select structural	2—4 in. thick	2,450	2,800	1,400	95	455	1,850	1,900,000
Select structural	2—4 in. wide	2,100	2,400	1,200	95	385	1,600	1,800,000
Dense No. 1		2,050	2,400	1,200	95	455	1,450	1,900,000
No. 1		1,750	2,050	1,050	95	385	1,250	1,800,000
Dense No. 2		1,700	1,950	1,000	95	455	1,150	1,700,000
No. 2		1,450	1,650	850	95	385	1,000	1,700,000
No. 3		800	925	475	95	385	600	1,500,000
Appearance		1,750	2,050	1,050	95	385	1,500	1,800,000
Stud		800	925	475	95	385	600	1,500,000
Construction	2—4 in. thick	1,050	1,200	625	95	385	1,150	1,500,000
Standard	4 in. wide	600	675	350	95	385	925	1,500,000
Utility		275	325	175	95	385	600	1,500,000
Dense select structural	2—4 in. thick	2,100	2,400	1,400	95	455	1,650	1,900,000
Select structural	5 in. and wider	1,800	2,050	1,200	95	385	1,400	1,800,000
Dense No.1		1,800	2,050	1,200	95	455	1,450	1,900,000
No. 1		1,500	1,750	1,000	95	385	1,250	1,800,000
Dense No. 2		1,450	1,700	775	95	455	1,250	1,700,000

Grade	Size classification						
No. 2		1,250	1,450	650	95	385	1,050	1,700,000
No. 3		725	850	375	95	385	675	1,500,000
Appearance		1,500	1,750	1,000	95	385	1,500	1,800,000
Stud		725	850	375	95	385	675	1,500,000
Dense select structural	Beams and stringers	1,900	—	1,100	85	455	1,300	1,700,000
Select structural		1,600	—	950	85	385	1,100	1,600,000
Dense No. 1		1,550	—	775	85	455	1,100	1,700,000
No. 1		1,300	—	675	85	385	925	1,600,000
Dense select structural	Posts and timbers	1,750	—	1,150	85	455	1,350	1,700,000
Select structural		1,500	—	1,000	85	385	1,150	1,600,000
Dense No. 1		1,400	—	950	85	455	1,200	1,700,000
No. 1		1,200	—	825	85	385	1,000	1,600,000
Dense select structural	Beams and stringers	1,900	—	1,250	85	455	1,300	1,700,000
Select structural		1,600	—	1,050	85	385	1,100	1,600,000
Dense No. 1		1,550	—	1,050	85	455	1,100	1,700,000
No. 1		1,350	—	900	85	385	925	1,600,000
Dense select structural	Posts and timbers	1,750	—	1,150	85	455	1,350	1,700,000
Select structural		1,500	—	1,000	85	385	1,150	1,600,000
Dense No. 1		1,400	—	950	85	455	1,200	1,700,000
No. 1		1,200	—	825	85	385	1,000	1,600,000

From Design Values for Wood Construction, National Forest Products Association, Washington, D.C., 1978. With permission.

Table 3
1982 ALLOWABLE STRESSES IN TENSION
PARALLEL TO THE GRAIN (PSI)

| | | Lumber width | | |
| | | | | 10 in. or |
Species	Grade	5, 6 in.	8 in.	larger
Douglas fir	Select struct	1200	1080	960
	No. 1	1000	800	600
	No. 2	650	520	390
	Stud	375	—	—
Southern pine	Select struct	1150	1025	920
	No.1	975	780	585
	No. 2	625	500	375
	Stud	350	—	—

Note: Lumber 2 to 4 in. thick, greater than 4 in. wide, and used at 19% m.c. max.

From Structures and Environment Handbook, Midwest Plan Services, Ames, Iowa, 1983, 405.2. With permission.

$$F'_c = 0.3E/(L/d)^2 \qquad \text{if } K < L/d \leq 50 \qquad (29)$$

$$F'_c = 0 \qquad \text{if } L/d > 50 \qquad (30)$$

where L = effective length of column, d = critical dimension of column, K = $0.671\sqrt{E/F_c}$, and L/d = maximum slenderness ratio.

In Equations 28 and 29 the slenderness term, L/d, should be the largest L/d of the strong and weak axes. Thus, use the larger value of L_x/b or L_y/d where L_x and L_y are the effective lengths with respect to buckling about the strong and weak axes, respectively.

Equation 31 is the NDS interaction equation for designing a solid sawn timber beam column subjected to any combination of axial compression loads, flexural loads, and eccentricity of axial loads:

$$\frac{f_c}{F'_c} + \frac{f_b + f_c(6 + 1.5J)\,(e/d)}{F_b - J\,f_c} \qquad (31)$$

where

$$J = \frac{L/d - 11}{K - 11}$$

but J should not be less than 0 nor greater than 1.0, f_b = flexural stress induced by lateral loads, and F'_c = allowable axial stress in pure compression based upon the maximum L/d ratio.

SPECIAL CONSIDERATIONS FOR STEEL

Structural steel sections are classified according to the method of fabrication. Hot rolled sections are rolled from hot bar stock while cold-formed sections are formed by cold rolling, press braking, or bend braking the sections from cold flat stock. The elements — webs and flanges — of hot rolled sections are generally, but not always, thicker than those of cold-

Table 4
SECTION PROPERTIES OF STANDARD DRESSED RECTANGULAR STRUCTURAL LUMBER

Nominal size b (in.) d	Standard dressed size (S4S) b (in.) d	Area of section A (in.²)	Moment of inertia I (in.⁴)	Section modulus S (in.³)	Mass in pounds per linear foot of piece when mass of wood per cubic foot equals	
					30 lb	35 lb
1 × 3	³/₄ × 2 ¹/₂	1.875	0.977	0.781	0.391	0.456
1 × 4	³/₄ × 3 ¹/₂	2.625	2.680	1.531	0.547	0.638
1 × 6	³/₄ × 5 ¹/₂	4.125	10.398	3.781	0.859	1.003
1 × 8	³/₄ × 7 ¹/₄	5.438	23.817	6.570	1.133	1.322
1 × 10	³/₄ × 9 ¹/₄	6.938	49.466	10.695	1.445	1.686
1 × 12	³/₄ × 11 ¹/₄	8.438	88.989	15.820	1.758	2.051
2 × 3	1 ¹/₂ × 2 ¹/₂	3.750	1.953	1.563	0.781	0.911
2 × 4	1 ¹/₂ × 3 ¹/₂	5.250	5.359	3.063	1.094	1.276
2 × 5	1 ¹/₂ × 4 ¹/₂	6.750	11.391	5.063	1.406	1.641
2 × 6	1 ¹/₂ × 5 ¹/₂	8.250	20.797	7.563	1.719	2.005
2 × 8	1 ¹/₂ × 7 ¹/₄	10.875	47.635	13.141	2.266	2.643
2 × 10	1 ¹/₂ × 9 ¹/₄	13.875	98.932	21.391	2.891	3.372
2 × 12	1 ¹/₂ × 11 ¹/₄	16.875	177.979	31.641	3.516	4.102
2 × 14	1 ¹/₂ × 13 ¹/₄	19.875	290.775	43.891	4.141	4.831
3 × 1	2 ¹/₂ × ³/₄	1.875	0.088	0.234	0.391	0.456
3 × 2	2 ¹/₂ × 1 ¹/₂	3.750	0.703	0.938	0.781	0.911
3 × 4	2 ¹/₂ × 3 ¹/₂	8.750	8.932	5.104	1.823	2.127
3 × 5	2 ¹/₂ × 4 ¹/₂	11.250	18.984	8.438	2.344	2.734
3 × 6	2 ¹/₂ × 5 ¹/₂	13.750	34.661	12.604	2.865	3.342
3 × 8	2 ¹/₂ × 7 ¹/₄	18.125	79.391	21.901	3.776	4.405
3 × 10	2 ¹/₂ × 9 ¹/₄	23.125	164.886	35.651	4.818	5.621
3 × 12	2 ¹/₂ × 11 ¹/₄	28.125	296.631	52.734	5.859	6.836
4 × 1	3 ¹/₂ × ³/₄	2.625	0.123	0.328	0.547	0.638
4 × 2	3 ¹/₂ × 1 ¹/₂	5.250	0.984	1.313	1.094	1.276
4 × 3	3 ¹/₂ × 2 ¹/₂	8.750	4.557	3.646	1.823	2.127
4 × 4	3 ¹/₂ × 3 ¹/₂	12.250	12.505	7.146	2.552	2.977
4 × 5	3 ¹/₂ × 4 ¹/₂	15.750	26.578	11.813	3.281	3.828
4 × 6	3 ¹/₂ × 5 ¹/₂	19.250	48.526	17.646	4.010	4.679
4 × 8	3 ¹/₂ × 7 ¹/₄	25.375	111.148	30.661	5.286	6.168
4 × 10	3 ¹/₂ × 9 ¹/₄	32.375	230.840	49.911	6.745	7.869
4 × 12	3 ¹/₂ × 11 ¹/₄	39.375	415.283	73.828	8.203	9.570
6 × 1	5 ¹/₂ × ³/₄	4.125	0.193	0.516	0.859	1.003
6 × 2	5 ¹/₂ × 1 ¹/₂	8.250	1.547	2.063	1.719	2.005
6 × 3	5 ¹/₂ × 2 ¹/₂	13.750	7.161	5.729	2.865	3.342
6 × 4	5 ¹/₂ × 3 ¹/₂	19.250	19.651	11.229	4.010	4.679
6 × 6	5 ¹/₂ × 5 ¹/₂	30.250	76.255	27.729	6.302	7.352
6 × 8	5 ¹/₂ × 7 ¹/₂	41.250	193.359	51.563	8.594	10.026
6 × 10	5 ¹/₂ × 9 ¹/₂	52.250	392.963	82.729	10.885	12.700
6 × 12	5 ¹/₂ × 11 ¹/₂	63.250	697.068	121.229	13.177	15.373
8 × 1	7 ¹/₄ × ³/₄	5.438	0.255	0.680	1.133	1.322
8 × 2	7 ¹/₄ × 1 ¹/₂	10.875	2.039	2.719	2.266	2.643
8 × 3	7 ¹/₄ × 2 ¹/₂	18.125	9.440	7.552	3.776	4.405
8 × 4	7 ¹/₄ × 3 ¹/₂	25.375	25.904	14.803	5.286	6.168
8 × 6	7 ¹/₂ × 5 ¹/₂	41.250	103.984	37.813	8.594	10.026

Table 4 (continued)
SECTION PROPERTIES OF STANDARD
DRESSED RECTANGULAR STRUCTURAL LUMBER

Nominal size b (in.) d	Standard dressed size (S4S) b (in.) d	Area of section A (in.²)	Moment of inertia I (in.⁴)	Section modulus S (in.³)	Mass in pounds per linear foot of piece when mass of wood per cubic foot equals	
					30 lb	35 lb
8 × 8	7 ½ × 7 ½	56.250	263.672	70.313	11.719	13.672
8 × 10	7 ½ × 9 ½	71.250	535.859	112.813	14.844	17.318
8 × 12	7 ½ × 11 ½	86.250	950.547	165.313	17.969	20.964
10 × 1	9 ¼ × ¾	6.938	0.325	0.867	1.445	1.686
10 × 2	9 ¼ × 1 ½	13.875	2.602	3.469	2.891	3.372
10 × 3	9 ¼ × 2 ½	23.125	12.044	9.635	4.818	5.621
10 × 4	9 ¼ × 3 ½	32.375	33.049	18.885	6.745	7.869
10 × 6	9 ½ × 5 ½	52.250	131.714	47.896	10.885	12.700
10 × 8	9 ½ × 7 ½	71.250	333.984	89.063	14.844	17.318
10 × 10	9 ½ × 9 ½	90.250	678.755	142.896	18.802	21.936
10 × 12	9 ½ × 11 ½	109.250	1204.026	209.396	22.760	26.554
12 × 1	11 ¼ × ¾	8.438	0.396	1.055	1.758	2.051
12 × 2	11 ¼ × 1 ½	16.875	3.164	4.219	3.516	4.102
12 × 3	11 ¼ × 2 ½	28.125	14.648	11.719	5.859	6.836
12 × 4	11 ¼ × 3 ½	39.375	40.195	22.969	8.203	9.570
12 × 6	11 ½ × 5 ½	63.250	159.443	57.979	13.177	15.373
12 × 8	11 ½ × 7 ½	86.250	404.297	107.813	17.969	20.964
12 × 10	11 ½ × 9 ½	109.250	821.651	172.979	22.760	26.554
12 × 12	11 ½ × 11 ½	132.250	1457.505	253.479	27.552	32.144

From National Design Specification for Wood Construction, National Forest Products Association, Washington, D.C., 1986, 77. With permission.

formed sections. Thus, local buckling of compression elements is a more common problem in cold-formed sections than in hot rolled sections.

Hot rolled and cold rolled steels used in structural framing are ductile materials with ductilities in the range of 15 to 30%. Yield strengths range from 36,000 to 100,000 psi and the stiffness of most structural steels is approximately 29×10^6 psi. The most commonly used structural steel in agricultural framing systems is A36 steel with a yield stress, F_y, of 36,000 psi.

HOT ROLLED STEEL

Hot rolled sections are characterized by a relatively few standard size sections. The shapes readily available are wide flange, channel, angle, square, and rectangular structural tubing and round pipe sections. The most commonly used shapes in agricultural applications are wide flange, rectangular tubing and round pipe. Properties of some typical hot rolled sections are summarized in Table 5.

Most standard shapes are proportioned so that local buckling does not occur before the section develops its full elastic capacity. The criterion (F_y in ksi) for development of the full elastic capacity is that the width to thickness ratio, b/t, of compression elements not exceed the following limits for unstiffened compression elements:

Table 5
SECTION PROPERTIES FOR DESIGNING

Designation	Area A (in.²)	Depth d (in.)	Flange Width bf	Flange Thickness tf	Web thickness tw (in.)	k (in.)	Axis X-X I (in.⁴)	Axis X-X S (in.³)	Axis X-X r (in.)	Axis Y-Y I (in.⁴)	Axis Y-Y S (in.³)	Axis Y-Y r (in.)	rT (in.)	d/Af	b/2tf	d/tw
W shapes																
W 12 × 36	10.6	12.24	6.565	0.540	0.305	1.06	281	46.0	5.15	25.5	7.77	1.55	1.77	3.45	6.08	40.1
× 31	9.13	12.09	6.525	0.465	0.265	1.00	239	39.5	5.12	21.6	6.61	1.54	1.75	3.98	7.02	45.6
× 27	7.95	11.96	6.497	0.400	0.237	0.93	204	34.2	5.07	18.3	5.63	1.52	1.74	4.60	8.12	50.5
W 12 × 22	6.47	12.31	4.030	0.424	0.260	0.93	156	25.3	4.91	4.64	2.31	0.847	1.03	7.20	4.75	47.3
× 19	5.59	12.16	4.007	0.349	0.237	0.87	130	21.3	4.82	3.76	1.88	0.820	1.01	8.70	5.74	51.3
× 16.5	4.87	12.00	4.000	0.269	0.230	0.81	105	17.6	4.65	2.88	1.44	0.770	0.975	11.2	7.43	52.2
× 14	4.12	11.91	3.968	0.224	0.198	0.75	88.0	14.8	4.62	2.34	1.18	0.754	0.957	13.4	8.86	60.2
W 10 × 45	13.2	10.12	8.022	0.618	0.350	1.18	249	49.1	4.33	53.2	13.3	2.00	2.21	2.04	6.49	28.9
× 39	11.5	9.94	7.990	0.528	0.318	1.12	210	42.2	4.27	44.9	11.2	1.98	2.19	2.36	7.57	31.3
× 33	9.71	9.75	7.964	0.433	0.292	1.00	171	35.0	4.20	36.5	9.16	1.94	2.16	2.83	9.20	33.4
W 10 × 29	8.54	10.22	5.799	0.500	0.289	1.06	158	30.8	4.30	16.3	5.61	1.38	1.57	3.52	5.80	35.4
× 25	7.36	10.08	5.762	0.430	0.252	1.00	133	26.5	4.26	13.7	4.76	1.37	1.56	4.07	6.70	40.0
× 21	6.20	9.90	5.750	0.340	0.240	0.87	107	21.5	4.15	10.8	3.75	1.32	1.53	5.06	8.46	41.3
W 10 × 19	5.61	10.25	4.020	0.394	0.250	0.93	96.3	18.8	4.14	4.28	2.13	0.874	1.05	6.47	5.10	41.0
× 17	4.99	10.12	4.010	0.329	0.240	0.87	81.9	16.2	4.05	3.55	1.77	0.844	1.03	7.67	6.09	42.2
× 15	4.41	10.00	4.000	0.269	0.230	0.81	68.9	13.8	3.95	2.88	1.44	0.809	1.00	9.29	7.43	43.5
× 11.5	3.39	9.87	3.950	0.204	0.180	0.75	52.0	10.5	3.92	2.10	1.06	0.787	0.975	12.2	9.68	54.8
W 8 × 35	10.3	8.12	8.027	0.493	0.315	1.00	126	31.1	3.50	42.5	10.6	2.03	2.22	2.05	8.14	25.8
× 31	9.12	8.00	8.000	0.433	0.288	0.93	110	27.4	3.47	37.0	9.24	2.01	2.21	2.31	9.24	27.8
× 28	8.23	8.06	6.540	0.463	0.285	0.93	97.8	24.3	3.45	21.6	6.61	1.62	1.80	2.66	7.06	28.3
× 24	7.06	7.93	6.500	0.398	0.245	0.87	82.5	20.8	3.42	18.2	5.61	1.61	1.78	3.07	8.17	32.4
W 8 × 20	5.89	8.14	5.268	0.378	0.248	0.87	69.4	17.0	3.43	9.22	3.50	1.25	1.42	4.09	6.97	32.8
× 17	5.01	8.00	5.250	0.308	0.230	0.81	56.6	14.1	3.36	7.44	2.83	1.22	1.40	4.95	8.52	34.8
W 8 × 15	4.43	8.12	4.015	0.314	0.245	0.81	48.1	11.8	3.29	3.40	1.69	0.876	1.04	6.44	6.39	33.1
× 13	3.83	8.00	4.000	0.254	0.230	0.75	39.6	9.90	3.21	2.72	1.36	0.842	1.02	7.87	7.87	34.8

Table 5 (continued)
SECTION PROPERTIES FOR DESIGNING

Designation	Area A (in.²)	Depth d (in.)	Flange Width b_f	Flange Thickness t_f	Web thickness t_w (in.)	k (in.)	Axis X-X I (in.⁴)	Axis X-X S (in.³)	Axis X-X r (in.)	Axis Y-Y I (in.⁴)	Axis Y-Y S (in.³)	Axis Y-Y r (in.)	r_T (in.)	d/A_f	b/2t_f	d/t_w
× 10	2.96	7.90	3.940	0.204	0.170	0.68	30.8	7.80	3.23	2.08	1.06	0.839	1.00	9.83	9.66	46.5
W 6 × 25	7.35	6.37	6.080	0.456	0.320	0.93	53.3	16.7	2.69	17.1	5.62	1.53	1.69	2.30	6.67	19.9
× 20	5.88	6.20	6.018	0.367	0.258	0.87	41.5	13.4	2.66	13.3	4.43	1.51	1.66	2.81	8.20	24.0
× 15.5	4.56	6.00	5.995	0.269	0.235	0.75	30.1	10.0	2.57	9.67	3.23	1.46	1.53	3.72	11.1	25.5
W 6 × 16	4.72	6.25	4.030	0.404	0.260	0.87	31.7	10.2	2.59	4.42	2.19	0.967	1.10	3.84	4.99	24.0
× 12	3.54	6.00	4.000	0.279	0.230	0.75	21.7	7.25	2.48	2.98	1.49	0.918	1.07	5.38	7.17	26.1
× 8.5	2.51	5.83	3.940	0.194	0.170	0.68	14.8	5.08	2.43	1.98	1.01	0.889	1.04	7.63	10.2	34.3
W 5 × 18.5	5.43	5.12	5.025	0.420	0.265	0.81	25.4	9.94	2.16	8.89	3.54	1.28	1.40	2.43	5.98	19.3
× 16	4.70	5.00	5.000	0.360	0.240	0.75	21.3	8.53	2.13	7.51	3.00	1.26	1.39	2.78	6.94	20.8
W 4 × 13	3.82	4.16	4.060	0.345	0.280	0.81	11.3	5.45	1.72	3.76	1.85	0.991	1.11	2.97	5.88	14.9
M shapes																
M 14 × 17.2	5.05	14.00	4.000	0.272	0.210	0.62	147	21.1	5.40	2.65	1.33	0.725	0.925	12.8	7.34	66.7
M 12 × 11.8	3.47	12.00	3.065	0.225	0.177	0.56	71.9	12.0	4.55	0.980	0.639	0.532	0.690	17.4	6.81	67.8
M 10 × 29.1	8.56	9.88	5.937	0.389	0.427	0.87	131	26.6	3.92	11.2	3.76	1.14	1.40	4.28	7.63	23.1
× 22.9	6.73	9.88	5.752	0.389	0.242	0.87	117	23.6	4.16	10.0	3.48	1.22	1.40	4.42	7.39	40.8
M 10 × 9	2.65	10.00	2.690	0.206	0.157	0.50	38.8	7.76	3.83	0.609	0.453	0.480	0.616	18.0	6.53	63.7
M 8 × 34.3	10.1	8.00	8.003	0.459	0.378	1.06	116	29.1	3.40	34.9	8.73	1.86	2.08	2.18	8.72	21.2
× 32.6	9.58	8.00	7.940	0.459	0.315	1.06	114	28.4	3.44	34.1	8.58	1.89	2.08	2.20	8.65	25.4
× 6.5	1.92	8.00	2.281	0.189	0.135	0.50	18.5	4.62	3.10	0.343	0.301	0.423	0.535	18.6	6.03	59.3
M 7 × 5.5	1.62	7.00	2.080	0.180	0.128	0.43	12.0	3.44	2.73	0.249	0.239	0.392	0.493	18.7	5.78	54.7
M 6 × 22.5	6.62	6.00	6.060	0.379	0.372	0.81	41.2	13.7	2.49	12.4	4.08	1.37	1.55	2.61	7.98	16.1
× 20	5.89	6.00	5.938	0.379	0.250	0.81	39.0	13.0	2.57	11.6	3.90	1.40	1.54	2.66	7.82	24.0
× 4.4	1.29	6.00	1.844	0.171	0.114	0.37	7.20	2.40	2.36	0.165	0.179	0.358	0.444	19.0	5.39	52.6
M 5 × 18.9	5.55	5.00	5.003	0.416	0.316	0.87	24.1	9.63	2.08	7.86	3.14	1.19	1.32	2.40	6.01	15.8
M 4 × 13	3.81	4.00	3.940	0.371	0.254	0.81	10.5	5.24	1.66	3.36	1.71	0.939	1.04	2.73	5.30	15.7

Abridged from Steel Construction Manual, 8th ed., American Institute for Steel Construction, New York, 1980. With permission.

Table 6
ALLOWABLE COMPRESSIVE STRESS FOR A36 STEEL COLUMNS

Main and secondary members KL/r not over 120						Main members KL/r 121 to 200				Secondary members[a] L/r 121 to 200			
$\frac{KL}{r}$	F_a (ksi)	$\frac{KL}{r}$	F_a (ksi)	$\frac{KL}{r}$	F_a (ksi)	$\frac{KL}{r}$	F_a (ksi)	$\frac{KL}{r}$	F_a (ksi)	$\frac{L}{r}$	F_{as} (ksi)	$\frac{L}{r}$	F_{as} (ksi)
5	21.39	45	18.78	85	14.79	125	9.55	165	5.49	125	9.80	165	7.08
10	21.16	50	18.35	90	14.20	130	8.84	170	5.17	130	9.30	170	6.89
15	20.89	55	17.90	95	13.60	135	8.19	175	4.88	135	8.86	175	6.73
20	20.60	60	17.43	100	12.98	140	7.62	180	4.61	140	8.47	180	6.58
25	20.28	65	16.94	105	12.33	145	7.10	185	4.36	145	8.12	185	6.46
30	19.94	70	16.43	110	11.67	150	6.64	190	4.14	150	7.81	190	6.36
35	19.58	75	15.90	115	10.99	155	6.22	195	3.93	155	7.53	195	6.28
40	19.19	80	15.36	120	10.28	160	5.83	200	3.73	160	7.29	200	6.22

[a] K taken as 1.0 for secondary members.

From Steel Construction Manual, 8th ed., American Institute of Steel Construction, Chicago, 1980. With permission.

$b/t < 76/\sqrt{F_y}$ for single- and double-angle struts

$b/t < 95/\sqrt{F_y}$ for columns and compression flanges of beams

$b/t < 127/\sqrt{F_y}$ for stems of tees

For stiffened compression elements, the b/t limits are

$$b/t < 238/\sqrt{F_y} \quad \text{for box sections}$$

$$b/t < 253/\sqrt{F_y} \quad \text{for uniformly compressed elements}$$

Equations 32 to 34 define the allowable compressive stresses set forth by AISC for centrically loaded hot rolled steel compression members:

$$F_c = \frac{[1 - (KL/r)^2/2C_c^2] \, F_y}{5/3 + \dfrac{3KL/r}{8C_c} - \dfrac{(KL/r)^3}{C_c^3}} \tag{32}$$

if $KL/r < C_c = \sqrt{2\pi^2 E/F_y}$

$$F_a = \frac{12 \, \pi^2 E}{23(KL/r)^2} \tag{33}$$

if $C_c < KL/r \leq 200$

$$F_a = 0 \tag{34}$$

if $KL/r > 200$

Table 6 summarizes the variation of allowable compressive stress with slenderness ratio for A36 steel. Tables 7 to 9 list the allowable compressive load for extra-strong pipe, standard pipe, and square structural tubing fabricated from A36 steel.

The wide-flange section is the most commonly used steel beam section for agricultural

Table 7
ALLOWABLE COLUMN LOAD IN KIPS FOR A36 EXTRA-STRONG PIPE

Effective Length, KL, in Feet with Respect to Radius of Gyration

Diameter (in.)	6	5	4	$3^{1}/_{2}$	3
Thickness (in.)	0.432	0.375	0.337	0.318	0.300
lb/ft	28.5	20.7	14.9	12.5	10.2
6	166	118	81	66	52
7	162	114	78	63	48
8	159	111	75	59	45
9	155	107	71	55	41
10	151	103	67	51	37
11	146	99	63	47	33
12	142	95	59	43	28
13	137	91	54	38	24
14	132	86	49	33	21
15	127	81	44	29	18
16	122	76	39	25	16
17	116	71	34	23	14
18	111	65	31	20	12
19	105	59	28	18	11
20	99	54	25	16	
22	86	44	21		
24	73	37	17		
26	62	32			
28	54	27			
30	47	24			

Properties

Area A, in.^{2}	8.40	6.11	4.41	3.68	3.02
I, in.^{4}	40.5	20.7	9.61	6.28	3.89
r, in.	2.19	1.84	1.48	1.31	1.14
B, in.$^{-1}$	0.688	0.822	1.03	1.17	1.36
a[a]	6.00	3.08	1.44	0.941	0.585

[a] Multiply values by 10^6.

Abridged from Steel Construction Manual, 8th ed., American Institute of Steel Construction, New York, 1980. With permission.

applications. Most of the standard wide-flange sections are compact if adequate lateral support is provided; that is, the sections are proportioned such that the cross section can support its full plastic moment capacity without buckling locally.

If a wide-flange section is compact and adequately supported laterally, the allowable flexural stress is

$$F_b = 0.66 \, F_y \tag{35}$$

The section is adequately supported if the unbraced length of the compression flange, L_b, is less than both L_c and L_u where:

$$L_c = 76 \, b_f / \sqrt{F_y} \tag{36}$$

Table 8
ALLOWABLE COLUMN LOAD IN KIPS FOR A36 STANDARD PIPE

Effective Length, KL, in Feet with Respect to Radius of Gyration

Diameter (in.)	6	5	4	$3^1/_2$	3
Thickness (in.)	0.280	0.258	0.237	0.226	0.216
lb/ft	18.9	14.6	10.7	9.1	7.5
6	110	83	59	48	38
7	108	81	57	46	36
8	106	78	54	44	34
9	103	76	52	41	31
10	101	73	49	38	28
11	98	71	46	35	25
12	95	68	43	32	22
13	92	65	40	29	19
14	89	61	36	25	16
15	86	58	33	22	14
16	82	55	29	19	12
17	79	51	26	17	11
18	75	47	23	15	10
19	71	43	21	14	9
20	67	39	19	12	
22	59	32	15	10	
24	51	27	13		
26	43	23			
28	37	20			
30	32	17			

Properties

Area A, in.^{2}	5.58	4.30	3.17	2.68	2.23
I, in.^{4}	28.1	15.2	7.23	4.79	3.02
r, in.	2.25	1.88	1.51	1.34	1.16
B, in.$^{-1}$	0.657	0.789	0.987	1.12	1.29
a[a]	4.21	2.26	1.08	0.717	0.447

[a]   Multiply values by 10^6.

Abridged from Steel Construction Manual, 8th ed., American Institute of Steel Construction, New York, 1980.

$$L_u = \frac{20,000}{\dfrac{d}{A_f}(F_y)} \tag{37}$$

where b_f = flange width, d = depth of the beam, A_f = flange area, and F_y = yield stress in ksi.

If the wide-flange section is compact except that $L_c < L_b < L_u$, then

$$F_b = 0.60 \, F_y \tag{38}$$

If $L_b > L_u$ and L_c, then $F_b < 0.60 \, F_y$ and is defined as the larger of the stresses evaluated from Equations 39 or 40, and 41:

Table 9

AXIAL COLUMN LOAD IN KIPS FOR A36 SQUARE STRUCTURAL TUBING

Effective Length, KL, in Feet with Respect to Radius of Gyration

Size (in.)	6 × 6			5 × 5			4 × 4		
Thickness (in.)	3/8	5/16	1/4	3/8	5/16	1/4	3/8	5/16	1/4
lb/ft	27.0	23.0	18.8	21.9	18.7	15.4	16.8	14.5	12.0
6	157	134	110	124	107	88	91	79	66
7	154	132	108	121	104	86	87	76	63
8	151	129	106	117	101	83	83	72	60
9	147	126	103	113	98	81	79	69	57
10	144	123	101	109	94	78	74	65	54
11	140	120	98	105	91	75	69	61	51
12	136	116	95	101	87	72	64	57	48
13	131	113	93	96	83	69	59	52	44
14	127	109	90	91	79	66	53	47	40
15	122	105	87	86	75	62	47	42	36
16	118	101	83	81	70	59	42	37	32
17	113	97	80	75	66	55	37	33	29
18	108	93	77	69	61	51	33	30	25
19	102	88	73	63	56	47	29	27	23
20	97	84	69	57	51	43	27	24	21
22	85	74	62	47	42	35	22	20	17
24	73	64	54	40	35	30	18	17	14
26	62	54	46	34	30	25			
28	54	47	39	29	26	22			
30	47	41	34	25	22	19			
32	41	36	30						
34	36	32	27						
36	32	28	24						
38		25	21						

Properties

Area A, in.^{2}	7.95	6.77	5.54	6.45	5.52	4.54	4.95	4.27	3.54
I, in.^{4}	40.5	35.5	29.9	22.0	19.5	16.6	10.2	9.23	8.00
r, in.	2.26	2.29	2.32	1.85	1.88	1.91	1.44	1.47	1.50
B, in.$^{-1}$	0.589	0.572	0.556	0.773	0.708	0.684	0.971	0.925	0.885
a[a]	6.05	5.29	4.44	3.29	2.91	2.47	1.53	1.37	1.19

[a] Multiply values by 10^6.

Abridged from Steel Construction Manual, 8th ed., American Institute of Steel Construction, New York, 1980. With permission.

$$F_b = \left[\frac{2}{3} - \frac{(L_b/r_T)^2 \, F_y}{1530 \times 10^3} \right] F_y$$

$$\text{if} \quad \left(\frac{102 \times 10^3}{F_y} \right)^{1/2} \leqslant L_b/r_T \leqslant \left(\frac{510 \times 10^3}{F_y} \right)^{1/2} \tag{39}$$

$$\text{or} \quad F_b = \frac{170 \times 10^3}{(L_b/r_T)^2}$$

$$\text{if} \quad L_b/r_T > \left(\frac{510 \times 10^3}{F_y} \right)^{1/2} \tag{40}$$

$$\text{and} \quad F_b = \frac{12 \times 10^3}{(d/A_f) \, L_b} \tag{41}$$

where L_b/r_T = slenderness ratio of the compression flange plus $\frac{1}{6}$ of the web area about the y-axis of the section (Table 5), and F_y = yield stress in ksi.

The allowable flexural stresses may be used conservatively to design steel flexural members that are compact in shape and are loaded through the shear center of the cross section. The allowable stresses for several wide-flange sections are plotted in Figure 6 for a wide range of unsupported lengths for A36 steel.

In structural steel beams, the allowable shear stress is $0.4 \, F_y$. Allowable bearing stresses at end reactions and concentrated loads equal $0.75 \, F_y$.

Steel beam-columns must be proportioned to satisfy one or two of the following interaction equations. If $f_a/F_a < 0.15$,

$$f_a/F_a + \frac{f_{bx}}{F_{bx}} + \frac{f_{by}}{F_{by}} \leqslant 1.0 \tag{42}$$

If $f_a/F_a > 0.15$, then both Equations 43 and 44 must be satisfied:

$$f_a/F_a + \frac{C_{mx} \, f_{bx}}{F_{bx}(1 - f_a/F'_{ex})} + \frac{C_{my} \, f_{by}}{F_{by}(1 - f_a/F'_{ey})} \leqslant 1.0 \tag{43}$$

$$\frac{f_a}{0.6 \, F_y} + \frac{f_{bx}}{F_{bx}} + \frac{f_{by}}{F_{by}} \leqslant 1.0 \tag{44}$$

where f_a, f_{bx}, f_{by} = actual compressive and flexural stresses; F_a = allowable axial stress for pure compression based upon the maximum slenderness ratio; F_{bx}, F_{by} = allowable pure flexural stresses about the x- and y-axes; F'_{ex}, F'_{ey} = Euler buckling load based upon KL/r about the axis of bending; C_m = 0.85 for members in frames subject to sidesway or in frames without sidesway with transverse loads between the ends in the plane of bending; C_m = 1.00 for frames with sidesway and transverse loads between the ends; and C_m = 0.6 − 0.4 $M_1/M_2 \geqslant 0.40$ for members in frames restrained from sidesway and not subject to transverse loading between supports in the plane of bending. M_1/M_2 is the ratio of the smaller to larger end moment and is positive for reverse-curvature bending and negative for single-curvature bending.

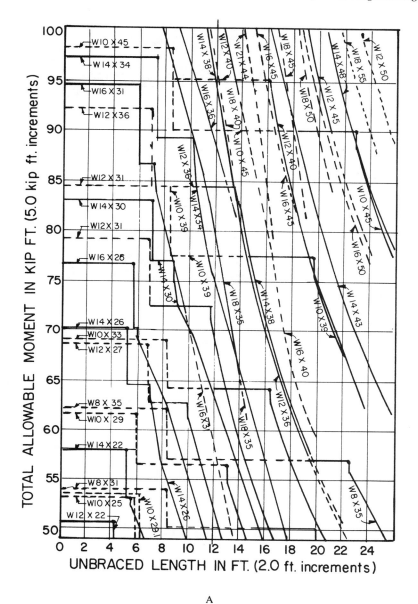

FIGURE 6. Allowable moments in beams, F_y = 36 ksi. (From Steel Construction Manual, 8th ed., American Institute of Steel Construction, New York, 1980. With permission.)

COLD FORMED STEEL

Many cold formed sections are custom fabricated for a specific job. There are few standard shapes. The elements of the sections are quite thin and local buckling of compression elements usually occurs before the elastic capacity is developed. Thus, reduced compressive stress levels or reduced section properties need to be used to design either columns or beams.

The procedures for designing cold-formed sections are beyond the scope of this handbook. For information on cold-formed design, consult the AISI Specification and Yu.[7]

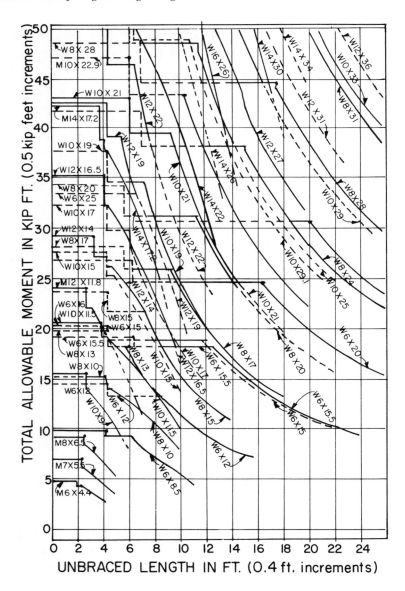

FIGURE 6B.

SPECIAL CONSIDERATIONS FOR REINFORCED CONCRETE

Concrete is much stronger in compression than in tension. Typically, the tensile strength of concrete is only 10% as great as the compressive strength.

The compressive strength of concrete is dependent upon many factors. Chief among them are the water to cement ratio of the mix, the length of cure, and the method of curing. Generally, the strength of concrete increases with decreasing water/cement ratios and increasing length of cure. Moist cured concrete is also stronger than dry cured concrete. Typical strengths of concrete mixes and the influence of water/cement ratios, cure length, and type are summarized in Figures 7 and 8. Recommended water/cement ratios and mix designs are presented in Table 10.

Reinforcing steel is added to the tension zones of concrete columns and beams. The steel

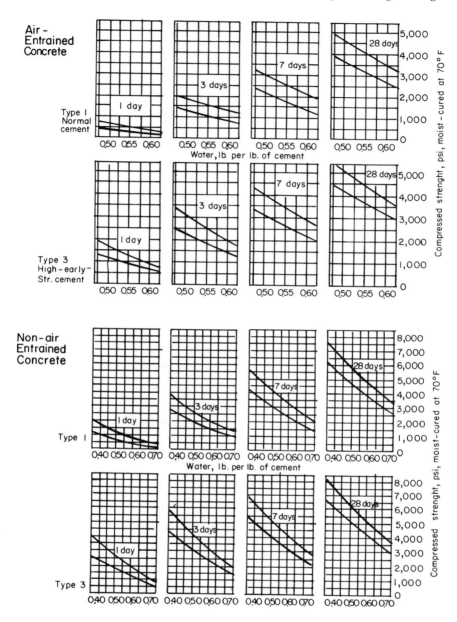

FIGURE 7.   Relationship between water-cement ratio and compressive strength for Portland cements at different ages. (These relationships are approximate and should be used only as a guide in lieu of data on job materials.) (From Design and Control of Concrete Mixtures, 11th ed., Portland Cement Association, Skokie, Ill., 1968. With permission.)

carries the tensile loads and compensates for the inherent weakness of the concrete. Typical reinforcing bar grades and sizes are given in Tables 11 and 12. Grade 40 is the most commonly used steel in agricultural applications. Reinforcing steel needs to have adequate cover to be effective. Recommended covers for typical construction are given in Table 13.

Many agricultural applications of reinforced concrete are singly reinforced beams or short centrically loaded columns. These are the only type members that are considered herein. The reader is directed to ACI 318-83 Specification, publications of ACI and PCA, and standard textbooks on reinforced concrete design for other applications.

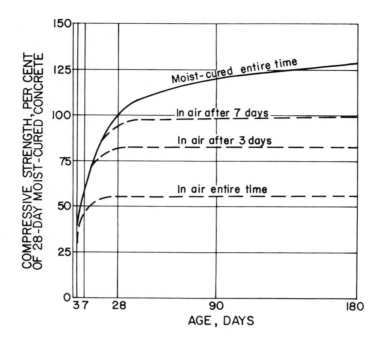

FIGURE 8.   Effect of cure length and type upon compressive strength of concrete. (From Design and Control of Concrete Mixtures, 11th ed., Portland Cement Association, Skokie, Ill., 1968. With permission.)

The ultimate design approach is the most commonly used approach in modern reinforced concrete practice. The design equations presented herein are based on ultimate strength criteria.

Reinforced concrete sections must be proportioned to carry factored service loads. The ACI code requires the following loads, U:

1.   For dead loads, D, plus live loads, L

$$U = 1.4D + 1.7L \tag{45}$$

2.   For dead, live, and wind load combinations the greater of

$$U = 0.75(1.4D + 1.7L + 1.7W)$$
$$U = 0.9D + 1.3W$$
$$U = 1.4D + 1.7L \tag{46}$$

3.   For dead, live, and lateral earth load combinations, H, the greater of

$$U = 1.4D + 1.7L + 1.7H \quad \text{if} \quad D \text{ and } L \text{ in direction of } H$$
$$U = 0.9D + 1.7H \quad \text{if} \quad D \text{ or } L \text{ oppose } H$$
$$U = 1.4D + 1.7L \tag{47}$$

In each of Equations 45 to 47, U represents the required load or moment the section must carry, and D, L, W, L, and H represent the service loads or moments.

Table 10
SUGGESTED TRIAL MIXES FOR CONCRETE

| | | Water added, gal (ℓ) | | | | | | |
| | | Gal (ℓ) of water for each sack of cement, using[a] | | | Suggested mixture for 1-sack trial batches,[b] ft^3 (m^3) | | | |
	Max size aggregate, in. (mm)	Damp[c] sand	Wet[d] (average) sand	Very[e] wet sand	Cement, sacks	Aggregates Fine	Aggregates Coarse	Ready-mix sacks cement, yr^3 (m^3)[f]
5-gal mix; use for concrete subjected to severe wear, weather, or weak acid and alkali solutions	$^3/_4$ in. (19)	4 $^1/_2$ (17)	4 (15)	3 $^1/_2$ (13)	1 (0.028)	2 (0.057)	2 $^1/_4$ (0.064)	7 $^3/_4$ (10 $^1/_4$)
6-gal mix; use for floors (home, barn), driveways, walks, septic tanks, storage tanks, structural concrete	1 in. (25)	5 $^1/_2$ (21)	5 (19)	4$^1/_2$ (17)	1 (0.028)	2 $^1/_4$ (0.064)	3 (0.085)	6 $^1/_4$ (8 $^1/_4$)
	1 $^1/_2$ in. (38)	5 $^1/_2$ (21)	5 (19)	4 $^1/_2$ (17)	1 (0.028)	2 $^1/_2$ (0.071)	3 $^1/_2$ (0.099)	6 (7 $^7/_8$)
7-gal mix; use for foundation walls, footings, mass concrete, etc.	1 $^1/_2$ in. (38)	6 $^1/_4$ (24)	5 $^1/_2$ (21)	4 $^3/_4$ (18)	1 (0.028)	3 (0.085)	4 (0.113)	5 (6 $^1/_2$)

[a] Increasing the proportion of water to cement reduces the strength and durability of cement. Adjust the proportions of trial batches without changing the water-cement ratio. Reduce gravel to improve smoothness; reduce both sand and gravel to reduce stiffness. One sack equals 94 lb (43 kg).

[b] Proportions will vary slightly depending on gradation of aggregates.

[c] Damp sand will fall apart after being squeezed in the palm of the hand.

[d] Wet sand will ball in the hand when squeezed, but leaves no moisture on the palm.

[e] Very wet sand has been recently rained on or pumped.

[f] Medium consistency [3 in. (76 mm) slump]. Order air-entrained concrete for outdoor use.

From Structures and Environment Handbook, MWPS-1, 11th ed., Midwest Plan Service, Ames, Iowa, 1983. With permission.

Table 11
GRADES AND STRENGTH OF REINFORCING STEEL

Grade	Design strength, f_y (psi)
40	40,000
50	50,000
60	60,000
80	80,000
100	100,000

Table 12
REINFORCING STEEL BAR SIZES

Bar no.[a]	Size (diam)	Area (in.^{2})
2	$^1/_4$	0.05
3	$^3/_8$	0.11
4	$^1/_2$	0.20
5	$^5/_8$	0.31
6	$^3/_4$	0.44
7	$^7/_8$	0.60
8	1	0.79

[a] Bar no. indicates the number of eights of an inch diameter.

Table 13
MINIMUM COVER REQUIREMENTS FOR STEEL

Placement type	Construction type	Cover (in.)
Cast-in-place	Cast against and exposed to earth	3
	Concrete exposed to earth and weather	
	$d_b \geq 3/4$ in.	2
	$d_b < 3/4$ in., W31 or D31 wire or smaller	1 1/2
	Concrete not exposed to weather or in contact with ground	
	Slabs, walls, joists ($d_b \leq 1 3/8$ in.)	3/4
	Primary reinforcement, ties, stirrups, spirals in beams and columns	1 1/2
Precast	Concrete exposed to earth and weather	
	Wall panels ($d_b \leq 1 3/8$ in.)	3/4
	Other members (3/4 in. $\leq d_b \leq 1 3/8$ in.)	1 1/2
	$d_b \leq 5/8$ in.)	1 1/4
	Concrete not exposed to weather or in contact with ground	
	Slabs, walls, joists ($d_b \leq 1 3/8$ in.)	5/8
	Primary reinforcement in beams	Greater of d_b or 5/8 in.
	Ties, stirrups, or spirals in beams or columns	3/8

The theoretical moment capacity of the rectangular reinforced beam section of Figure 9 is

$$M_n = T(d - a/2) = A_s f_y(d - a/2) \tag{48}$$

or

$$M_n = C(d - a/2) = 0.85 f_c'(ab)(d - a/2) \tag{49}$$

provided the section is underreinforced.

If the section is underreinforced

$$a = \frac{A_s f_y}{0.85 f_c' b} \tag{50}$$

where A_s = area of reinforcing steel, f_y = yield stress of steel, a = depth of concrete compression block, f_c = 28-day compressive strength of concrete, b = width of the rectangular section, and d = distance from compression edge to the reinforcing steel.

A rectangular section is underreinforced if $A_s < A_{sb}$ where

$$A_{sb} = \frac{0.85 B_1 f_c'}{f_y} \left(\frac{87}{87 + f_y} \right) bd \tag{51}$$

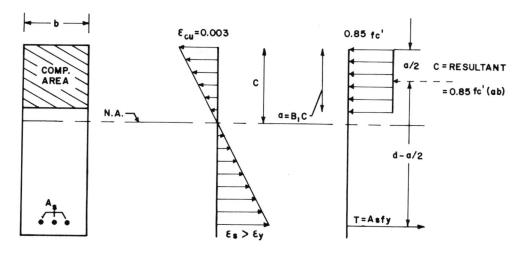

FIGURE 9.   Stress and strain distributions in an underreinforced rectangular beam at ultimate capacity.

or when the steel ratio $\rho = A_s/bd < \rho_b$ where

$$\rho_b = \frac{0.85 \; B_1 \; f_c'}{f_y} \left(\frac{87}{87 + f_y} \right) \tag{52}$$

In Equations 51 and 52 f_c' and f_y both have units of ksi and $B_1 = 0.85$ for $f_c' \leq 4$ ksi.

To assure underreinforced behavior, the ACI code requires that the maximum allowable steel ratio, ρ_{all}, be less than $0.75 \; \rho_b$. Thus,

$$\rho_{all} = \frac{(A_s)all}{bd} \leq \frac{0.638 \; B_1 \; f_c'}{f_y} \; \frac{87}{87 + f_y} \tag{53}$$

To prevent cracking of lightly loaded beams, ACI requires a minimum steel ratio, ρ_{min}.

$$\rho_{min} = \frac{200}{f_y} \tag{54}$$

where f_y is in psi.

Equation 55 is the design criterion to be satisfied for underreinforced and singly reinforced rectangular beams:

$$\phi M_n \geq M_u \tag{55}$$

where ϕ = capacity reduction factor for flexure (0.90), and M_n = theoretical moment capacity from Equation 48 or 49.

If the reinforced beam depth is less than 10 in., shear reinforcements (stirrups) are not required. This covers the vast majority of agricultural applications. If the beam depth is greater than 10 in., shear reinforcement is necessary and the reader is directed to any standard text on reinforced concrete and the ACI 318-83 Specifications for details. Other common situations where shear reinforcement is generally not required are footers and slabs.

Deflection analysis of concrete beams is beyond the scope of this text. Deflections need not be computed, however, if the beam depth, h, is at least as great as the values given in Table 14.

Table 14
MINIMUM THICKNESS OF NONPRESTRESSED BEAMS OR ONE-WAY SLABS UNLESS DEFLECTIONS ARE COMPUTED[a]

| | Minimum thickness, h | | | |
Member[b]	Simply supported	One end continuous	Both ends continuous	Cantilever
Solid one-way slabs	$\ell/20$	$\ell/24$	$\ell/28$	$\ell/10$
Beams or ribbed one-way slabs	$\ell/16$	$\ell/18.5$	$\ell/21$	$\ell/8$

Note: Values given shall be used directly for members with normal weight concrete (w_c = 145 pcf) and Grade 60 reinforcement. For other conditions, the values shall be modified as follows. (a) For structural lightweight concrete having unit weights in the range 90 to 120 lb/ft^3, the values shall be multiplied by 1.65 to 0.05 w_c but not less than 1.09, where w_c is the unit weight in lb/ft^3. (b) For f_y other than 60,000 psi, the values shall be multiplied by $(0.4 + f_y/100,000)$.

[a]   Span length ℓ is in inches.
[b]   Members not supporting or attached to partitions or other construction are likely to be damaged by large deflections.

From Building Code Requirements for Reinforced Concrete, ACI-318-77, American Concrete Institute, Detroit, 1977, 30. With permission.

To be effective, reinforcing steel must be embedded in concrete at least a distance l_d before encountering the design moment it is required to carry. For rebars No. 11 and smaller in size, required development length, l_d, is the larger of

$$l_d = 0.04 \, A_b f_y / \sqrt{f_c'}$$

$$\text{or} \quad l_d = 0.004 \, d_b f_y$$

$$\text{or} \quad l_d \geq 12 \text{ in.} \tag{56}$$

where A_b = area of 1 rebar; f_y = steel yield stress, psi; f_c' = concrete compressive strength, psi; d_b = rebar diameter, in.; and l_d = development length, in.

Short Centrically Loaded Columns

A centrically loaded reinforced concrete column is classified as short if (1) KL/r < 34 in braced frames without sidesway, and (2) KL/r < 22 in unbraced frames with sidesway where K = effective length factor for column end conditions, L = unsupported column length, $r = \sqrt{I_g/A_g}$, I_g = moment of inertia of the gross column area, and A_g = gross column area.   The theoretical load capacity of the short centrically loaded reinforced column of Figure 10 is

$$P_n = (0.85 \, f_c')(A_g - A_s) + f_y A_s \tag{57}$$

After consideration of accidental eccentricity of loading and capacity reduction factors, the maximum allowable load for spiral columns is

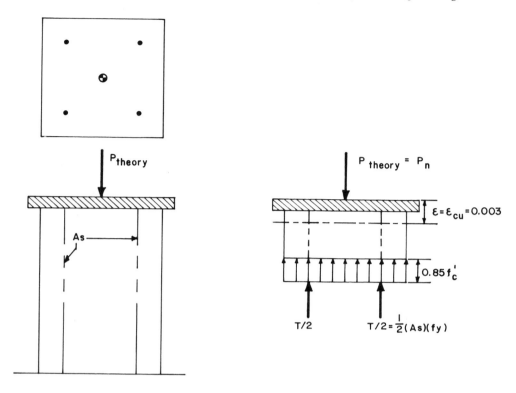

FIGURE 10.   Theoretical axial ultimate load capacity of a short reinforced concrete column.

$$(\phi\ P_n)_{max} = 0.85\ \phi\ [0.85\ f_c'(A_g - A_s) + A_s f_y] \tag{58}$$

and for tied columns is

$$(\phi\ P_n)_{max} = 0.80\ \phi\ [0.85\ f_c'(A_g - A_s) + A_s f_y] \tag{59}$$

where $\phi = 0.75$ for spiral columns, $\phi = 0.70$ for tied columns, A_g = gross column area, and A_s = total steel area. The criterion for short column design for a dead plus live load combination is

$$(\phi\ P_n)_{max} \geqslant P_u = 1.4 P_D + 1.7 P_L \tag{60}$$

The ties in tied columns (Figure 11a) must be No. 3 rebars or larger and spaced less than or equal to the smaller of 16 longitudinal bar diameters, 48 tie bar diameters, or the least dimension of the column. There should be at least four longitudinal bars in a tied column, and every corner and alternate longitudinal bar needs to be supported by a tie. The maximum allowable distance between a corner tied bar and an alternate bar is 6 in. on both sides.

Lateral reinforcement in spiral columns (Figure 11b) should be at least

$$\rho_s \geqslant 0.45\left(\frac{A_g}{A_c} - 1\right)\frac{f_c'}{f_y} \tag{61}$$

where $\rho_s = \dfrac{\text{volume of spiral steel/loop}}{\text{volume of concrete enclosed by spiral loop}}$, A_c = concrete area enclosed by outside of spiral, and A_g = gross column area.

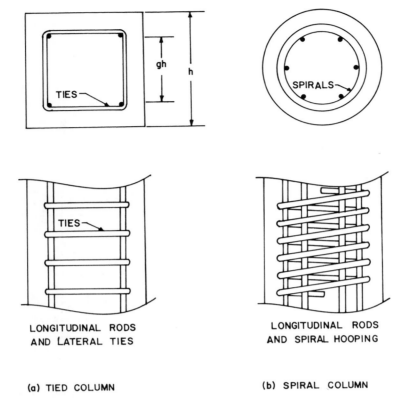

(a) TIED COLUMN   (b) SPIRAL COLUMN

FIGURE 11.   Reinforced column types.

Spirals must be No. 3 bars or greater and spiral spacings should be between 1 and 3 in., but greater than (4/3) × (maximum aggregate size). At least six longitudinal bars are required for spiral columns.

The limits on longitudinal reinforcement for all reinforced columns are

$$0.01 \leq \rho \leq 0.08 \qquad (62)$$

where

$$\rho = \frac{A_s}{A_g}$$

Consideration of eccentrically loaded columns, long columns, and beam-columns is beyond the scope of this text. Details for any of these situations may be found in standard texts on reinforced concrete design and ACI and PCA publications.

REFERENCES

1. Building Code Requirements for Reinforced Concrete, ACI 318-83, American Concrete Institute, Detroit, 1983.
2. Design and Control of Concrete Mixtures, 11th ed., Portland Cement Association, Skokie, Ill., 1968.
3. Design and Control of Concrete Mixtures, 12th Ed., Portland Cement Association, Skokie, Ill., 1978.
4. Design Specification for Cold-Formed Steel Design Manual, American Iron and Steel Institute, Washington, D.C., 1986.
5. **Higdon, A. et al.,** *Mechanics of Materials,* 3rd ed., John Wiley & Sons, New York, 1976.
6. National Design Specification for Wood Construction, National Forest Products Association, Washington, D.C., 1986.
7. Steel Construction Manual, 8th ed., American Institute of Steel Construction, New York, 1980.
8. Structures and Environment Handbook, 9th ed., Midwest Plan Service, Ames, Iowa, 1983.
9. **Yu, W. W.,** *Cold-Formed Steel Structures,* McGraw-Hill, New York, 1973.

STUD FRAMES, POLE FRAMES, STEEL FRAMES, AND TILT-UP CONCRETE

I. L. Winsett

INTRODUCTION

General-purpose agricultural buildings are usually less than 30 ft in height and are designed to carry their own weight plus the load of people, animals, fixtures, furniture, and equipment. This is relatively light loading limited to 50 to 100 lb/ft². A typical example is a residence or a shop.

Raised floors are structural designs that utilize floor framing systems, usually constructed of wood. An important factor in the design of a wood floor system is to equalize shrinkage and expansion at the outside wall and the interior supports. This is accomplished by having the same total depth of wood at the interior supports as at the outside framing. Thus, as beams and joists approach moisture equilibrium, there are only small differences in the amount of shrinkage.

Grades of dimension lumber vary considerably by species. A sequence of first-, second-, third-, fourth-, and sometimes fifth-grade material is used. In general, the first grade is for a high or special use, the second for better than average, the third for average, and the fourth and fifth for more economical construction. Joists and girders are usually second-grade material of a species while sills and posts are usually of third or fourth grade. Specific recommendations for each species are available from the organizations listed in ASAE Standard S288.3, Section 8.[1]

The strength of wood framing materials varies with the defects, their size, and location. The strength of wood varies greatly between species, with southern pine and Douglas fir being the strongest and most commonly used framing materials in the U.S.

Typically, floor framing is usually either 2 × 8, 2 × 10, or 2 × 12 placed 16 in. on center. The floor joists are selected to meet strength and stiffness requirements. Strength requirements depend upon the loads to be carried. Stiffness requirements place limits on deflection (sag) under load. The grades usually used for joists are "standard" for Douglas fir, "number two or number two KD" for southern pine, and comparable grades for other sociation (SFPA).

SAGGING OR DEFLECTION AS A DESIGN GUIDE

When engineering analysis or the strength of the materials available is not known, the trial loading on an individual member may be attempted. An adequate wood floor framing member, 10 ft long, should sag or deflect no more than $1/_2$ in. when loaded. A 16-ft span should deflect no more than $3/_4$ in. If excessive deflection is encountered when loading the framing, the size of the framing member can be increased, spacing of the framing can be decreased, or the span can be reduced.

Floors for stacking of fertilizer, grain, cement, or other heavy agricultural crops should be limited to a maximum depth of 4 ft unless a special engineering design is undertaken. Even these loads can overload a floor framing system. The guide for safety is to watch the deflection of the floor joists as the load is applied. The deflections mentioned earlier should not be exceeded.

WALL FRAMING

Walls are usually framed of either wood or light gage steel sections. Typical lightweight structures use 2 × 4 studs placed 2 ft on centers. The purpose of the sidewall framing is to withstand the wind loads and to transmit the roof load to the foundation, except in pole construction. Therefore, loading of the sidewall from the inside with grain or other product is generally not permitted unless the foundation, anchorages, and sidewalls are significantly strengthened.

Wind blowing on the side or end of the structure will tend to make the framing collapse unless diagonal bracing can be provided in some manner. Pole framing is an exception as it transfers the wind forces directly into the ground and consequently does not require the degree of diagonal bracing required in other framing methods. However, diagonal wind bracing is required at the top of the structure when tying rafters and trusses to the poles. The typical bracing requirements are listed in the Midwest Plans Service sketches and Structures and Environment Handbook.[2] Where walls are to be covered with plywood, the plywood can be expected to carry the diagonal loading and no additional wind bracing for the walls is required. Roofing systems on all structures require special attention to wind bracing.

Agricultural buildings are generally less than 30 ft in height and can generally be provided with a wind velocity design pressure of 15 lb/ft². There are corrections for height and wind exposure as noted in ASAE S228.3. The 15-lb wind loading is normally considered adequate for winds in the 80-mi/hr range.

ROOFS

Many materials are utilized for roofing systems requiring many different designs. A typical wood framing system uses 2 × 4s set on edge for spans up to 10 ft. Typically, 2 × 6s are used up to 16 ft. They are usually placed 2 ft on centers.

Typical of roof loadings are 2 to 10 lb/ft² for dead load. The live load for roofs would consist of the wind and snow. A roof will have to be checked for snow with no wind and wind on one side plus snow on the appropriate side. The typical weight attributed to snow is that 10 in. of snow approximates 1 in. of water, which is about 5 lb/ft². Therefore, areas receiving accumulations of 10 in. of snow or less would have only 5 lb/ft² live load plus the dead load. Most roof designs provide a minimum of 15 to 20 lb/ft² for wood framing systems, which is quite adequate across the southern U.S. Heavier designs are required in areas receiving heavier snow.

Some light steel framed structures designed for only 5 lb/ft² live roof load have failed under relatively small accumulations of snow. However, if the design allowed for the dead load, live load, and wind, this should not be a problem in southern U.S. Failures can usually be attributed to the lack of a brace or poor construction practices.

Wooden roof trusses are now used extensively for economy of materials, as heavier roof loads can be carried over longer spans than would be possible with individual framing members. Spans from 20 to 60 ft are typical. The trusses are usually rated in pounds per linear feet of truss and require the user to determine the spacing of the trusses. Of course, in colder climates, with large buildups of snow, live loads can go up to 35 or 40 lb ft², which requires closer truss spacing. The same approach would apply to steel trusses or framing members as well.

JOINTS

Fasteners of wood joints are critical. In fact, most structural failures of wood buildings involve inadequate fastening of the members. Nails have a limited load-carrying ability, require adequate spacing to distribute their load, and must not split the framing materials. The nail holding power of smooth shanked nails in treated wood is drastically reduced. Therefore special ring shanked nails must be used for good joints in treated wood. The various sized nails are rated in their ability to resist withdrawal and to withstand lateral loads. These values are available from the various wood and forest product associations.

Steel framing is available for many designs. The spacing of the frames is limited by the size and type of the materials provided by the manufacturer and the particular wind and snow loadings to be applied to the structure. Therefore manufacturer's recommendations must be used for all steel-framed buildings.

TILT-UP CONCRETE

Tilt-up concrete sidewalls can be cast with the simpliest of materials and tools. A pipe tilting frame can be used to pull the panel erect. This construction is practical where a sand floor covered with plastic or a concrete slab is available. The panels must be tilted up on the footings and not relocated unless heavy equipment is available for moving the panels.

The concrete panels are usually 4 in. thick and must be reinforced with steel to provide the needed strength for tilting up. A 25-hp or heavier tractor is adequate to tilt the panels up. The footing of concrete or slab is poured first. Vertical reinforcing rods are placed at the time of the footing. After the panel is raised into place, forms are placed on the inside and outside of the adjacent panels. This forms a concrete column reinforced with the vertical steel when the void is filled with concrete. Note that a very small aggregate must be used for the concrete to fit into the small cavities and close tolerances of the column. The panel is usually limited to an 8- or 10-ft height. Of course, larger panels can be utilized with heavy construction equipment. Specific designs and typical reinforcement for concrete tilt-up sidewalls and construction techniques are available from the Portland Cement Association.

Sweating of the concrete walls can be a serious problem in some applications and locations. Surfaces can be insulated and treatment of inside walls can help to minimize this particular problem. Conventional roofing systems are used for tilt-up construction.

POLE FRAMES

Treated wood poles are excellent construction materials that are low in cost and quite durable. When used, each pole is placed a minimum of 3 ft 6 in. into the ground. In large buildings, the pole may extend as much as 5 ft into the ground and require a concrete pad under the base of the pole, or it may be back-filled with gravel. Pole spacings can be determined from the spacing charts from the Midwest Plan Service[2] as a guide.

Roof trusses are often used for the framing of pole buildings. Ideally, a truss is placed on two poles. However, closer truss spacings require carrying the load of intermediate trusses to the poles. This is done by lintel beams placed between the poles. The sizes of these lintels or top plates can be determined from Lytle's text in the Midwest Plan Service.[2]

Wall framing for doors and windows usually consists of two framing members placed around the opening. Avoid placing window openings where framing members are located. If an opening must cut a framing member, care must be taken to transmit the load to another piece of framing to insure that there is no weak point in the structure. Framing the opening is usually adequate when only one framing member is cut. Cutting of trusses requires special considerations.

SIDING

There are many choices of siding materials available including aluminum, galvanized steel, wood, exterior plywood, tempered hardboard, and concrete blocks.

The wood siding should be kept 6 in. or more above ground level. Using treated wood for siding is always a good practice. Proper nailing is always necessary for all materials, but it is expecially important in attaching aluminum, where at least 100 nails should be used per 100 ft^2.

PLANS

There are many economical designs for pole buildings and other wooden structures available from the Midwest Plan Service, Iowa State University, Ames, Iowa 50010. They have plans for most common agricultural buildings in a large variety of sizes.

ROOF

Common roofing materials include asbestos-cement, galvanized steel, aluminum, wood shingles, asphalt shingles, selvage-edge roll roofing, and smooth-surfaced roll roofing. The asphalt shingles, selvage-edge, and smooth-surfaced roll roofing require a solid deck underneath, while asbestos-cement, galvanized steel, aluminum, and wood shingles use spaced decking.

Twenty-eight-gage or heavier galvanized-steel roofing is recommended for farm buildings. Most aluminum is lightweight (0.019 in. thick); however, a heavier gage, if available, would add strength.

Some metal roofs have a tendency to leak at the joints when the joints are filled with water for various reasons. This sets up a siphon and water continues to flow through the joint; however, metal roofing with joints designed to prevent siphoning is now available.

Spacing of the framing for metal roofing may go up to 3 ft, but individual manufacturer specifications should be followed. Use only ring shank or screw shank nails equipped with a neoprene gasket to secure a metal roof.

Plywood decking is an excellent choice for the roofing materials requiring a solid deck. Plywood $^3/_8$ in. thick will usually span up to 24 in. and $^3/_4$ in. plywood can span up to 48 in. when equipped with metal edge clips.

REFERENCES

1. ASAE Standard S288.3, Agricultural Engineers Yearbook, American Society of Agricultural Engineers, St. Joseph, Mich. 1982—83.
2. Midwest Plan Service, Structures and Environment Handbook, 10th ed., Iowa State University, Ames, 1977.

FOUNDATIONS*

C. Harold Moss

INTRODUCTION

Any properly constructed building must be supported by an adequate foundation that will support the weight of the building. The foundation must maintain the integrity of the structure through all weather conditions and, depending on the soil conditions, be proportioned so as to prevent any uneven settlement that would limit or restrict use of or in any way make a portion of the building unserviceable to the occupants. For a structure of substantial investment or where human welfare is a consideration, soil load carrying capacities and characteristics should be detemined by a soils testing laboratory. Likewise, foundations of such buildings should be designed by qualified engineers according to recognized engineering practices. The following should in no way be used in lieu of adequate soil investigations and foundation designs.

TYPES OF MATERIALS

Man has used and still uses many materials to provide foundations for buildings. Mats of stone and masonry units have been laid to provide a suitable base for building. Also, wooden mats and flat steel plates have sometimes been used. Timber poles driven into the ground have been used since man's early beginning. Today's modern structures may be founded on concrete, timber or steel piles, drilled and poured caissons, or soil with some highly specialized stabilization process.

The nature of this chapter is not to provide design procedures, but rather to offer guidelines in selecting and constructing foundations for light-frame and one- to two-story farm structures. Therefore, discussion will be primarily limited to wall, column, and floor foundations constructed of poured-in-place concrete, with some consideration given to masonry and concrete foundation walls (Tables 1 and 2).

LOCATION OF PLACEMENT

Footings should be placed on undisturbed soil if at all possible. If a building must be placed on a fill, the soil should be well tamped. Hand-tamped soil (tamping with a block of stone or steel) should be tamped in 1- to 2- in. layers, whereas the gasoline-driven ''jumping tampers'' can usually provide adequate compaction in layers from 6 to 10 in.

Before any footing is placed, it must be determined that the soil has the ability to support the load of the structure without excessive or uneven settlement. The supporting capability of a soil is called the soil bearing capacity. Safe bearing capacities for different types of soil are usually given in pounds per square foot (psf). The bearing capacity can be obtained by established laboratory testing procedures.

Some factors that affect the load carrying capacity of the soil are its composition, amount of organic material, amount of water, confinement, and depth below surface. Some generally accepted safe soil bearing capacities are listed in Table 3. These values should not be used in lieu of laboratory analysis.

When a load such as the weight of a structure is applied to soil, the soil compacts, thus

---

* All tables, sizes, and other data specifications throughout this chapter are approximate only and of a general nature. All data are for information only and are not intended to be used for design purposes.

Table 1
GUIDELINES FOR READY-MIXED CONCRETE

	Minimum 28-day compressive strength (psi)	Slump (in.)	Maximum large aggregate (in.)	Minimum cement (bags/yd)	Water (gal/bag)
Footings and slabs	2000	2—4	$1^{1}/_{2}$	5	7
Walls and piers	2500	3—5	$^{3}/_{4}$	6	6

Table 2
GUIDELINES FOR FIELD-MIXED CONCRETE

Maximum course aggregate size (in.)	Cement (sacks/yd³)	Water (gal/sack)	Approx. proportions by volume per sack of cement		
			Cement	Sand	Course aggregate
$^{3}/_{4}$	6.0	5	1	$2^{1}/_{2}$	$2^{3}/_{4}$
$1^{1}/_{2}$	5.5	5	1	$2^{1}/_{2}$	$3^{1}/_{2}$

Table 3
SAFE SOIL BEARING CAPACITIES

Type of soil	Bearing capacity (psf)
Loose fill	Unstable
Soft clay, loam	2,000
Dry, firm sand or clay	4,000
Compacted sand	6,000
Compact course gravel	8,000
Hardpan	12,000
Limestone	20,000
Granite	60,000 +

causing settlement. The amount or rate of settlement can be determined only from laboratory analysis of the particular soil. In general, clays usually result in higher settlement, with less in compacted sand and even less in coarse compacted gravel. The pressure acting between the bottom of the footing and the soil can be called "contact pressure". Sometimes it is necessary to place a foundation in an area where it will bear on soils with different settlement values. In the areas where high settlement is expected, the size of the foundation should be increased so that the contact pressure for the entire foundation is equalized. Likewise, where high loads occur, the foundation size should be increased to provide a more uniform contact pressure. Generally in light construction, settlement becomes a problem when one portion of the structure settles more or at a different rate than some other portion. Uneven settlement can cause cracking in foundations and walls, binding of doors and windows, sagging of floors, and possibly limit the intended use of the structure. The total foundation of a structure can be made more rigid by increasing the size or adding reinforcing steel. This increase in rididity helps distribute the load, thus increasing the chance for uniform settlement throughout the structure.

Another consideration in placement of a foundation is that the bottom of the footing must be below the frost line. The frost line is the depth below the ground surface where subfreezing

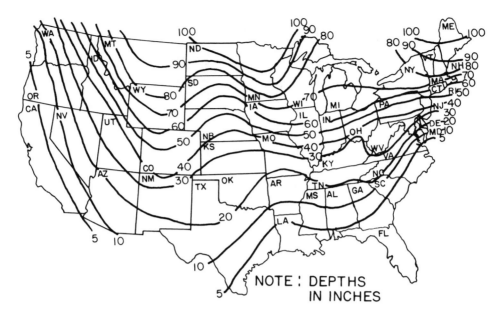

FIGURE 1.   Maximum depth of frost penetration in the U.S.

temperatures occur during winter months. Freezing temperature in the upper layers of the soil draws water up from the ground water table and forms ice. This formation of ice in the upper layers of the soil can cause a heaving action that will damage or even destroy building foundations. During warmer months, the excess water in the upper layers of the soil caused by frost action can cause excessive settlement and foundation problems. Foundations are placed below the frost line, usually as specified by local building codes. Figure 1 shows the maximum depth of frost penetration expected within the U.S. Clayey soils are subject to frost action more than coarse sands and gravels. Therefore an alternate method of eliminating frost action would be to remove all clayey or frost-susceptible soils down to the maximum depth of frost penetration and replace them with coarse sand or gravel.

Frost action in soils can also be induced from low temperatures inside a building such as cold storage and freezer buildings. In these type buildings, adequate insulation must be provided between the floor and the supporting soil. It is also advisable to found the floor slabs on 8 to 12 in. of gravel fill.

Whereas it is not a part of foundation design, it does deserve mention that during the preparation of soil for foundation work is a good time to treat the soil for termite protection, especially in the southern part of the U.S.

FOOTING DESIGN AND CONSTRUCTION

Wall footings are relatively long narrow footings placed directly below foundation walls. They support the loads from walls, floor, ceilings, and roofs. Wall footings should be formed with the sides vertical and the bottom flat and horizontal. The size of the footing must always be determined by local codes and designed to suit local soil conditions. As a general rule, however, the width of the footing should be not less than twice the width of the foundation wall and the footing thickness should be the same as the wall thickness but not less than 8 in. (see Figure 2). The footing should be centered on the wall with equal projection on each side of the wall. The footing should not project more than 6 in. beyond the face of the wall unless it is reinforced.

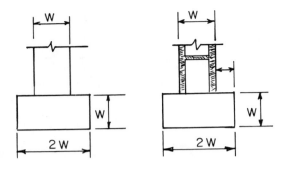

FIGURE 2.   Typical wall footing dimension ratios.

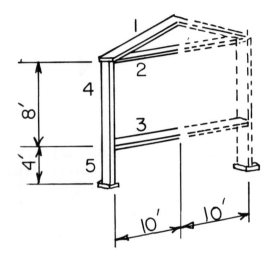

FIGURE 3.   Determining the load on a footing by ex—amining a strip of the building 1 ft wide.

Calculation

(1)	Roof live load	20 psf
	Roof dead load	15 psf
(2)	Ceiling live load	10 psf
	Ceiling dead load	10 psf
(3)	Floor live load	50 psf
	Floor dead load	15 psf
Subtotal		120 psf

The footing supports 10 ft of roof, ceiling, and floor; thus:

$$10 \text{ ft} \times 120 \text{ psf} = 1200 \text{ lb/ft}$$

(4)   Wall dead load

$$25 \text{ psf} \times 8 \text{ ft high} = 200 \text{ lb/ft}$$

(5)   Wall dead load

$$150 \text{ psf} \times 4 \text{ ft high} = 600 \text{ lb/ft}$$

Total load to footing = 2000 lb/ft

If soil bearing capacity is 1500 psf, then the minimum width of footing is

$$\frac{2000}{1500} = 1.33 \text{ ft}$$

When soil conditions are not uniform and load-bearing qualities are questionable, it is advisable to place 2- or $3^{1}/_{2}$ in. reinforcing rods longitudinally in the footing.

When footings are subjected to relatively heavy loads, calculations should be made to determine their correct size. It is also advisable to size footings such that all parts of the footings have the same pressure in contact with the soil. The load on a continuous wall footing is the sum of both the dead load and live load of the portions of any wall, floor, or roof it supports. The dead load refers to the weight of the structure, and the live load refers to the load to which the building is subjected, such as: people, stored items, furnishings, livestock, wind, snow, etc. Both dead load and live load are designated in pounds per square foot (psf). Local building codes and various handbooks usually specify minimum allowable live loads, depending on the use of the structure. To determine the load on a footing, examine a strip of the building 1-ft wide (see Figure 3).

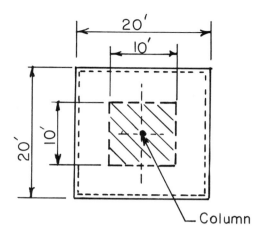

FIGURE 4.   Example of calculations involving columns.

Calculation

(1)	Roof dead load	15 psf
	Roof live load	20 psf
(2)	Ceiling dead load	10 psf
	Ceiling live load	10 psf
(3)	Floor dead load	15 psf
	Floor live load	50 psf
Total		120 psf

Area supported by the column is

$$10 \text{ ft} \times 10 \text{ ft} = 100 \text{ sf}$$

Load to column $= 100 \times 120 = 12,000$ lb

For an allowable soil capacity of 3000 psf the footing area required is

$$\frac{12000}{3000} = 4 \text{ ft}^2$$

Size of footing is 2 ft $\times$ 2 ft

Footings that support columns carry much higher load concentrations than wall footings. Calculations should be made to determine the soil contact area required to suit the load and soil capacity. The load is determined by finding the square footage of building times the dead load and live load applied to that area. The dead loads and live loads are determined in the same manner as described previously. An example is given in Figure 4.

For unreinforced concrete footings that support columns, the thickness should not be less than one half the length of the longest side. (For above example, thickness would be 1 ft.) Concrete footings larger than 2 ft² should have reinforcing bars (re-bars) placed 3 in. from the bottom. The determination of size and quantity of re-bars should be made by engineering calculations. There are many variables to consider in determining the reinforcing required, such as strength of concrete, re-bar and soil, load, and geometry of footing. However, below is an example for 3000-psi concrete, 60,000-psi re-bars and 3000-psf soil bearing capacity with a 1-ft-thick footing:

Footing size square	Reinforcing bars equally spaced (place each way in footing bottom)
2 ft 6 in.	4 #4 bars (8 required)
3 ft 0 in.	4 #4 bars (8 required)
3 ft 6 in.	4 #4 bars (8 required)
4 ft 0 in.	5 #4 bars (10 required)
4 ft 6 in.	6 #4 bars (12 required)
5 ft 0 in.	6 #5 bars (12 required)

Concrete slabs are often used for the ground floor of buildings (Figure 5) and serve as both the floor and foundation. The slab can be placed as an integral part of the wall and column foundations or poured separately and allowed to "float" or settle independently of wall foundations. If walls or columns have high loads, it is advisable to separate the footings and slab, thus preventing cracks in the slab.

An important consideration for slabs on grade is the preparation of the soil below the slab. The soil should be undisturbed or well compacted as previously discussed. A layer of gravel approximately 4 in. thick or more should be added for slabs that might require drainage

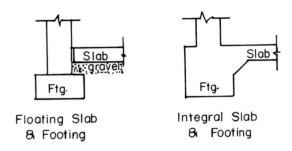

Floating Slab
& Footing

Integral Slab
& Footing

FIGURE 5.   Concrete slabs as the ground floor of buildings.

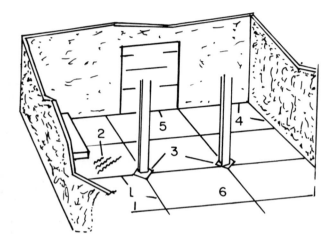

FIGURE 6.   Construction measures which help prevent cracking: (1) control joints, (2) reinforcing steel at protruding corner, (3) isolation joint at column, (4) isolation joint at wall and slab, (5) construction joint at doorway, and (6) prevention of rapid moisture loss during finishing.

or if soil might be subject to expansion or shrinkage when moisture content changes. Vapor barriers (usually 4 or 6 mil polyethylene), although not always required, are effective in preventing moisture from migrating through slabs. Care should be used to prevent puncture of vapor barrier during construction. A layer of sand should be placed between gravel and vapor barrier where both are used.

The primary object in the construction of concrete slabs on grade is to reduce and control the cracks. Many volumes have been written on controlling cracking in concrete floors. In this chapter, we will discuss only a few of the basic measures used to prevent cracking, (Figure 6). Rapid loss of moisture during placing and finishing can cause random shrinkage cracks. The conditions that cause these shrinkage cracks are combinations of low humidity, high wind velocity, and higher temperature in the concrete than in the air. A concrete mix design with a lower water requirement is less susceptible to shrinkage cracks. However, water reducing admixtures to the concrete are not recommended. It is more important to cover the concrete with wet burlap immediately after screeding and floating and keep covered until ready for troweling, and then exposing only enough area to allow work to proceed. A curing material (wet burlap, wet sand, or sprayed compound) should be applied as soon as possible after finishing. Since cracking cannot be prevented completely, joints should be provided in the slab to avoid cracks from developing where they are not wanted. Joints can be formed by limiting the size of the concrete pour or providing a tooled or saw cut joint

Table 4
WALL BEARING CAPACITIES

	Allowable compressive load (lb/ft)	
Wall thickness (in.)	Poured concrete (2500 psi)	Concrete block
8	60,000	6,800
10	75,000	8,600
12	90,000	10,000

Table 5
BEARING WALL REINFORCING

	Reinforcing in center of wall	
Wall thickness (in.)	Horizontal reinf.	Vertical reinf.
6	#4 @ 12 in.	#3 @ 12 in.
8	#4 @ 10 in.	#3 @ 9 in.
10	#5 @ 12 in.	#4 @ 12 in.
12	#4 @ 12 in.[a]	#3 @ 12 in.[a]

[a] Bars to be in each face of wall for walls 12 in. thick or more.

approximately one fourth of slab thickness. Joints should be located no more than 25 ft apart (15 ft is preferable) each way. The distance between joints in one direction should not be more than one and a half times the distance in the other direction. Where column footings are below the slab, columns should be isolated from the slab by providing a pocket around the column and filling it with a secondary pour. Construction joints should always be provided at doorways. For concrete and block walls, a pocket for the full thickness of the wall should be left in the doorway and filled later with a secondary pour. Cracks in slabs also tend to occur where wall corners protrude into the slab. If construction joints cannot be provided at these corners, reinforcing steel (two or three pieces of $1/2$-in. diameter by 4 ft long) should be placed in the top of the slab at the corner apex. The preceding measures will not eliminate all cracks in floor slabs, but will limit them and provide a more serviceable slab.

FOUNDATION WALLS

Foundation walls for light construction are generally built of poured concrete or concrete block (Table 4). Poured concrete walls are usually more reliable, but involve considerable labor and time. Concrete block is more easily constructed and generally is sufficient for most loads encountered in light construction.

The thickness of the wall is often determined by the thickness of the above walls rather than the load to be supported. A general rule of thumb, however, is that the ratio of height to thickness should not be more than 25 for unreinforced concrete, (i.e. 8-in.-thick wall not more than 25 × 8 in. = 200 in. or 16.7 ft). Walls less than 6 in. thick should not be used for load bearing foundations. For walls greater in height than 25 times the thickness, reinforcing bars should be added according to Table 5.

Concrete walls should be tied to the footing by either extending reinforcing vertically out of the footing into the wall or providing a key in the footing by placing a 2 × 4 in the top of the footing, then removing before pouring the wall.

Because of their economy and speed of erection, concrete block foundation walls are

Table 6
MAXIMUM UNREINFORCED BASEMENT WALL HEIGHT

Wall thickness (in.)	No surcharge (ft)		Surcharge[a] (ft)	
	Concrete	Conc. blk.	Concrete	Conc. blk.
8	8	4.75	7	—
10	9.5	5.6	8	4.75
12	11	6.5	9.5	5.6

[a]   Surcharge is light traffic or parking adjacent to wall.

Table 7
REINFORCED BASEMENT WALLS

Wall height (ft)	Concrete				Concrete block	
	8 in. thick		12 in. thick		8 in. thick	12 in. thick
	Vertical	Horiz.	Vertical	Horiz.	Vertical	Vertical
8	#4 @ 16 in.	#4 @ 10 in.	#4 @ 12 in.	#5 @ 12 in.	#5 @ 16 in.	#5 @ 16 in.
9	#4 @ 12 in.	#4 @ 10 in.	#4 @ 12 in.	#5 @ 12 in.	#5 @ 16 in.	#5 @ 16 in.
10	#4 @ 9 in.	#4 @ 10 in.	#4 @ 12 in.	#5 @ 12 in.	#5 @ 8 in.	#5 @ 16 in.
12	#5 @ 9 in.	#4 @ 10 in.	#4 @ 9 in.	#5 @ 12 in.	#5 @ 8 in.	#5 @ 8 in.

[a]   Provide horizontal trussed reinforcing in alternate courses.

widely used. Load bearing block walls should not be higher than 20 times their thickness. For high walls, it is recommended that the top course be a lintel block filled with concrete with a reinforcing bar continuous. Block walls can be reinforced by placing vertical bars in the block cells and filled with concrete or placing trussed wire reinforcing (i.e., "Dur-o-wall") horizontally in the mortar bed. A general rule of thumb is to provide a #5 bar at 32 in. on center for 8-in. block and at 16 in. on center for 12-in. block. Wire trussed reinforcing should be used on every third joint in reinforced block walls. Both concrete and concrete block walls must be laterally braced until the vertical load from the structure above is added. An unbraced wall becomes a retaining wall and is subject to easier cracking or failure.

Where earth is only on one side of a foundation wall, the wall becomes a basement wall and should be designed to resist lateral pressures. Due to the many variables, the design of basement walls cannot be completely covered in this paper. Of first importance is to provide drain tile to eliminate any water pressure against the wall. Reinforcing bars are normally provided $1\frac{1}{2}$ in. from the inside face of the wall to resist the lateral pressure. Table 6 shows the limits of unreinforced basement walls.

When the depth of earth against a basement wall exceeds that in Table 6, the wall should be designed by a qualified engineer for the conditions as they exist. Table 7 gives some general guidelines for reinforced basement walls.

Foundation wall and footing design depend on many variables. The preceding tables and charts should not in any way be used in place of adequate design. This information may be helpful, however, in general planning and estimating overall requirements for light farm structures.

REFERENCES

1. **Badzinski, S., Jr.,** *Carpentry in Residential Construction,* Prentice Hall, Englewood Cliffs, N.J., 1972.
2. *CRSI Handbook,* Concrete Reinforcing Steel Institute, Chicago, 1963 and 1975.
3. **Lytle, R. J.,** *Farm Builder's Handbook,* Structures Publishing Co., Farmington, Mich., 1969.
4. **Muller, E. J.,** *Architectural Drawing and Light Construction,* Prentice Hall, Englewood Cliffs, N.J., 1967.
5. **Sowers, G. B. and Sowers, G. F.,** *Introductory Soil Mechanics and Foundations,* MacMillan, New York, 1961.

INSULATION AND VAPOR BARRIERS

Cecil Hammond

INTRODUCTION

Thermal insulation decreases energy waste and reduces energy cost associated with heating and cooling structures. Insulation reduces heat transfer, keeping heat inside during the winter and outside during the summer.

Heat flows from the warm interior of a structure to the cooler exterior; this is called "heat loss". The reverse occurs during the summer when heat flows from the exterior to the cooler interior; this is called "heat gain".

TYPES

Insulation can be loose-fill, flexible, rigid boards, foamed-in-place, or reflective type (see Figure 1). Insulating materials usually contain tiny air pockets and must be kept dry to perform properly.

VAPOR BARRIERS

Moisture deteriorates insulation, causes paint to peel, and produces mildew and rotting in wood. Always place vapor barriers against the warm side of the wall, floor, or ceiling (see Figure 2). Continuous barriers are the most effective way of preventing moisture movement. Vapor barriers can be of aluminum foil, kraft paper, or plastic sheeting. Aluminum paint, applied in two layers under the covering paint on inside walls, is sometimes used when existing walls have insulation blown in. When foam materials are pumped into walls, they usually make an effective vapor barrier. If new insulation is added to existing insulation, do not install a vapor barrier between layers of insulation.

HEAT TRANSFER

Heat is transferred in three ways:

1. Conduction-heat is transferred through dense materials readily, such as heat moving from a cook stove through a skillet. The vast majority of heat transfer through structural surfaces such as ceiling, walls, and floors occurs through conduction.
2. Convection-heat is transferred by air movement over a hot surface such as a radiator. This process is aided by air movement due to density changes (warm air is lighter).
3. Radiation-heat is transferred through air and transparent materials, such as occurs in heat radiation from a fireplace to persons or objects nearby. Heat normally transferred by radiation can be reflected by silver, copper, and aluminum or their alloys.

HEAT LOSS OR GAIN

The rate of heat conduction through a surface can be predicted from the relation

$$Q = A\Delta t/R$$

where Q = heat flow (steady state) in Btu/hr, Δt = temperature difference across the

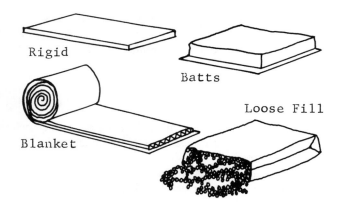

FIGURE 1.   Insulating materials.

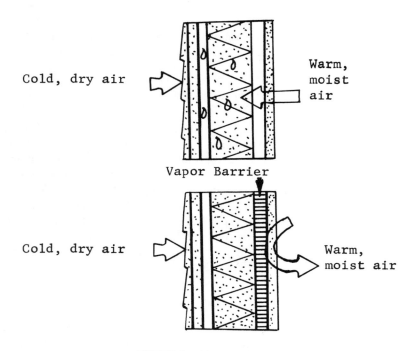

FIGURE 2.   Vapor barriers.

conducting surfaces in °F, R = overall thermal resistance in (hr) (ft²) °F/Btu, and A = area in ft².

The overall resistance to heat flow can be obtained by adding the resistances of the individual materials (see Figure 3).

Another term commonly used in heat transfer is called thermal transmittance or conductance and is represented by U, in Btu/(hr) (ft²) (°F). The expression for heat transfer as before is

$$Q = UA\Delta t$$

where U = thermal transmittance in Btu/(hr) (ft²) (°F). U, therefore, equals 1/R or $1/U_i = R_i$

Tables 1 to 5 give the thermal transmittance (U) and resistance (R) values for some building materials.

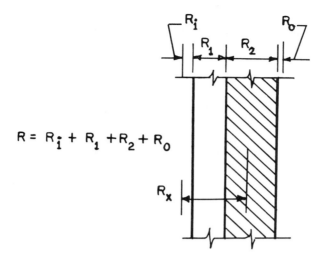

FIGURE 3.   Individual resistances and total R-values for a non-homogeneous wall.

Table 1
RESISTANCES FOR AIR ON INTERIOR AND EXTERIOR SURFACES

Position of surface	Heat flow	R
Still air (interior)		
Horizontal (ceiling)	Upward	0.61
Sloping — 45°	Upward	0.62
Vertical (wall)	Horizontal	0.68
Sloping — 45°	Downward	0.76
Horizontal (Floor)	Downward	0.92
Moving air (exterior) any position		
15-mi/hr wind (for winter)		0.17
$7^1/_2$-mi/hr wind (for summer)		0.25

From *ASHRAE Handbook of Fundamentals,* American Society of Heating, Refrigerating, and Air Conditioning Engineers, New York, 1977. With permission.

Table 2
THERMAL RESISTANCE (R) VALUES FOR AIR SPACES

Position of air space	Direction of heat flow	Thickness of air space (in.)			
		$^1/_2$	$^3/_4$	$1^1/_2$	$3^1/_2$
Horizontal	Up	0.84	0.87	0.89	0.93
45°	Up	0.90	0.94	0.91	0.96
Vertical	Horizontal	0.91	1.01	1.02	1.01
Horizontal	Down	0.92	1.02	1.14	1.21
45°	Down	0.92	1.02	1.09	1.05

From *ASHRAE Handbook of Fundamentals,* American Society of Heating, Refrigerating, and Air Conditioning Engineers, New York, 1977. With permission.

Table 3
RESISTANCES (R) OF BUILDING AND INSULATING MATERIALS

Material	Description		Resistance per inch thickness	(R) for thickness listed
Building board	Asbestos-cement board		0.25	—
Boards, panels	Asbestos-cement board	$^1/_8$ in.	—	0.03
Subflooring	Asbestos-cement board	$^1/_4$ in.	—	0.06
Sheathing	Gypsum or plaster board	$^3/_8$ in.	—	0.32
Woodbased panel	Gypsum or plaster board	$^1/_2$ in.	—	0.45
Products	Gypsum or plaster board	$^5/_8$ in.	—	0.56
	Plywood		1.25	—
	Plywood	$^1/_4$ in.	—	0.31
	Plywood	$^3/_8$ in.	—	0.47
	Plywood	$^1/_2$ in.	—	0.62
	Plywood	$^5/_8$ in.	—	0.77
	Plywood or wood panels	$^3/_4$ in.	—	0.93
	Vegetable fiber board			
	Sheathing, regular density	$^1/_2$ in.	—	1.32
	Sheathing (18 lb/ft³)	$^{25}/_{32}$ in.	—	2.06
	Sheathing, intermediate density	$^1/_2$ in.	—	1.22
	Nail-base sheathing	$^1/_2$ in.	—	1.14
	Shingle backer	$^3/_8$ in.	—	0.94
	Shingle backer	$^5/_{16}$ in.	—	0.78
	Sound deadening board	$^1/_2$ in.	—	1.35
	Tile and lay-in panels, plain or acoustic		2.50	—
	—	$^1/_2$ in.	—	1.25
	—	$^3/_4$ in.	—	1.89
	Laminated paperboard		2.00	—
	Homogeneous board from repulped paper		2.00	—
	Hardboard			
	Medium density (50 lb/ft³)		1.37	—
	High density, service temp service, underlay (55 lb/ft³)		1.22	—
	High density, std. tempered		1.00	—
	Particleboard			
	Low density (37 lb/ft³)		1.85	—
	Medium density (50 lb/ft)		1.06	—
	High density (62.5 lb/ft³)		0.85	—
	Underlayment	$^5/_8$ in.	—	0.82
	Wood subfloor	$^3/_4$ in.	—	0.94
Building membrane	Vapor-permeable felt		—	0.06
	Vapor-seal, 2 layers of mopped 15-lb felt		—	0.12
	Vapor-seal, plastic film		—	Negl.
Finish flooring materials	Carpet and fibrous pad		—	2.08
	Carpet and rubber pad		—	1.23
	Cork tile	$^1/_8$ in.	—	0.28
	Terrazzo	1 in.	—	0.08
	Tile-asphalt, linoleum, vinyl, rubber		—	0.05
	Wood, hardwood finish	$^3/_4$ in.	—	0.68
Insulating materials	Mineral fiber, fibrous form processed from rock, slag, or glass			

Table 3 (continued)
RESISTANCES (R) OF BUILDING AND INSULATING MATERIALS

Material	Description		Resistance per inch thickness	(R) for thickness listed
Blanket and batt	Approx. 2—2³/₄ in.		—	7
	Approx. 3—3¹/₂ in.		—	11
	Approx. 5¹/₂ in.		—	19
	Approx. 6—7 in.		—	22
	Approx. 8¹/₂ in.		—	30
Board and slabs	Cellular glass		2.63	—
	Glass fiber, organic bonded		4.00	—
	Expanded rubber (rigid)		4.55	—
	Expanded polystyrene extruded, plain		4.00	—
	Expanded polystyrene extruded, (R-12 exp.)		5.00	—
	Expanded polystyrene extruded, (R-12 exp.) (thickness 1 in. and greater)		5.26	—
	Expanded polystyrene, molded bead		3.57	—
	Expanded polyurethane (R-11 exp.) (Thickness 1 in. or greater)		6.25	—
	Mineral fiber with resin binder		3.45	—
	Mineral fiberboard, wet felted			
	Core or roof insulation		2.94	—
	Acoustical tile		2.86	—
			2.70	—
	Mineral fiberboard, wet molded			
	Acoustical tile		2.38	—
	Wood or cane fiberboard			
	Accoustical tile	¹/₂ in.	—	1.25
		³/₄ in.	—	1.89
	Interior finish (plank, tile)		2.86	—
	Wood shredded (cemented in preformed slabs)		1.67	—
Loose fill	Cellulose insulation (milled paper or wood pulp)		3.13—3.70	—
	Sawdust or shavings		2.22	—
	Wood fiber, softwoods		3.33	—
	Perlite, expanded		2.70	—
	Mineral fiber (rock, slag, or glass)			
	Approx. 3³/₄—5 in.		—	11
	Approx. 6¹/₂—8³/₄ in.		—	19
	Approx. 7¹/₂—10 in.		—	22
	Approx. 10¹/₄—13³/₄ in.		—	30
	Vermiculite (expanded)			
	7.0—8.2 lb/f²		2.13	—
	4.0—6.0 lb/f²		2.27	—
Roof insulation	Preformed, for use above deck			
	Approx.	¹/₂ in.	—	1.39
	Approx.	1 in.	—	2.78
	Approx.	1¹/₂ in.	—	4.17
	Approx.	2 in.	—	5.56
	Approx.	2¹/₂	—	6.67
	Approx.	3 in.	—	8.33

Table 3 (continued)
RESISTANCES (R) OF BUILDING AND INSULATING MATERIALS

Material	Description		Resistance per inch thickness	(R) for thickness listed
Masonry materials	Cement mortar		0.20	—
Concrete	Gypsum-fiber concrete 87$\frac{1}{2}$% gypsum, 12$\frac{1}{2}$% wood chips		0.60	—
	Lightweight aggregates including	120 lb/f^2		0.19
	expanded shale, clay or slate; ex-	100		0.28
	panded slags: cinders; pumice;	80		0.40
	vermiculite; also cellular concrete	60		0.59
		40		0.86
		30		1.11
		20		1.43
	Perlite			
	40 lb/ft^3		1.08	—
	32 lb/ft^3		1.41	—
	20 lb/ft^3		2.00	—
	Sand gravel or stone aggregate (oven dried)		0.11	—
	Sand and gravel or stone aggretate (not dried)		0.08	—
	Stucco		0.20	—
Masonry units	Brick, common		0.20	—
	Brick, face		0.11	—
	Clay tile, hollow			
	1 cells deep	3 in.	—	0.80
	1 cells deep	4 in.	—	1.11
	2 cells deep	6 in.	—	1.52
	2 cells deep	8 in.	—	1.85
	2 cells deep	10 in.	—	2.22
	3 cells deep	12 in.	—	2.50
	Concrete blocks, three oval core			
	Sand and gravel aggregate	4 in.	—	0.71
		8 in.	—	1.11
		12 in.	—	1.28
	Cinder aggregate	3 in.	—	0.86
		4 in.	—	1.11
		8 in.	—	1.72
		12 in.	—	1.89
	Lightweight aggregate (expanded	3 in.	—	1.27
	shale, clay, slate or slag;	4 in.	—	1.50
	pumice)	8 in.	—	2.00
		12 in.	—	2.27
	Concrete blocks, rectangular core			
	Sand and gravel aggregate			
	2 core, 8 in. 36 lb		—	1.04
	Same with filled cores		—	1.93
	3 core, 6 in. 19 lb		—	1.65
	Same with filled cores		—	2.99
	2 core, 8 in. 24 lb		—	2.18
	Same with filled cores		—	5.03
	3 core, 12 in. 38 lb		—	2.48
	Same with filled cores		—	5.82
	Stone, lime, or sand		0.08	—

Table 3 (continued)
RESISTANCES (R) OF BUILDING AND INSULATING MATERIALS

Material	Description		Resistance per inch thickness	(R) for thickness listed
	Gypsum partition tile			
	3 × 12 × 30 in. solid		—	1.26
	3 × 12 × 30 in. 4-cell		—	1.35
	4 × 12 × 30 in. 3-cell		—	1.67
Plastering materials	Cement plaster, sand aggregate		0.20	—
	Sand aggregate	$^3/_8$ in.	—	0.08
		$^3/_4$ in.	—	0.15
	Gypsum plaster			
	Lightweight aggregate	$^1/_2$ in.	—	0.32
		$^5/_8$ in.	—	0.39
	Lightweight aggregate on metal lath	$^3/_4$ in.	—	0.47
	Perlite aggregate		0.67	—
	Sand aggregate		0.18	—
		$^1/_2$ in.	—	0.09
		$^5/_8$ in.	—	0.11
	Sand aggregate on metal lath $^3/_4$ in.		—	0.13
	Vermiculite aggregate		0.59	—
Roofing	Asbestos-cement shingles		—	0.21
	Asphalt roll roofing		—	0.15
	Asphalt shingles		—	0.44
	Built-up roofing	$^3/_8$ in.	—	0.33
	Slate	$^1/_2$ in.	—	0.05
	Wood shingles, plain and plastic film faced		—	0.94
Siding materials (on flat surface)	Shingles			
	Asbestos-cement		—	0.21
	Wood, 16 in., 7$^1/_2$ exposure		—	0.87
	Wood, double, 16 in., 12 in. exposure		—	1.19
	Wood, plus insul. backer board	$^5/_{16}$ in.	—	1.40
	Siding			
	Asbestos-cement, $^1/_4$ in. lapped		—	0.21
	Asphalt roll siding		—	0.15
	Asphalt insulating siding $^1/_2$ in. board		—	1.46
	Wood, drop, 1 × 8 in.		—	0.79
	Wood, bevel, $^1/_2$ × 8 in., lapped		—	0.81
	Wood, bevel, $^3/_4$ × 10 in., lapped		—	1.05
	Wood, plywood, $^3/_8$ in., lapped		—	0.59

Table 3 (continued)
RESISTANCES (R) OF BUILDING AND INSULATING MATERIALS

Material	Description		Resistance per inch thickness	(R) for thickness listed
	Aluminum or steel, over sheathing, hollow-backed		—	0.61
	Insulating-board backed nominal, $^3/_8$ in.		—	1.82
	Insulating-board backed nominal, $^3/_8$ in. foil backed		—	2.96
	Architectural glass		—	0.10
Woods	Maple, oak, and similar hardwoods		0.91	—
	Fir, pine, and similar softwoods		1.25	—
	Fir, pine, and similar softwoods	$^3/_4$ in.	—	0.94
		$1^1/_2$ in.	—	1.89
		$2^1/_2$ in.	—	3.12
		$3^1/_2$ in.	—	4.35
		$5^1/_2$ in.	—	6.88
		$7^1/_2$ in.	—	9.38
		$9^1/_4$ in.	—	11.56
		$11^1/_4$ in.	—	14.06

From *ASHRAE Handbook of Fundamentals,* American Society of Heating, Refrigerating, and Air Conditioning Engineers, New York, 1977. With permission.

Table 4
THERMAL TRANSMITTANCE (U) OF WINDOWS AND SKYLIGHTS

Part A: Exterior Vertical Panels

Description	Winter no shade	Summer no shade	Winter indoor shade[a]	Summer indoor shade[a]
Flat Glass				
Single glass	1.10	1.04	0.83	0.81
Insulating glass — double				
$3/16$ in. air space	0.62	0.65	0.52	0.58
$1/4$ in. air space	0.58	0.61	0.48	0.55
$1/2$ in. air space	0.49	0.56	0.42	0.52
$1/2$ in. air space, low emmitance coating				
e = 0.20	0.32	0.38	0.30	0.37
e = 0.40	0.38	0.45	0.36	0.44
e = 0.60	0.43	0.51	0.38	0.48
Insulating glass — triple				
$1/4$ in. air space	0.39	0.44	0.31	0.40
$1/2$ in. air space	0.31	0.39	0.26	0.36
Storm windows	0.50	0.50	0.42	0.48

Part B: Exterior Horizontal Panels (Skylights)

Description	Winter	Summer
Flat glass		
Single glass	1.23	0.83
Insulating glass — double		
$3/16$ in. air space	0.70	0.57
$1/4$ in. air space	0.65	0.54
$1/2$ in. air space	0.59	0.49
$1/2$ in. air space, low emittance coating		
e = 0.20	0.48	0.36
e = 0.40	0.52	0.42
e = 0.60	0.56	0.46
Plastic domes		
Single walled	1.15	0.80
Double walled	0.70	0.46

Part C: Adjustment Factors for Various Windows and Sliding Patio Door Types (Multiply U-Values in Parts A and B by Factors Below)

Description	Single glass	Double or triple glass	Storm windows
Windows			
All glass	1.00	1.00	1.00
Wood sash: 80% Gl.	0.90	0.95	0.90
Wood sash: 60% Gl.	0.80	0.85	0.80
Metal sash: 80% Gl.	1.00	1.20	1.20
Sliding patio doors			
Wood frame	0.95	1.00	—
Metal frame	1.00	1.10	—

[a] Values apply to tightly closed venetian and vertical blinds, draperies, and roller shades.

From *ASHRAE Handbook of Fundamentals*, American Society of Heating, Refrigerating and Air Conditioning Engineers, New York, 1977. With permission.

Table 5
THERMAL TRANSMITTANCE (U) OF SLAB DOORS

	Winter			Summer
		Storm door		
Material and thickness	No storm door	Wood	Metal	No storm door
Solid wood				
1 in.	0.64	0.30	0.39	0.61
$1^1/_4$ in.	0.55	0.28	0.34	0.61
$1^1/_2$ in.	0.49	0.27	0.33	0.47
2 in.	0.43	0.24	0.29	0.42
Steel Door				
$1^3/_4$ in. mineral fiber core	0.59	—	—	0.58
$1^3/_4$ in. urethane foam core	0.19	—	—	0.18
$1^3/_4$ in. polystyrene foam core	0.47	—	—	0.46

From *ASHRAE Handbook of Fundamentals,* American Society of Heating, Refrigerating, and Air Conditioning Engineers, New York, 1977. With permission.

REFERENCES

1. *ASHRAE Handbook of Fundamentals,* American Society of Heating, Refrigerating and Air Conditioning Engineers, New York, 1977.

PIT AND SLATTED FLOOR CONSTRUCTION

Cecil Hammond

SLATTED FLOORS

Slatted floors allow a convenient way of handling animal waste. Simplified waste handling, which has eliminated the need to scrape or wash floors, has made confinement housing acceptable.

Partially slatted and partially solid floors have worked well for most livestock and have generally shown less foot and claw damage. Flooring materials and arrangement of pens is a factor; for example, research has shown that a long narrow pen 8 to 10 ft wide and 22 to 24 ft long is preferred for swine when floors are partially slatted. Pen shape is not a factor on totally slatted floors. Narrow slats have a tendency to have more claw problems than wider slats. The surface of some slat materials can be quite slick, particularly when wet.

Design Loads

The recommended design loads on slatted floors in livestock pens are given in Table 1. Gang slats or slots that are interconnected between supports require three or more slats to deflect together. In this case, design individual slats to carry the recommended load per unit of length using the distance between interconnections, and design the full span of individual slats for one half the recommended load per unit of length.

Shape and Spacing of Slats

The size of the slat opening and slat width depends upon the composition of the manure and experience with foot injury, animal response, and traction.

Slats can be made of many materials. Concrete is durable, heavy, and requires stronger support. Wood warps, wears, and can be chewed by some livestock. Metal, plastic, or a combination of materials are also supplied commercially. Important considerations are cost, life expectancy, corrosion, strength, noise, and surface as it affects traction characteristics.

Slats greater than 1 in. in depth, as most materials are, tend to pass wastes better when they have a tapered cross section as shown in Figure 1.

Some general recommendations for slat size and spacing are given in Table 2. If flooring is purchased commercially, follow manufacturer's instructions. Wide slats in farrowing stalls should be covered to protect small pigs.

MANURE PITS

Waste storage is closely tied to the type of livestock housing. Requirements vary for minimum size of waste storage depending upon climates, as do requirements for construction, maintenance, use, and location.

Storage capacity may be regulated and varies with the type, size, and number of animals and amount of dilution water. Handling of runoff water and the storage time also determine storage necessary. If manure is to be field spread, 3 to 6 months storage capacity is usually required to allow spreading when crop and field conditions allow. Cleaning with high-pressure water may increase storage requirements significantly, as will the amount of dilution water required if the manure is to be pumped through an irrigation system. In general, storage capacity = manure production per animal per day × number of animals × storage interval in days + dilution water + freeboard of 1 ft minimum.

Table 1

DESIGN LOADS FOR GROUPS OF ANIMALS ON SLOTTED FLOORS

Animal	Live load in pounds per linear foot of slat	Live load in pounds per square foot of floor (slat support)
Cattle, dairy and beef	250	100
Calves, dairy and beef	150	50
Sheep	120	50
Swine[a]		
0—50 lb.	50	35
50—200 lb.	100	50
200—400lb.	150	65
400—500 lb.	170	70

[a]   Slats support for small pigs (up to 50 lb) should meet an alternate requirement of supporting a concentrated load of 220 lb moved to produce maximum stress in bending or shear. Slats in farrowing pens should be designed to support 250 lb moved to produce maximum stress in bending or shear. From Structures and Environment Handbook (MWPS - 1), 9th ed., Midwest Plan Service, Ames, Iowa, 1977. With permission.

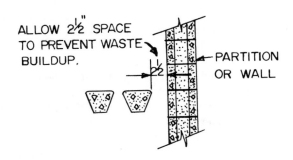

FIGURE 1.   Tapered cross section cleans effectively.

Estimates of daily waste production for various animals are given in Table 3. Actual waste production will vary and can easily be ±20% of values given.

Filling Liquid Manure Pits

Before liquid manure pits are filled with manure, water should be added to keep solids submerged and maintain a liquid state. Add 3 to 4 in. of water to pits under slatted floors and 6 to 12 in. if pit is to be loaded with batches of scraped waste. Dilution of waste to over 90 per liquid is common. Even with adequate liquid, pits under slatted floors produce noxious gases and should be ventilated.

Manure Pit Design

Manure pits or tanks fall into three general groups: (1) shallow tanks of relatively simple design, (2) large tanks above or below grade requiring careful design, and (3) covered tanks that require complete design based upon all loads which occur. Of primary concern are the requirements from management practices, safety, and structural integrity.

Pit depth may be limited by the type of pump, soil depth above bedrock, and level of water table. Some pits will float out of the groung in periods of high water table. Pumps

Table 2
GENERAL RECOMMENDATIONS FOR SLAT SIZE AND SPACING

Animal	Slat width	
	$1^1/_4$—3 in.	4—8 in.
Cattle		
Beef and dairy cows	—	$1^1/_2$—$1^3/_4$ in.
Dairy calves	$^3/_4$ in.	$1^1/_4$ in.
Sheep		
Ewes	1 in.	1 in.
Lambs	$^5/_8$—$^3/_4$ in.	$^3/_4$—1 in.
Feeders	$^3/_4$—1 in.	1 in.
Swine[a]		
Farrowing[b]	$^3/_8$ in.(1 in. behind sow)	$^3/_8$—1 in.
25—40 lb	$^1/_2$ in.	$^3/_4$—1 in.
40—220 lb	—	1—$1^1/_4$ in.
Sows	—	1—$1^1/_4$ in.

[a] Firm footing required during breeding.
[b] Wide slats back of sows should be covered during farrowing.

From Structures and Environment Handbook (MWPS - 1), 9th ed., Midwest Plan Service, Ames, Iowa, 1977, 241. With permission.

Table 3
ESTIMATES OF DAILY WASTE PRODUCTION FOR VARIOUS ANIMALS

Animal	Wt (lb)	Total waste production		
		lb/day	ft^3/day	% water
Beef cattle	500	30	0.5	88.4
	750	45	0.75	88.4
	1000	60	1.0	88.4
	1250	75	1.2	88.4
Dairy cattle	150	12	0.19	87.3
	250	20	0.32	87.3
	500	41	0.66	87.3
	1000	82	1.32	87.3
	1400	115	1.85	87.3
Poultry				
Layers	4	0.21	0.0035	74.8
Broilers	2	0.14	0.0024	74.8
Sheep	100	4.0	0.062	75.0
Horse	1000	45.0	0.75	79.5
Swine				
Nursery pig	35	2.3	0.038	90.8
Growing pig	65	4.2	0.070	90.8
Feeder pig	150	9.8	0.16	90.8
	200	13.0	0.22	90.8
Gestate sow	275	8.9	0.15	90.8
Sow and litter	375	33.0	0.54	90.8
Boar	350	11.0	0.19	90.8

From Structures and Environment Handbook, (MWPS - 1), 9th ed., Midwest Plan Service, Ames, Iowa, 1977, 242. With permission.

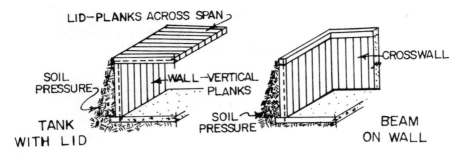

FIGURE 2. Manure pit design. (From Structures and Environment Handbook (MWPS - 1), 9th ed., Midwest Plan Service, Ames, Iowa, 1977, 72. With permission.)

Table 4
PIT AND TANK DESIGN
LOADS ON WALLS[a]

Soil condition	Pressure inward for design (lb/ft³)
Saturated or high water table	60
Moderately drained	30
Well drained	15

[a] Loads on partitions occur in either direction and outward for walls above ground.

From Structures and Environment Handbook (MWPS - 1), 9th Ed., Midwest Plan Service, Ames, Iowa, 1977, 72. With permission.

are usually effective in agitation up to 40 ft. Agitation access is recommended at least every 40 ft.

Pits and tanks should be designed to withstand all loads from earth pressures, live loads, uplift from water table, and hydrostatic effects. Walls and lids are usually designed as strips as shown in Figure 2, with a top beam often required if no lid is used.

Hydrostatic effects on the walls and floors from liquid manure should be figured at 60 lb/ft³. Earth pressures from backfilling are given in Table 4. Adequate provisions must be made to carry lateral loads to partitions, endwalls, cross beams, lids, or pilasters.

A surcharge of 100 lb/ft² should be added to the hydrostatic wall load where heavy equipment is operated within 5 ft of walls.

Open pits should be enclosed with a fence at least 5 ft high for safety. If pits are within a building, locate electrical entrances outside the building and, when possible, place electrical wiring and fixtures behing the vapor barrier.

REFERENCES

1. Livestock Waste Facilities Handbook (MWPS-18), 1st ed., Midwest Plan Service, Ames, Iowa, 1975.
2. Structures and Environment Handbook (MWPS-1), 9th ed., Midwest Plan Service, Ames, Iowa, 1977.

Electrical Systems and
Appliances for
Agricultural Structures

ELECTRIC LOADS, CIRCUITS, AND WIRES

Robert H. Brown

INTRODUCTION

Proper operation and service of an electric circuit depends upon proper selection of electric wires and cables. Safety, economy, and circuit performance are involved. Normally either copper (mostly) or aluminum wires are used due to their excellent conductivity and durability. This treatment of the topic is restricted to these two metals.

RESISTANCE

An ideal wire for an electric circuit would have zero resistance to the flow of electrons (current). Absolute zero is impossible, but one should keep the resistance as small as practical. The resistance of a wire is calculated from the relation

$$R = \rho \frac{\ell}{A}$$

where R is the resistance in ohms, A is the cross-sectional area of the material in circular mils, and ρ is the resistivity of the material in ohm circular mils per foot of length. For copper $\rho = 10.8$, and for aluminum $\rho = 17.0$.

As seen from the equation, R will be smallest when ℓ is small and A is large.

Example — Calculate the resistance of 1000 ft of copper wire with a diameter of 100 mils.

$$A = dia^2 = 100^2 = 10,000$$

$$\rho = 10.8$$

$$R = \frac{(10.8)(1000)}{10,000} = 1.08 \ \Omega$$

Resistance values for all standard sizes of copper and aluminum wires are furnished in tabulated form in textbooks[1,2] and in the N.E.C.[3] (see properties of conductors). The above equation holds for any material whose value for ρ is known.

AMPACITY

Another important property of electric wires is current carrying capacity — how many amperes can be safely conducted through the wire. This is appropriately termed "ampacity". The rated (maximum safe) value of amperes permitted to flow in a given wire is highly dependent on temperature. It is the concern about the insulation surrounding the wire and its being damaged by high temperatures that places a limit (ampacity) for the wire. A melted or damaged insulation would make the wire unsafe and probably would trip the protective devices of the circuit.

When an electric current flows in the wire, heat is produced. This is the major source of heat that could cause insulation damage. Also, the proximity of other wires, the location of the wire, and ambient temperature have a bearing on the situation. Accordingly, the tables containing "ampacity" ratings of wires have headings showing temperatures and have notes about the wire being in "free air" in a cable, buried in the earth, etc. The principal guide is that the insulation will melt long before any change occurs to the metal; therefore, insulation is the key.

Insulation types given in Table 1 are type R,T,V, etc.; those having heat-resistive properties contain the letter H, while those permitted in wet locations contain the letter W. A type RHW insulation would be rubber material with heat tolerance and permitted in wet locations.

DESIGNATIONS

When ordering an electric wire, specify: American Wire Gage (AWG), size (No.12, No. 8, etc.), metal (copper, etc.), and insulation (type TW, R, etc.).

Example — 500 ft of AWG No. 12 copper wire with type TW insulation.

SELECTING WIRES

Electric circuit conductors are chosen after consideration of ampacity, voltage drop, and mechanical strength.

1.   Short wires — Wires and cables less than 30 ft in length are considered short. Select on the basis of ampacity and choose appropriate wet or dry type of insulation, etc.
2.   Long wires — Lengths of 30 ft and greater are considered long, and consideration must be given to voltage drop as well as to ampacity. Voltage drops of 1, 2, or 3% of the source voltage are usual considerations. The same percentage of money spent for electricity will be wasted as is designated for the voltage drop design. Two percent is a good choice. Next, calculate the required circular mil area of wire to limit the voltage drop to the chosen percentage and select the appropriate wire size from a table containing wire properties.

Example 1

Select and designate the copper wire for serving a 4800-W 240-V electric load located 100 ft from the source. This is to be an underground cable service.

$$cm = \frac{22^* \times A \times \text{length one way}}{\text{allowable voltage drop}}$$

$$cm = \frac{22 \times \dfrac{4800}{240} \times 100}{2\%(240)}$$

$$cm = \frac{22 \times 20 \times 100}{4.8} = 9166$$

From Table 2 it is determined that No. 10 wire has a cross-sectional area of 10,380 circular mils, which is sufficient, and No. 12 has 6530, which is too small; therefore, No. 10 copper with TW insulation on wires should be selected. A two-wire, underground cable is needed and a grounding conductor should be included: therefore, type UF cable 10-2 copper with ground (UF 10-2 WG) should be specified.

Example 2

Choose for the same load as for Example 1, except provide the service overhead (amperes

---

*   Use 34 if consideration is for aluminum wires.

Table 1

ALLOWABLE AMPACITIES OF INSULATED CONDUCTORS RATED 0-2000 V, 60 TO 90°C

Not More Than Three Conductors in Raceway or Cable or Earth (Directly Buried), Based on Ambient Temperature of 30°C (86°F)

Temperature rating of conductor

Copper

Size AWG MCM	60°C (140°F) Types RUW, T, TW, UF	75°C (167°F) Types FEPW, RH, RHW, RUH, THW, THWN, XHHW, USE, ZW	85°C (185°F) Types V, MI	90°C (194°F) Types TA, TBS, SA, AVB, SIS, FEP, FEPB, RHH, THHN, XHHW
14	15	15	25	25
12	20	20	30	30
10	30	30	40	40
8	40	45	50	50
6	55	65	70	70
4	70	85	90	90
3	80	100	105	105
2	95	115	120	120
1	110	130	140	140
0	125	150	155	155
00	145	175	185	185
000	165	200	210	210
0000	195	230	235	235

Aluminum or Copper-clad aluminum

Size AWG MCM	60°C (140°F) Types RUW, T, TW, UF	75°C (167°F) Types RH, RHW, RUH, THW, THWN, XHHW, USE	85°C (185°F) Types V, MI	90°C (194°F) Types TA, TBS, SA, AVB, SIS, RHH, THHN, XHHW[a]
14	—	—	—	—
12	15	15	25	25
10	25	25	30	30
8	30	40	40	40
6	40	50	55	55
4	55	65	70	70
3	65	75	80	80
2	75	90	95	95
1	85	100	110	110
0	100	120	125	125
00	115	135	145	145
000	130	155	165	165
0000	155	180	185	185

Note: The load current rating and the overcurrent protection for these conductors should not exceed 15 A for 14AWG, 20 A for 12AWG, and 30 A for 10 AWS copper, or 15 A for 12 AWG and 25 A for 10 AWG aluminum and copper-clad aluminum. For larger sizes, see Table 310-16, National Electrical Code.[3]

a For dry locations only. See 75°C column for wet locations.

= 20). No. 10 wire is again the minimum size, but it must be supported at frequent intervals. No. 8 wire is the smallest size for practical outside overhead use. It would more than suffice for voltage drop and for ampacity. Therefore, No. 8 with TW or RW insulation individual conductors should be selected. Run three wires, one being for grounding.

Example 3

Select the cable to use for proper installation of a 4500-W, 240-V water heater for the milk handling area of a dairy barn,[4] 60 ft from the main service entrance. Because this is a damp location, choose UF cable. This is a long run, so calculate

$$cm = \frac{22^* \times \dfrac{4500}{240} \times 60}{4.8} = 5156$$

No. 12 has ample circular mil cross section (6530) and has an ampacity of 20 A. The calculated amps value is $\dfrac{4500}{240} = 18.75$ which appears to be adequate. However, there is an additional safety guideline calling for "no continuous single appliance load which exceeds 80% of the rating of the wire (or circuit)" — 20 × 80% = 16 A. Accordingly, choose 10-2 WG UF cable.

Example 4

The required power capacity of the service entrance for a dairy barn complex has been determined as 34,800 W 120/240 V (three-wire single phase). Determine the proper size service entrance switch and service entrance conductors and the correct size wire to use for the service-drop wires for this dairy structure. The length of run is 16 ft for the entrance and 28 ft for the service drop.

It is recognized that these lengths are classed as "short" wires; therefore, voltage drop is not a problem. The wires (and/or cable) will be located in a wet location. Type SE style U unarmored service entrance cable is a possibility, or if an installation of rigid conduit is desired, the three individual wires could be run through the conduit. Both specifications will be identified:

$$P = E \, I \, \text{Cos} \, \theta$$

Assume that the power factor (Cos θ) is approximately 1.0; then 34,800 W ÷ 240 V = 145 A. Select the next largest standard size switch as 150-A** 120/240-V single-phase, heavy-duty, fusible main, service entrance switch.

The wire must have an ampacity of 150 A, and from Table 1 (copper wire section) it is determined that 2/0 type TW or 1/0 type THW wires would be adequate.

Type SE cable with three AWG 2/0 type TW copper wires or three type THW AWG 1/0 wires in 1¹/₂-in. diameter rigid conduit are recommended. Refer to Table 2 or Table 3, Chapter 9, NEC[3] where it can be seen that three 1/0 wires are permitted in 1¹/₂-in. conduit or three 2/0 wires with insulation type (TW) wires could be selected for the 1¹/₂-in. diameter conduit, whichever is least expensive.

The service drop wires are single conductors in free air. Note that left-hand columns of

---

* Use 34 if consideration is for aluminum wires.

** The "80%-capacity" limit is not applicable due to diversity of load.

Table 2
PROPERTIES AND DATA OF COPPER CONDUCTORS[a]

Wire size AWG	Cross-sectional area in circular mils	Ampacity single conductor in free air		Maximum number of conductors having type T, TW THW[b] insulation permitted in standard trade sizes of conduit or tubing					
		60°C T, R	75°C THW, RH	½ in.	¾ in.	1 in.	1¼ in.	1½ in.	2 in
14	4,110	20	20	9	15	25	44	60	99
12	6,530	25	25	7	12	19	35	47	78
10	10,380	40	40	5	9	15	26	36	60
8	16,510	55	65	2	4	7	12	17	28
6	26,240	80	95	1	2	4	7	10	16
4	41,740	105	125	1	1	3	5	7	12
3	52,620	120	145	1	1	2	4	6	10
2	66,360	140	170	1	1	2	4	5	9
1	83,690	165	195		1	1	3	4	6
0	105,600	195	230		1	1	2	3	5
00	133,100	225	265		1	1	1	3	5
000	167,800	260	310		1	1	1	2	4
0000	211,600	300	360			1	1	1	3

[a] Refer to Tables 3A and 8 of Chapter 9 and Table 310-17 of the National Electric Code[3] for additional data.
[b] RHW also if without outer covering.

Table 3
PROPERTIES AND DATA OF ALUMINUM CONDUCTORS[a]

Wire size AWG	Cross-sectional area in circular mils	Ampacity single conductor in free air		Maximum number of conductors having type T, TW, THW[b] insulation permitted in standard trade sizes of conduit or tubing					
		60°C T, R	75°C THW, RH	½ in.	¾ in.	1 in.	1¼ in.	1½ in.	2 in.
14	4,110	—	—	9	15	25	44	60	99
12	6,530	20	20	7	12	19	35	47	78
10	10,380	30	30	5	9	15	26	36	60
8	16,510	45	55	2	4	7	12	17	28
6	26,240	60	75	1	2	4	7	10	16
4	41,740	80	100	1	1	3	5	7	12
3	52,620	95	115	1	1	2	4	6	10
2	66,360	110	135	1	1	2	4	5	9
1	83,690	130	155		1	1	3	4	6
0	105,600	150	180		1	1	2	3	5
00	133,100	175	210		1	1	1	3	5
000	167,800	200	240		1	1	1	2	4
0000	211,600	230	280			1	1	1	3

[a] Refer to Tables 3A and 8 of Chapter 9 and Table 310-17 of the National Electric Code[3] for additional data.
[b] RHW also if without outer covering.

Table 2 show the ampacity of single conductors in free air. Either No. 2 type THW copper wires or No. 1 type TW copper has sufficient ampacity and would be satisfactory.

Example 5

Feeder lines to serve electric motors are frequent choices that must be made. Several appropriate guidelines should be followed.

Select the proper feeder line size to serve a 5-hp 240-V single-phase electric motor located 100 ft from the service entrance. From Table 1 of the chapter on Guidelines For Installation of Motors, the full load current of a 5-hp 240-V motor is 28 A. Also from this chapter, the minimum ampacity for the feeder lines (branch circuit) of this installation must be at least:

$$125\% \times 28 \text{ A} = 35 \text{ A}$$

This indicates a temporary choice of No. 8 copper wire having an ampacity of 40 A with Type T insulation.

To allow for the surge current at starting and to keep voltage drop manageable during this period, a very good procedure is to assume 2000 W/hp and compute circular mil area required to limit the voltage drop to 2% under the assumed condition. Thus,

$$cm = \frac{22 \times \dfrac{5 \times 2000}{240} \times 100}{2\% \times 240}$$

$$cm = \frac{22 \times 41.6 \times 100}{4.8} = 19{,}097$$

No. 8 wire has 16,510-circular mil cross-sectional area and No. 6 has 26,240. These values are furnished in Table 2. Therefore, select No. 6 type T (or TW if a conduit run is desired) copper wires.

From Table 2, the thin wall rigid conduit size required for two No. 6 wires is $^3/_4$ in. Run the wires to a junction box or to a controller near the motor; then use flexible conduit or type S cord to the motor itself (allows for vibration). The conduit wall serves as the grounding conductor. With type S cord, add another wire (6-3, type S). Also, type SE cable is an alternate choice to the conduit method. For this wiring method, specify two No. 6 insulated copper conductors and one No. 8 bare conductor in the cable. The No. 8 bare wire is used as the grounding conductor for non current-carrying metal parts.

COMPUTATION OF ELECTRICAL-DEMAND LOADS FOR FARMS

The procedure for calculating the voltamps-demand of various farm buildings is described in ASAE Engineering Practice, EP294.1,[4] and in Article 220 of the National Electrical Code.[3] The essentials of these procedures are outlined below. The calculations form the basis of selection for size of wire for feeder lines and service entrance conductors and for the size (rating) of service entrance equipment.

Dwelling (100 A or Greater Service)

Complete the total voltamps as appropriately determined for each of the following three categories:

1. Select the largest load from A, B, C, and D; then enter value: _____
    A. Air Conditioning and cooling including heat pump compressors ×
       100 % (use nameplate volts × amps in the case of electric motors)
    B. Central electric space heating × 65%
    C. Less than four separately controlled space heaters × 65%
    D. Four or more separately controlled space heaters × 40%
2. Determine the total wattage of all other loads, described below, and then
    enter up to 10,000 W as contributing toward the dwelling total: _____

A. 1500 VA for each small appliance branch circuit (two minimum)
B. 1500 VA for each laundry branch circuit
C. 3 VA/ft² for general lighting and general use receptacles
D. Total of the nameplate ratings of all fixed appliances such as water heater, range, wall-mounted oven, counter-mounted cooking unit, clothes dryer, washer motors (multiply nameplate volts × amps)

3. Subtract 10,000 from the sum of the voltamp obtained from A + B + C + D in category 2. Multiply the remainder by 40%, and enter value:

Total
_____

Other Farm Buildings

The procedure outlined below is recommended for determining the wattage demand of all farm buildings other than the dwelling. Assume no difference in voltamps and watts:

1. List all of the types of electrical loads in the building — voltamps of lights, heaters, motors, appliances, etc. Determine the voltamps required for each load. In the case of electric motors, multiply the nameplate volts by amps and use this product as the voltamps of the motor. Increase the voltamp value of the largest motor by 25% and use it as the rating of this motor.

2. Identify from the list in No. 1 all of the loads that are expected to be _____
    "on" at the same time (operating without diversity). The sum of these loads must be considered at full value. Enter as voltamps for No. 2. Additionally, if the largest motor is not a part of the loads expected to operate without diversity and if the voltamps of that motor is greater than the full-value sum, then use the voltamps of the motor on the line for No. 2. Also, if any other single load, such as a space heater, is not a part of the "without diversity" group and its wattage is larger than the value determined for No. 2, then use it on line No. 2. Still further, if the load entered on line No. 2 is less than 13,800 VA, then add other loads from the list until the total is approximately 13,800 (230 × 60).

3. Add the voltamps of loads not used in No. 2 up to a sum of 13,800 W. Enter voltamps × 50%   _____

4. Add voltamps of all remaining loads. Enter value × 25%.   _____

Total
_____

Computing Total Load for the Farm

The total farm load includes loads from all buildings and the dwelling. The loads are combined in a special way, taking diversity into consideration.

1. Voltamp total for farm dwelling   _____
2. Total voltage value for largest building load   _____
3. Next largest building load × 75%   _____
4. Next largest building load × 65%   _____
5. All remaining × 50%   _____

Total
_____

Determining Service Ratings and Wire Sizes

The wire size and size of service entrance equipment can be determined after the appropriate load has been calculated in accordance with the procedure described in the previous section.

Unless otherwise specified, a three-wire, single-phase service is assumed for the farmstead. This means that both 115 and 230 V are available for the circuit. Calculate the value of amperes required for the load in question by using the equation:

$$\text{Amperes} = \frac{\text{Voltamp required for the load}}{230 \text{ V}}$$

(this assumes PF = 1.00)

If the voltamps for the dwelling = 21,000, the ampere value is $\dfrac{21000}{230} = 91.3$. Always select wires and switches with an ampere rating at least as large as calculated; therefore, choose a 100-A service entrance switch and, from Table 1, AWG No. 2 copper wire size for the service entrance cable or conduit wires. Other farm buildings would be handled in a similar manner. Also, as is sometimes the case, if a single, central meter pole serving the entire farmstead is used, the procedure is to determine the total load for the farm; then (if the total is 42,000 VA)

$$\frac{42000}{230} = 183 \text{ A}$$

Select the matching or next largest standard size (60, 100, 125, 150, 200, and 400 are standard size switches) such as: 200-A service disconnect switch and, from Table 1, AWG 3/0 copper wires with type RHW or THW insulation in 2-in. conduit size (Table 2) as the service entrance conductor.

The feeder lines from meter pole to the various buildings should be determined on the basis of 2% voltage drop and in accordance with procedures outlined in Example 2.

EQUATIONS FOR ELECTRIC CIRCUITS

DC Series Circuits

$$I_{in} = \frac{E_{in}}{R_T}$$

$$R_T = R_{Total} = R_1 + R_2 + \text{---}$$

I is in amperes, E is in volts, and R is in ohms

$$E_{in} = E_{input} = E_1 + E_2 + \text{---}$$

$$I_{in} = I \text{ of the circuit}$$

$$P_{in} = E_{in}I_{in} = \frac{E_{in}^2}{R_T} = I_{in}^2 R_T$$

DC Parallel Circuits

$$I_{in} = \frac{E_{in}}{R_e}$$

$$R_e = \text{equivalent } R = \frac{1}{\dfrac{1}{R_1} + \dfrac{1}{R_2}} + \text{---}$$

E_{in} = E of the parallel branches

$$I_{in} = I_1 + I_2 + \text{---}$$

$$P_{in} = E_{in}I_{in} = \frac{E_{in}^2}{R_e} = I_{in}^2 R_e$$

AC Series

$$I_{in} = \frac{E_{in}}{Z}$$

$$Z = \sqrt{R_T^2 + (X_{L_T} - X_{c_T})^2}$$

E_{in} = Vector sum of $E_1 + E_2 + \text{---}$ $\qquad X_L = 2\pi fL$

I_{in} = I of circuit $\qquad X_C = \dfrac{1}{2\pi fC}$

$$P = E_{in}I_{in}\cos\theta = I^2 R_T = \frac{E_{R_T}^2}{R_T}$$

$$\theta = \text{Tan}^{-1}\frac{X_{L_T} - X_{c_T}}{R_T}$$

AC Parallel

$$I_{in} = EY_T$$

where Y_T = vector sum of $\dfrac{1}{Z_1} + \dfrac{1}{Z_2} + \text{---}$, E_{in} = E of parallel branches, I_{in} = vector sum of $I_1 + I_2 + \text{---}$, $P = E I \cos\theta = I_{R_1}^2 R_1 + I_{R_2}^2 R_2 + \text{---}$, and $\theta = \text{Tan}^{-1}\dfrac{\text{Vert}}{\text{Horz}}$ components of Y_{Total} or $\dfrac{\text{Vert}}{\text{Horz}}$ components of I_{in}.

Three-Phase Delta Circuits

$$E_L = E_{Phase}$$

$$I_L = \sqrt{3}I_{Phase}$$

$$P = \sqrt{3}E_L I_L \cos\theta_{phase}$$

$\cos\theta_{phase}$ assumes all phases balanced

and $\theta_{phase} = \text{Tan}^{-1}\dfrac{X_{phase}}{R_{phase}}$

If unbalanced phases, use:

$$P = P_{phase\ 1} + P_{ph\ 2} + P_{ph\ 3}$$

Three-Phase Wye Circuits

$$E_L = \sqrt{3}E_{phase}$$

$$I_L = I_{phase}$$

$$P = \sqrt{3}E_L I_L \cos\theta_{phase}$$

$$P = P_{phase\ 1} + P_{ph\ 2} + P_{ph\ 3}$$

REFERENCES

1. **Brown, R. H.,** *Farm Electrification,* McGraw-Hill, New York, 1956.
2. **Gray, A. and Wallace, G. A.,** *Principles and Practice of Electrical Engineering,* 6th ed., McGraw-Hill, New York, 1948.
3. National Electrical Code, National Fire Protection Association, Quincy, Mass., 1984.
4. Computing Electrical Demands for Farms, ASAE EP 294.1, American Society of Agricultural Engineers, St. Joseph, Mich., 1984.

GUIDELINES FOR INSTALLATION AND OPERATION OF SINGLE-PHASE ELECTRIC MOTORS

Robert H. Brown

INTRODUCTION

Five decisions must be made prior to the preparation of specifications for installing electric motors. These involve determination of the proper size or rating of electric circuit components, namely: (1) motor running overcurrent protection, (2) motor branch circuit short-circuit and ground-fault protection, (3) size of wire for the feeder and/or branch circuit, (4) type and rating of the motor controller, and (5) size or rating of the disconnecting means.

The regulations and guidelines described below are based on electric circuit theory[1] and Article 430 of the National Electric Code entitled, "Motors, Motor Circuits and Controllers".[2]

INSTALLATION GUIDELINES

Motor Running Overcurrent Protection

Each continuous-duty motor should be protected against overload (excessive input amperes) as follows:

1. If the motor is less than 1 hp, is manually started, is within sight from the starting location, and is not permanently installed, the branch circuit ground-fault and/or short-circuit protective device may serve to protect the motor. Additional protection may be added but is not mandatory.
2. For all other situations a separate overload device is the best choice, but a built-in thermal overload device may serve. Either of these must be selected to trip at or must be rated at not more than (A) 125% of motor nameplate current for motors with service factors not less than 1.15 and rated for a temperature rise not over 40°C (see the motor name-plate for these values) and (B) 115% of motor nameplate current for motors having service factors exceeding 1.15 or temperature rise rating over 40°C.

When designing circuits and choosing components for motor circuits, this is the only instance where the value of current printed on the nameplate of the motor is used. For all other determinations, use the value of current furnished in Tables 1 to 3.

Motor Branch Circuit Short-Circuit and Ground-Fault Protection

The branch circuit conductors serving the motor and the motor control apparatus are to be protected against short circuits and grounds by selecting overcurrent devices as follows:

1. Refer to Table 430-152, page 320, 1978 National Electrical Code, or refer to the guidelines shown in Table 4 which were taken from the data in Table 430-152.
2. Where the values in Table 4, when multiplied by current values in Table 1, yield an overcurrent device rating which will not start the motor, they may be increased to 400% of the motor current rating if it is not over 100 A. They may be increased to 300% if the motor current rating is over 100 A. They may be increased to the next largest standard size overcurrent device in all instances.

Protection of the branch circuit conductors as chosen by the above guidelines is also applicable to a circuit serving more than one motor, provided each motor has individual

Table 1

FULL-LOAD CURRENTS IN AMPERES OF SINGLE-PHASE ALTERNATING CURRENT ELECTRIC MOTORS

hp rating of motor	Voltage rating of motor		
	115 (110—120)	230 (220—240)	208
$^1/_6$	4.4	2.2	2.4
$^1/_4$	5.8	2.9	3.2
$^1/_3$	7.2	3.6	4.0
$^1/_2$	9.8	4.9	5.4
$^3/_4$	13.8	6.9	7.6
1	16.0	8.0	8.8
$1^1/_2$	20.0	10.0	11.0
2	24.0	12.0	13.2
3	34.0	17.0	18.7
5	56.0	28.0	30.8
$7^1/_2$	80.0	40.0	44.0
10	100.0	50.0	55.0

From Table 430-148, National Electrical Code, National Fire Protection Association, Quincy, Mass., 1984. With permission.

Table 2

FULL-LOAD CURRENTS IN AMPERES OF POLYPHASE INDUCTION MOTORS

hp rating	Voltage rating of motor		
	115	230	460
$^1/_2$	4	2.0	1
$^3/_4$	5.6	2.8	1.4
1	7.2	3.6	1.8
$1^1/_2$	10.4	5.2	2.6
2	13.6	6.8	3.4
3		9.6	4.8
5		15.2	7.6
$7^1/_2$		22	11
10		28	14
15		42	21
20		54	27
25		68	34
40		104	52
50		130	65

Note: For other values see Table 430-150, National Electrical Code.[2]

Table 3

FULL-LOAD CURRENTS IN AMPERES OF DIRECT CURRENT MOTORS

hp	Armature voltage rating			
	90	120	180	240
$^1/_4$	4.0	3.1	2.0	1.6
$^1/_2$	6.8	5.4	3.4	2.7
1	12.2	9.5	6.1	4.7
$1^1/_2$		13.2	8.3	6.6
2		17.0	10.8	8.5
3		25.0	16.0	12.2
5		40.0	27.0	20.0
10		76.0		38.0
15				55.0
25				89.0
40				140.0

Note: For other values see Table 430-147, National Electrical Code.[2]

motor running overcurrent protection as selected in step 1 and provided one of the following conditions exists: (1) no motor exceeds 1 hp and the rating of the motor controllers is not exceeded, or (2) the branch circuit protective device was selected using the smallest motor of the group.

Motor Circuit Conductors

Single motor — The minimum wire size for conductors serving a single motor, either

Table 4
GUIDELINES FOR SELECTING THE AMPERE RATING OF DEVICES FOR PROTECTING THE MOTOR BRANCH CIRCUIT AGAINST SHORT CIRCUITS AND GROUND FAULTS

Code letter on motor nameplate	Percentage of motor full-load current allowed for rating of the		
	Regular fuse	Time-delay fuse	Circuit breaker
A	150	150	150
B—E	250	175	200
F—V	300	175	250
None	300	175	250

feeder or branch circuit, is selected on the basis of having at least ampacity equal to 125% (motor full-load current shown in Table 1). The ampacity of insulated copper and aluminum conductors is given in Table 1 of the preceding chapter (Electric, Loads, Circuits, and Wires), and in Table 310-16 of the NEC. Note that this determination yields a minimum size; for a more adequate choice refer to Example 5 in the preceding chapter (Electric Loads, Circuits, and Wires).

More than one motor — The conductor ampacity should at least be sufficient to serve an ampere load equal to the sum of the full-load ampere rating of all of the motors plus 25% times the full-load ampere rating of the largest motor in the group.

Motor Controllers

The controller includes any switch or device normally used to start and stop a motor. Its size is selected as follows. The general guideline is that the controller should have a horsepower rating not less than the horsepower rating of the motor. The following exceptions may be used if desired:

1. $1/8$ hp and smaller — the branch circuit overcurrent device may serve.
2. $1/3$ hp and smaller and portable — the attachment plug cap may serve.
3. 2 hp and less and stationary — a general-use switch having an ampere rating not less than twice the full-load current rating of the motor is the minimum permitted controller.

DISCONNECTING MEANS

A disconnecting means is required to disconnect the motor, controllers, and branch circuit wires (or feeder) from the electric source.

The disconnect means must be in sight from the controller location. The means may be in the same enclosure with the controller or may be in a separate enclosure. The general guideline is that the disconnecting means must be a motor circuit switch rated in horsepower — not less than the horsepower rating of the motor. The disconnecting means must also have an ampere rating, in addition to the horsepower rating, which must be at least equal to 115% times the full-load current rating of the motor.

There are important *exceptions* to the general guideline as follows:

1. If $1/8$ hp and smaller, the branch circuit overcurrent device may serve as the disconnecting means.
2. If $1/3$ hp and smaller, the plug cap and receptable may serve.

3.   If a portable motor, the plug cap and receptable may serve for any size motor.
4.   A general-use switch rated in amperes (not necessarily in horsepower) may serve if the ampere rating of the switch is at least twice the full-load current rating of the motor.

If the disconnecting means serves for more than one motor, the ampere rating must be not less than the total locked-rotor amperes for all the motors and horsepower rating must be not less than the sum of the horsepower ratings of all of the motors being served.

REFERENCES

1. EEI, ASAE, *Agricultural Wiring Handbook,* Edison Electric Institute, New York, 1970.
2. National Electrical Code, National Fire Protection Association, Quincy, Mass., 1984.

ELECTRIC MOTOR CHARACTERISTICS AND SCHEMATIC DIAGRAMS

Robert H. Brown

DIRECT CURRENT MOTORS

Direct current (DC) motors operate from a DC supply such as batteries, DC generators, or rectified alternating current (AC). Cordless power tools, clocks, automobile starters, motors in electric vehicles, motors powered by photovoltaic cells, and the like are DC motors. Frequently, batteries are not used as the source of DC power, being replaced by an AC-operated power supply that uses electronic means (rectification) to convert AC to DC.

In past years ships, airplanes, trains, and street cars were major users of DC motors. Today the major application is in industries where efficient and automatic speed control is the dominant issue. These motors are more expensive than AC motors and have commutators and brushes that require maintenance and cause sparking. Nevertheless, when a motor must be battery-operated or when very excellent, low-cost speed variations and/or speed control are needed, the DC motor is the choice.

DC motors are classified as shunt, series, or compound.[1] The names arise from the method of connecting the field windings, as in parallel with the armature (shunt), in series with the armature (series), or in the case of two field windings, one in parallel and one in series with the armature (compound). The schematic diagrams for these motors are shown in Figure 1. To reverse the direction of rotation, one must reverse the direction of current flow in either the armature or the field, but not both. This is done by interchanging the connections of the field lead wires at the armature (if previously F_1 to A_1 and F_2 to A_2, change to F_1 to A_2 and F_2 to A_1).

The speed of these motors is normally adjusted by varying the current in the field circuit. An increase in field current results in a decrease in speed, as can be seen in the following equation:[2]

$$S = \frac{E_A - I_A R_a}{k\phi}$$

where S is speed in rpm, E_A is voltage applied to armature, ϕ is field flux and is directly related to field current, k is a constant of the units involved, and R_a is resistance of the armature winding.

It is also seen from the equation that speed is directly related to armature voltage. Wide variations in speed (zero to full load) can be obtained by varying this voltage, but after the motor is started it is not usually as convenient to vary armature voltage as it is to vary field current.

The torque of a DC motor is dependent upon field current (this current sets up the magnetic flux, ϕ and the armature current (I_A) as follows:

$$T = K^1 \phi I_A$$

where K^1 is a constant for the units involved. For shunt and compound motors, the starting torque is usually selected as 150% of full-load torque and is secured by making I_A 150% of rated (nameplate) full-load current. I_A is normally controlled by adjusting the external portion (R_{ext}) of the total armature circuit resistance, R_A, where $R_A = R_{ext} + R_a$. Reference to the speed equation and substituting $E_A = E_T - I_A R_{ext}$ and solving for I_A yields:

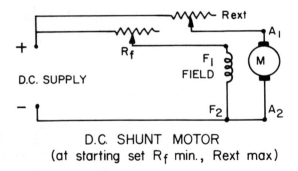

D.C. SHUNT MOTOR
(at starting set R_f min., Rext max)

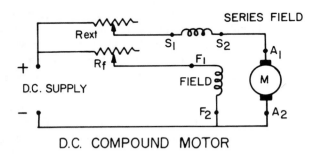

D.C. COMPOUND MOTOR

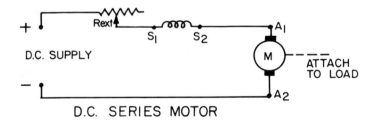

D.C. SERIES MOTOR

FIGURE 1. Schematic diagrams for DC motors.

$$I_A = \frac{E_T - k\phi S}{R_{ext} + R_a}$$

E_T is the voltage applied at the source.

At the instant of starting KϕS equals zero and $I_A = \dfrac{E_T}{R_{ext} + R_a}$. The series motor has the highest starting torque per input ampere since armature current and field current are the same value. Thus, both I_A and ϕ are maximum at starting. The shunt and compound motors are classed as "constant-speed" motors, but this does not mean a fixed speed. It is normal to expect a decrease in speed (probably around 10%) as load is increased from no load to full load. The speed of a series motor varies with load, dramatically, running very fast (such as 2000 rpm) at no load and dropping down to very slow (several hundred rpm) at full load. This motor exhibits the characteristic of a constant horsepower motor. As torque increases, speed decreases so that speed times torque is essentially constant.

The horsepower output of a DC motor (and an AC motor) is generally calculated using torque (T) and speed (S) values in lb-ft·ft and rpm, as follows:

$$hp = \frac{T \times S}{5250}$$

$$kW = hp(0.746)$$

and in SI units

The full-load values for hp and S are furnished on the nameplate of the motor. The torque at full load can thus be calculated for any particular motor. The starting torque for machines such as the shunt motor is probably 150% times the full-load torque calculated in the above manner. (Multiply lb-ft·ft by 1.355818 to obtain N·m.)

AC MOTORS

AC electric motors are by far the most popular and most important group of electric motors. Since the electric power distribution systems of today are AC systems, it is more energy efficient to connect a motor directly to the AC source than to convert to DC and then supply the motor. In addition, the AC induction motor, synchronous motor, and series motor offer a source of shaft torque and speed (rpm) that excels. The AC induction motor is by far the most popular motor of this group. It operates at essentially a constant speed and is available as a polyphase or single-phase motor. The synchronous motor operates at one fixed speed and also assists in power factor correction for an electric system.[2] The series motor is a constant-horsepower variable-speed motor, mainly used for portable power tools. The distinguishing characteristics of these motors, which are important in selecting the type of motor for a specific application and in understanding the motor's performance while in operation, are summarized in the next three sections.[3]

Induction Motors

This is the most important group of commercial electric motors. Their operating characteristics are constant speed, variable torque, variable horsepower, average power factors, and average efficiencies. These motors are so named because the input energy to the rotating member is accomplished through the process of induction.

Induction motors have two main parts: the stationary member (or stator) and the rotating member (or rotor). There are two types of rotors: the squirrel cage and the wound rotor. The squirrel-cage rotor appears outwardly as a solid mass, but actually consists of a series of bars placed in slots around the periphery of the rotor. The conductors are short circuited at each end by an end ring. This is usually the type of rotor found in AC motors and is a low-cost, rugged, reliable component part. The wound-rotor type of construction appears quite different and is used only for the three-phase wound-rotor induction motor. The windings are visible and are not short circuited, but are connected to slip rings located on one end of the shaft. The slip rings provide a means for connecting external resistance into the rotor circuit, which changes the electrical characteristics of the rotor, namely, power factor and impedance, and therefore varies the input current, torque, and speed of the motor. The wound-rotor induction motor has a greater initial cost than the squirrel cage.

Every induction motor rotates because the stator winding sets up a rotating magnetic field. Torque is developed when this magnetic field cuts across the rotor conductors. The rotor turns in the same direction as the rotating field and therefore must rotate more slowly. At full load, the rotor speed is usually about 97% of that of the rotating field. The speed of the rotating field, the synchronous speed, does not depend on voltage, but is a function of frequency and the number of poles of the stator winding (see formulas). Normally the AC supply frequency is fixed at 60 Hz. However, the new speed controllers for induction motors can supply power at various frequencies (up to 60 Hz) and use this means of causing the desired variation in rpm.

Induction-Motor Formulas

Some useful relationships for AC motors running with induction characteristics are

$$\text{rpm of rotating field} = \frac{120 \times \text{frequency}}{\text{number of poles}}$$

$$\text{Percent slip} = 100 \frac{\text{rpm of rotating field} - \text{rpm of rotor}}{\text{rpm of rotating field}}$$

$$\text{Torque, ft-lb} = \frac{\text{hp} \times 5250}{\text{rpm}}$$

$$\text{Watts input (single phase)} = \text{EI cos } \theta$$

$$\text{Watts input (three phase)} = 3 \text{ EI cos } \theta$$

$$\text{Watts input} = \frac{\text{hp output} \times 746}{\% \text{ efficiency}}$$

Approximations for Induction Motors

1. Rotor speed: ranges from 95 to 99% of rotating-field speed
2. Cos θ increases with load to about 83% at full load
3. Power input: 1200 W/hp up to $1/2$ hp, 1000 W/hp for motor sizes $1/2$ hp and larger
4. Torque: varies as the square of the ratio of applied voltage; thus, reducing voltage from 120 to 80 V reduces the torque to $(80/120)^2 = 44.4\%$ of its value with 120 V applied

Series Motors

The AC series motors operate with series characteristics and always have brushes, a commutator, and a DC-type wound rotor. The rotor winding either is connected in series with the stator winding or is arranged electrically so as to yield the same result. The AC series motor has the stator winding connected in series with the rotor (via the brushes) and rotor current is conducted into the rotor. This is a true series motor. On the other hand, in the case of repulsion-type motors, the rotor is not connected to the stator, but the brushes are connected together to provide a closed rotor circuit, and all rotor currents are induced currents.

Motors operating with series characteristics develop torque because of the interaction between rotor current and field flux. They do not depend upon a rotating magnetic field. These motors provide high starting torque, good running torque, variable speed, and approximately constant horsepower output. The no-load speed is very high, but the speed decreases rapidly as load is increased. They are best adapted for jobs such as food mixers, portable power tools, sewing machines, lawn mowers, and similar applications where an increase in load should be accompanied by a decrease in speed.

Synchronous Motors

The synchronous motor runs at an exact speed for all loads up to the pull-out point. If load is increased past this value, the motor will stop. In industry it is used because of this speed characteristic and because it can be made to take a leading current from the source, thereby improving the overall power factor of combined electrical loads.

The synchronous motor is usually of the revolving-field type, requires a DC source for exciting the field poles, and is provided with an amortisseur winding (in the pole faces) for self-starting. Other than the amortisseur winding, the synchronous motor is identical to the

AC generator, and it can therefore be used as a standby generator for emergency use. For this reason, it is adapted for use on certain types of farms.

Summary of Operating Performance

Operating performance can be predicted and evaluated after first identifying the type of motor involved.[4] It will have one of the three sets of characteristics described above, either induction, series, or synchronous. One group of motors, the repulsion type, has a combination of these characteristics such as starting series and running induction. A summary of performances is presented in Table 1.

In order to assist in identification and to appreciate the listing of major parts in Table 1, one should refer to the schematic diagrams given in Figure 2. The name of the motor, the major features, and the operating characteristics are all interrelated. For example, a two-value capacitor motor has two capacitors, a capacitor motor does not have a centrifugally operated switch, a shaded-pole motor has a shading coil, etc. Also note that all motors having a squirrel-cage rotor operate with induction characteristics.

MOTOR NAMEPLATES

The motor manufacturer furnishes nameplate data of significant importance to the specifier, the installer, and the user of an electric motor. Values specified on the nameplate are full-load values. The messages and values gained from motor nameplate data are as follows:[4]

1. hp (full-load output horsepower) — The motor develops horsepower as needed to drive the load, from no load to full load. The efficiency of the motor is poorest at no load and at partial loads. It is greatest at full load.
2. Amps (full-load input current to the motor) — Expect this value of input current when the motor is connected to a voltage source having a magnitude as given on the nameplate and when the motor is delivering its rated horsepower. The starting current (surge of current at the instant of starting) will be greater. The input current to induction motors does not vary as much from no load to full load as one would expect. The power factor of the motor varies with load (improves as load increases) and the power input can increase as load increases without much change in current. (Refer to the equations at the end of this chapter.)
3. Volts — For single-voltage motors, there will be one value of volts on the nameplate. This is the rated voltage for the motor. For dual-voltage motors there will be two values of volts given on the nameplate (also two values of amps will be furnished). The rated input volts times amps product is the same, regardless of which supply voltage is used, since the highest voltage corresponds to the lowest amps value. For proper connections of the internal windings of a dual-voltage motor, refer to the schematic diagrams for electric motors (Figure 3).
4. rpm — The full-load speed of the motor is the value given. It is also known as the rated speed of the motor. Expect the no-load speed of AC induction motors to be greater, by approximately 3 to 5%.
5. Temperature rise (DUTY), 40 or 50°C (CONT) or (INTER) — The duty will be either continuous or intermittent. For the motor rated ''CONT'' and ''40°C'' which is being operated continuously, the temperature will rise to a value of 40°C *above ambient* temperature. Intermittent-duty (INTER) motors are designed for periods of 5, 15, 30, or 60 min and will overheat if operated continuously.
6. Service factor — The manufacturer furnishes this value to affirm that the motor may be overloaded slightly. Numbers such as 1.15 or 1.25 indicate 15 or 25% overload permitted above the full-load values on the nameplate.

Table 1
AC MOTOR CHARACTERISTICS

Name of motor	Operating characteristics	Usual range hp-size available	Major parts	Starting torque % of full load	Electric power source
Split-phase	Induction	1/20—1/2	Squirrel-cage rotor, CS,[a] starting winding, main winding	175	Single phase
Capacitor	Induction	1/12—3/4	Squirrel-cage rotor, capacitor, main winding, second winding	70	Single phase
Capacitor-start	Induction	1/6—5	Squirrel-cage rotor, CS, capacitor, starting winding, main winding	375	Single phase
Two-value capacitor	Induction	1/2—10	Squirrel-cage rotor, CS, two capacitors, main winding, second winding	300	Single phase
Repulsion	Series	1/6—5	Main winding, brushes, commutator	450	Single phase
Repulsion-start	Starts series runs induction	1/6—10	Main winding, brushes, CS, commutator	450	Single phase
Repulsion-induction	Combination series/ induction	1/6—5	Main winding, brushes, commutator, dual winding on rotor	300	Single phase
Shaded pole	Induction	1/20—1/8	Squirrel-cage rotor, stator poles, shading coils	50	Single phase
AC series	Series	1/20—200	Brushes, commutator, stator poles	450	Single phase
Universal	Series	1/20—200	Brushes, commutator, stator poles	450	Direct current or single phase
Three-phase squirrel cage	Induction	1/6—200	Squirrel-cage rotor, main winding	185	Three phase
Three-phase wound rotor	Induction	1/6—200	Main winding, wound rotor, 3 slip rings, brushes	Varies with resistance, 100 minimum	Three phase

| Two-phase squirrel cage | Induction | — | Main winding, squirrel-cage rotor | 185 | Two phase |
| Three-phase synchronous | Synchronous | 5—200 | Main winding, rotor poles, 2 slip rings, brushes | 110 | Three phase and DC |

[a] Centrifugally operated switch.

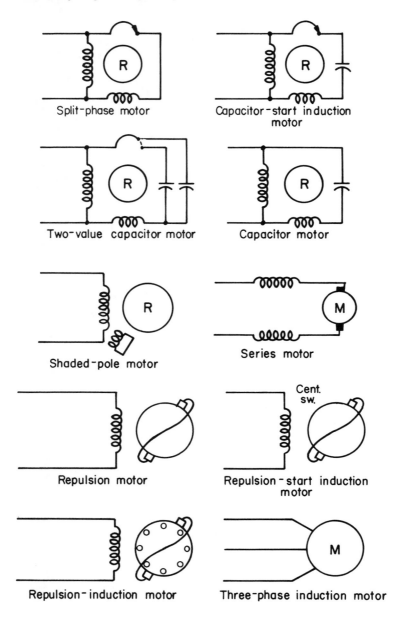

FIGURE 2.   Schematic diagrams for AC motors.

7.   Code — This nameplate entry, a letter of the alphabet, describes the locked-rotor kilovolt-amperes per horsepower for electric motors. (When the motor is stopped, and is switched ''on'', for an instant of time the full power is applied and the motor is not rotating, thus ''locked''.) This value gives an indication of how much surge current to expect when the motor starts. The first letters of the alphabet (A, B, C, etc.) convey the message that this motor has a very small value of surge current per horsepower when compared to code letters such as P, R, or S. Refer to Table 2 or to Table 430-7(b) of the National Electric Code for specific values. For example, assume that the magnitude of starting current for a particular motor is to be calculated using nameplate data and the data in Table 2 as follows. Given nameplate data is hp, 3; V, 230; rpm,

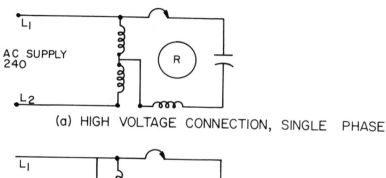

(a) HIGH VOLTAGE CONNECTION, SINGLE PHASE

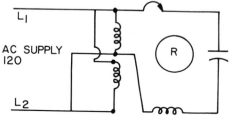

(b) LOW VOLTAGE CONNECTION, SINGLE PHASE

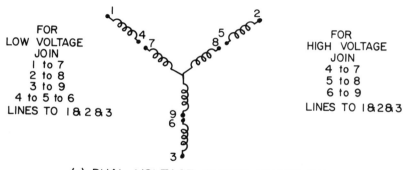

(c) DUAL VOLTAGE, THREE PHASE INDUCTION

FIGURE 3.   Schematic diagrams for dual-voltage connections of AC motors.

1725; amps, 17; code, G. Reference to Table 2 shows that code "G" includes motors having a kilovolt-ampere rating per horsepower with locked rotor in the range 5.6 to 6.29. One might use the low value or the high value, but an average value is suggested. The starting current would be

$$\text{amps} = \frac{\text{average kVA value} \times \text{hp} \times 1000}{V}$$

$$\text{amps} = \frac{\dfrac{5.6 + 6.29}{2} \times 3 \times 1000}{230}$$

$$\text{amps} = 77.5$$

8.   Other nameplate entries specify the style, model number, serial number, and manufacturer. These are self-explanatory and are useful in ordering spare parts and in determining the dimensions of importance when mounting and fastening the motor into position.

Table 2
LOCKED-ROTOR
INDICATING CODE
LETTERS FOR ELECTRIC
MOTORS[a]

Code letter	Kilovolt-amperes per horsepower with locked rotor
A	0—3.14
B	3.15—3.54
C	3.55—3.99
D	4.0—4.49
E	4.5—4.99
F	5.0—5.59
G	5.6—6.29
H	6.3—7.09
J	7.1—7.99
K	8.0—8.99
L	9.0—9.99
M	10.0—11.19
N	11.2—12.49
P	12.5—13.99
R	14.0—15.99
S	16.0—17.99
T	18.0—19.99
U	20.0—22.39
V	22.4—and up

[a] Refer to Table 430-7(b) of the National Electric Code.

AC SINGLE-PHASE MOTOR CIRCUIT

$$P = E\,I\,\cos\theta$$

E and I are nameplate values of the motor. I is the full load value, and $\cos\theta = PF =$ power factor of motor. $\cos\theta$ varies with load on motor ranging from about 0.2 to 0.9. For estimating the full-load power input, use 1000 W/hp or use PF -0.75 in the above equation along with nameplate values of E and I (Table 3).

AC THREE-PHASE MOTOR CIRCUIT

$$P = \sqrt{3}E\,I\,\cos\theta$$

E and I are nameplate values of the motor, with I being the rated full-load current of the motor. $\cos\theta$ varies with load. For estimating the full-load power input, use 900 W/hp or PF $= 0.83$ and the above equation.

Table 3
STANDARD HORSEPOWER SIZES
OF ELECTRIC MOTORS

Single-Phase AC Induction Motors

1/20	1/6	1/2	1 1/2	5	15
1/12	1/4	3/4	2	7 1/2	20
1/8	1/3	1	3	10	25

Polyphase Induction Motors (3-Phase)

1/8	3/4	5	25	75	250
1/6	1	7 1/2	30	100	etc.
1/4	1 1/2	10	40	125	
1/3	2	15	50	150	
1/2	3	20	60	200	

DC Motors

1/20 — 250 in sizes as listed above and

300	500	700
400	600	800

Note: 1 hp output = 746 W. Input watts = 746
÷ efficiency of motor.

REFERENCES

1. **Lister, E. C.,** *Electric Circuits and Machines,* 6th ed., McGraw-Hill, New York, 1983.
2. **Gray, A. and Wallace, G. A.,** *Principles and Practice of Electrical Engineering,* 6th ed., McGraw-Hill, New York, 1948.
3. **Veinott, C. G.,** *Fractional Horsepower Electric Motors,* McGraw-Hill, New York, 1940.
4. **Brown, R. H.,** *Farm Electrification,* McGraw-Hill, New York, 1956.

LIGHTING SYSTEM DESIGN

Robert H. Brown

INTRODUCTION

Graphical techniques along with some proven mathematical equations are used in the design of lighting systems. Procedures and accepted guidelines have been developed for both indoor and outdoor systems. The indoor special features are room shape and surface color, while the outdoor special features are mounting height and luminaire characteristics. The materials included in this chapter are for outdoor designs such as work areas, recreational areas, and roadways.

The basis for most lighting decisions arises from data presented in the Illuminating Engineering Society (IES) Lighting Handbook.[1] This is the standard for the industry and presents in detail all information needed for various lighting system designs.

When properly designed, lighting systems are intended to provide the recommended average illumination level (footcandles) in a space with a reasonable and acceptable degree of uniformity. This is true for a room or an outdoor space.

IES DEFINITIONS

Illumination (footcandle) meter — An instrument for measuring the illumination on a surface. Most such instruments consist of one or more barrier layer cells connected to a meter calibrated in footcandles.

Illumination — The density of the luminous flux incident on a surface; it is the quotient of the luminous flux by the area of the surface when the surface is uniformly illuminated. Level of illumination (not the act of illuminating) is the thrust of this definition.

Inverse-square law — A law stating that the illumination at a point on a surface varies directly with the candlepower of a point source and inversely as the square of the distance between the light source and the point.

Speed of light — The speed of all radiant energy, including light, is 2.997925×10^8 m/sec in a vacuum or approximately 186,000 mi/sec.

Street lighting luminaire — A complete lighting device consisting of a light source together with its direct appurtenances such as globe, reflectors, refractor, housing, and such support as is integral with the housing. The pole, post, or bracket is not considered part of the luminaire.

Luminous flux — The time rate of flow of light.

Luminous flux density at a surface — Luminous flux density per unit area of the surface. When referring to flux incident on a surface, it is identical to illumination.

Maintenance factor (MF) — A factor formerly used to denote the ratio of the illumination on a given area after a period of time to the initial illumination on the same area. Another term, light loss factor, is often used and like MF takes into account temperature, voltage variations, dirt accumulation on the luminaire, lamp depreciation, and atmospheric conditions.

Incandescent filament lamp — A lamp in which light is produced by a filament heated to incandescence by an electric current.

Mercury lamp — An electric discharge lamp in which the light is produced by the radiation from a mixture of a metallic vapor (such as mercury) and the products of the disassociation of halides (such as halides of indium or sodium).

High-pressure sodium lamp — A sodium vapor lamp in which the partial pressure of the vapor during operation is of the order of 10^4 N/m^2 (or 0.1 atm).

Table 1
CHARACTERISTICS OF LIGHT SOURCES

Source	Initial lumens	Rated life (hr)	lm/W	Remarks
Incandescent				
10 W	80		8.0	
25 W	230	750—1000	9.2	
60 W	860		14.3	
75 W	1180		15.7	
100W	1680		16.8	
150 W	2880		19.2	
Mercury		10,000—16,000	30—65	Greenish blue light; re-strike 3—7 min
Fluorescent	Wide range size and color	7,500—12,000	45—75	For interior applications
Metal halide		7,000—10,000	50—90	Changes color during warmup, somewhat like mercury color but improved color rendering; restrike 15 min
High-pressure sodium		10,000—15,000	110	Restrike 2 min; golden white color
Low-pressure sodium			170	Yellow color will restart immediately

Fluorescent lamp — A low-pressure mercury electric discharge lamp in which a fluorescing coating (phosphor) transforms some of the UV energy, generated by the discharge, into light.

LIGHT SOURCES

The IES definition for those light sources appropriate for outdoor lighting is given in the previous section. The characteristics and abilities of these sources to produce light flux are presented in Table 1. Note the significant difference in the lumens/watt of the various sources and also note the minimum time that must elapse before restroking (restarting) certain types of light sources. Generally speaking, fluorescent sources are not used in outdoor designs, but are a majority choice for interior systems.

The progress in improving the efficiency of light sources, as indicated by the lumens per watt value ranging from the 8 lm/W of Edison's early lamp to the 170 lm/W of low-pressure sodium sources, represents a giant step toward efficient lighting systems. The data of Table 1 also show that higher wattage incandescent lamps are more efficient than small wattage lamps.

FOOTCANDLE VALUES RECOMMENDED

The illumination levels recommended for various locations and activities are given in the IES Handbook.[1] The complete listing is extensive and is the values accepted as design standards in the lighting industry. For agricultural and rural area uses, the most frequently needed values are furnished in Tables 2 and 4.[1,2]

OUTDOOR LIGHTING

Isofootcandle Curves
These curves indicate the manner in which the illumination from a single light source (luminaire) is distributed on a horizontal plane.

Values for each point were obtained by:

$$\frac{\text{CP directed toward point}}{\text{Dist}^2 \text{ to that point from luminaire}} \times \text{Cos } \alpha$$

Typical curves are shown in Figure 1. The values given are in footcandles (fc) or lumens per square foot. Note that the mounting height (MH) used when obtaining the data is specified. For other MH use the curves as presented and obtain an initial value of footcandles; then convert this value as follows:

$$\text{fc at actual MH} = \text{fc}_{\text{chart}} \left(\frac{\text{Chart MH}}{\text{Actual MH}} \right)^2$$

Example 1

Use the photometric data[3] in Figure 1 and determine the footcandle value at a point on the pavement side which is 60 ft across from and 30 ft along (longitudinal) the side of the lamp position. MH = 50 ft. (The chart mounting height of Figure 1 is 40 ft.)

$$\text{Across ratio} = \frac{60}{50} = 1.2 \text{ (transverse)}$$

$$\text{Along ratio} = \frac{30}{50} = 0.6 \text{ (longitudinal)}$$

Chart value is 1.6 fc (intersect point of 1.2 on 0.6)

$$\text{Actual fc} = 1.6 \left(\frac{40}{50} \right)^2 1.6(0.64) = 1.02$$

Isofootcandle diagrams thus provide a means of evaluating the uniformity of illumination and determining the footcandles available at a specific point.

When seeking the footcandle value at a particular point, the footcandles from one luminaire must be added to the footcandles from other luminaires to gain the total footcandles. When more than one luminaire contributes illumination at any particular point in the lighted area, whether street side or curb side, add the values of footcandles from each luminaire, as determined separately, in order to determine the total footcandle value.

Example 2

Calculate the total footcandle value at point X in Figure 4a if all luminaires have the photometric data shown in Figure 1. (Curb side and house side are the same.)

fc from A curb side across 6/40 = 0.15 and along 40/40 = 1.00	2.00
fc from C curb side same as A	2.00
fc from B street side across 46/40 = 1.15 and along 0/40 = 0	2.40
Total	6.40 fc

Utilization Curves

Utilization curves for a particular luminaire are available from the lamp manufacturers.[3] These curves show the quantity of light striking the horizontal plane — in front of — and behind the luminaire. Utilization curves are the most widely used of photometric data because

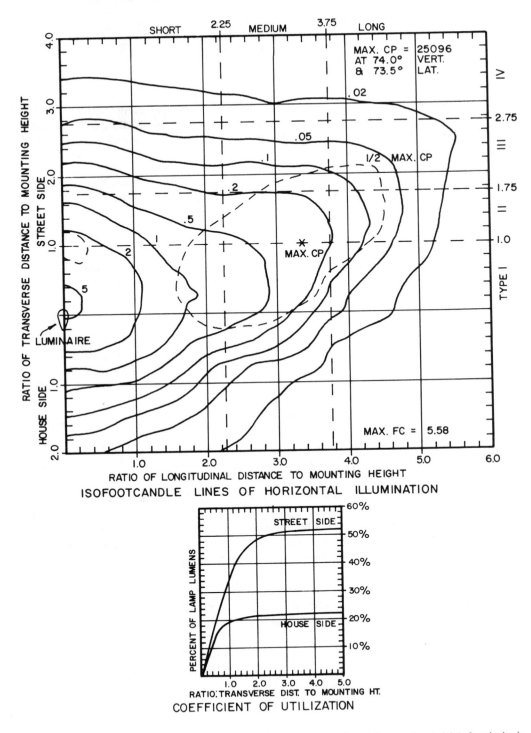

FIGURE 1.   Photometric report. Footcandle data for luminaire mounted at 40 ft (mounting height). Luminaire is ITT. 400 horizontal. Lamp is 400-W high-pressure sodium. Initial lumens 50,000.

they express *the percentage of light* falling on an area in terms of total light generated by the light source.

These curves will provide the very important term "coefficient of utilization" (CU), which must be used in all lighting design calculations.

Normally the photometric data such as shown in Figure 1 will include an isofootcandle diagram and utilization curve for the front (or street) side of the luminaire and another utilization curve for the back (or road) side of the luminaire. The total CU for the luminaire installation is the sum of the two coefficients.

To utilize the photometric data chart for determining the coefficients, one must first determine the ratio of distance to MH. Enter the graphical charts at points corresponding to these ratios and read the values at the appropriate intersection with the utilization curve.

Example 3

Use the photometric data in Figure 1 and determine the CU for the luminaire mounted as shown in Figure 4b.

$$\text{curb ratio} = \frac{10}{40} = 0.25 \qquad \text{Chart value is } 0.05$$

$$\text{Roadside ratio} = \frac{50}{40} = 1.25 \qquad \text{Chart value is } \underline{0.41}$$

$$\text{CU} = 0.46$$

Equations for Outdoor Lighting Designs

$$\text{fc} = \frac{\text{lm}}{\text{ft}^2} \text{ by definition}$$

therefore

$$\text{fc} = \frac{\text{Initial lumens} \times \text{CU} \times \text{Maint. factor}}{\text{Area}}$$

This equation will yield the average intensity. The maintenance factor is 0.8 for incandescent lamps and 0.9 for mercury lamps. It ranges in value from 0.74 to 1.00. It also follows that:

$$\text{fc} = \frac{\text{lm} \times \text{CU} \times \text{MF}}{\text{Width of section} \times \text{Spacing between luminaires}}$$

and

$$\text{Spacing} = \frac{\text{lm} \times \text{CU} \times \text{MF}}{\text{fc} \times \text{width}}$$

Example 4

Refer to the photometric data and utilization curves in Figure 1. Determine the average footcandles produced in the roadway with the following information:

Table 2
RECOMMENDED ILLUMINATION LEVEL OF FOOTCANDLES
(IES VALUES) FOR OUTDOOR APPLICATIONS

Application	Minimum average footcandles
Construction	
General	10.0
Excavation	2.0
Malls	5.0—10.0
Parks	0.2
Parking areas	
Industrial	1.0
Shopping center	2.5
Playgrounds	5.0
Recreation and sports	
Badminton	10.0
Baseball Class C	30.0 infield, 20.0 outfield
Softball recreation	10.0 infield, 7.0 outfield
Softball league	20.0 infield, 15.0 outfield
Tennis, recreation	10.0
Volleyball	10.0
Football Class III	30.0
Roadways (intermediate class)	
Expressway	1.4
Local	0.6
Alleys	0.4
Sidewalks	
Intermediate	0.6
Storage yards	
Inactive	1.0

Lumens   = 50,000
Width   = 50 ft (total roadway width)
Length   = 60 ft (spacing between luminaires)
Overhang = 10 ft (into the roadway)
MF   = 0.9
MH   = 40 ft
CU determined from the chart as $0.41 + 0.05 = 0.46$

$$fc = (50,000 \times 0.46 \times 0.9) \div 3000 = 6.9$$

Example 5
Using the curves in Figure 1 and the data in Example 4, determine the spacing if footcandles are specified as 2.0.

$$\text{Spacing} = \frac{\text{lm} \times \text{CU} \times \text{MF}}{\text{fc} \times \text{width}} = \frac{50,000 \times 0.46 \times 0.9}{2 \times 50}$$

Length (or spacing) = 207 ft

Design Procedure for Roadway Lighting

The planning and the design of a new outdoor lighting installation begins with selections from tables and calculations to determine the amount of light (lumens) required. The average footcandle value recommended for a particular area is shown in Table 2. The total lumens required for a given area ia calculated. Next is the selection of a luminaire and a choice of

MH. Then, calculate the spacing of luminaires.[4,5] Also involved at this point is the fact that providing an average value of footcandles in a space does not insure uniformity. The average could be gained via a luminaire at each end of the space where it would be very bright underneath, and yet there might be complete darkness in the middle of the area.

A uniformity ratio of 1 to 3 is the cutoff point. The design is acceptable when the lowest level of light intensity (in footcandles) is at least equal to one third of the average footcandle value in the space.

$$\text{Uniformity ratio} = \frac{\text{Minimum level of intensity}}{\text{Average footcandle value}} = 1/3 \text{ or greater}$$

Illustrative Example of Design

Example 6

Determine the spacing of luminaires along a 40-ft-wide local roadway if it is desired to provide 0.6 fc on the average and to use the luminaire with photometric data as presented in Figure 2 (assume MF = 0.9).

1. The footcandle value is selected from Table 2 (choose 0.6).
2. The luminaire mounting height is selected as 30 ft. (Note and choose about the same as the MH given on the data page, Figure 2.) Overhang is to be 6 ft (usually 4 to 10 ft). A staggered arrangement of luminaires is planned and these decisions lead to the layout of Figure 4c.
3. Determine the CU.

$$\text{Street side } \frac{\text{DIST}}{\text{MH}} = \frac{34}{30} = 1.13 \quad \text{Chart value is } 0.44$$

$$\text{Curb side } \frac{6}{30} = 0.2 \quad \text{Chart value is } \underline{0.04}$$

$$\text{Total } 0.48$$

4. Determine the spacing.

$$\text{fc} = \frac{\text{Initial lumens} \times \text{MF} \times \text{CU}}{\text{Width} \times \text{Spacing}}$$

$$\text{Spacing} = \frac{7900 \times 0.9 \times 0.48}{40 \times 0.6}$$

Spacing = 142 ft (use 140 ft for convenience)

This represents a tentative choice for spacing. If the uniformity ratio is acceptable, the spacing is to be 140 ft. If the ratio is not acceptable, the spacing should be decreased or the MH decreased or both.

5. Calculations for uniformity ratio
    A. Three test points, shown in Figure 4d, are normally used for uniformity ratio determinations — x_1 is directly across from a luminaire at the roadway curb, x_2 is in the middle of the roadway and one half way between adjacent luminaires, and x_3 is one fourth of the way between luminaires on the same side of the roadway.

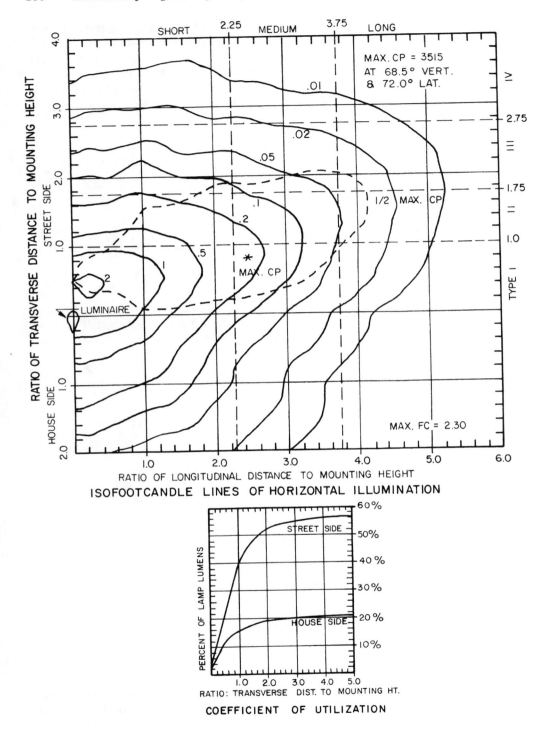

ISOFOOTCANDLE LINES OF HORIZONTAL ILLUMINATION

COEFFICIENT OF UTILIZATION

FIGURE 2.   Photometric report. Footcandle data for luminaire mounted at 25 ft (mounting height). Luminaire is ITT. 250 horizontal. Lamp is 175-W mercury vapor. Initial lumens 7900.

Table 3
TABULATIONS FOR DETERMINING UNIFORMITY RATIO

Luminaire	Longitudinal ratios			Luminaire contribution to fc values at:		
	x_1	x_2	x_3	x_1	x_2	x_3
A	280/30	350/30	350/30	0	0	0
Ratio of dist. to MH	9.33	11.67	11.67			
B	140/30	210/30	210/30	0	0	0
Ratio of dist to MH	4.67	7.00	7.00			
C	0/30	70/30	70/30	0.50	0.25	0.28
Ratio of dist. to MH	0.00	2.33	2.33			
D	140/30	70/30	70/30	0	0.25	0.09
Ratio of dist. to MH	4.67	2.33	2.33			
E	280/30	210/30	210/30	0	0	0
Ratio of dist. to MH	9.33	7.00	7.00			
Totals				0.50	0.50	0.37

B. Calculate the transverse (across) ratios of DIST/MH as follows:

A, C, and E to x_1 and x_3: $\dfrac{34}{30} = 1.13$

A, C, E, and D to x_2: $\dfrac{14}{30} = 0.47$

B and D curb side to x_1 and x_3: $\dfrac{\cdot 6}{30} = 0.20$

C. Draw these three (horizontal) lines on the chart of Figure 2.
D. Determine the values needed to provide the data as shown and as arranged (for convenience) in the tabulations of Table 3.
E. After calculating the longitudinal ratios, draw vertical lines on the chart corresponding to these ratios (along street). In this example draw 2.33 and 4.67; omit 7.00, 9.33, and 11.67 since they are off the chart.
F. Read the footcandle values of intensity where the vertical (transverse) ratio lines intersect the horizontal (longitudinal) ratio lines. Record in the tabulation under x_1, x_2, and x_3. Then total the intensity values for each test point.
G. Correct the total values of footcandles for MH if required. It is required in this example and the correction factor is 0.69. Corrected footcandle totals are $x_1 = 0.347$, $x_2 = 0.347$, $x_3 = 0.257$.
H. Calculate the uniformity ratio for the smallest value, namely, $x_3 = 0.257$.

$$\text{Uniformity ratio} = \frac{0.257}{0.60} = \frac{1}{2.3}$$

This is an acceptable design, as it is within the one third limit.

Lighting of Outdoor Work Areas

The general principles and design procedures in the case of lighting systems for roadways also apply to particular areas such as parking lots and storage, processing, and work areas. New considerations include two luminaires per pole (poles are very expensive) and in the case of recreational areas a much larger "bank" of luminaires. It is also likely that the designer would seek photometric data for luminaires adapted to these special areas rather than use the roadway luminaires. However, for illustrative purposes, the new consideration when using two luminaires per pole will be described while making use of the data presented in Figure 3.

Example 7

Assume that two Figure 3 luminaires are to be mounted on each pole of a proposed arrangement as shown in Figure 4e.

A holding area for crop drying wagons is to be constructed as indicated. To take advantage of using only three poles, there are to be two luminaires per pole for a total of six for the holding area. Determine the spacing of the poles that will provide an average footcandle value in the area of 2.0.

1.
$$\text{Spacing} = \frac{20,600 \times 0.8 \times \text{CU}}{2 \times 80} = 103(\text{CU})$$

2.   Determine CU using luminaires A and B mounted on the same pole:

#1   Street ratio is $\dfrac{72}{40} = 1.8$ Coef is   0.52

#1   Curb ratio is $\dfrac{8}{40} = 0.2$ Coef is   0.04

#2   Curb ratio is $\dfrac{88}{40} = 2.2$ Coef is   0.21

but 8 ft of the 88 for luminaire B is not in the area under consideration, so subtract curb ratio of 8/40 = 0.2.

Coef is $\underline{-0.04}$

Total   0.73

3.   The spacing is

$$103(\text{CU})$$

$$= 103(0.73)$$

$$= 75.2 \text{ ft} \approx 75 \text{ ft}$$

Choose 75 to 80 ft as spacing for the poles.

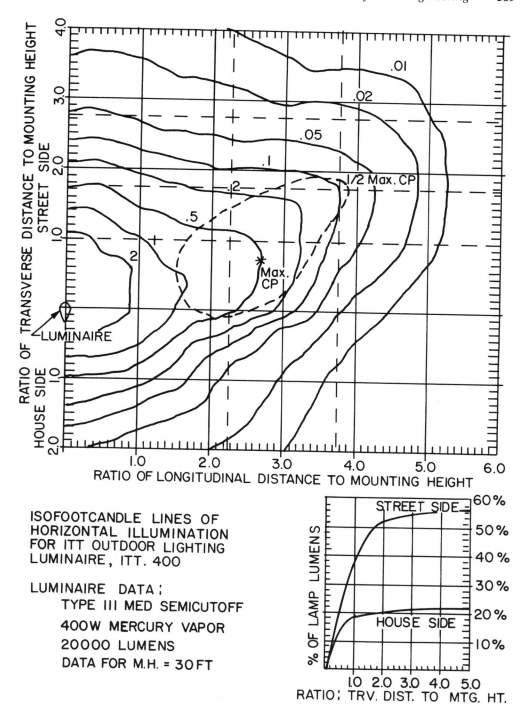

ISOFOOTCANDLE LINES OF
HORIZONTAL ILLUMINATION
FOR ITT OUTDOOR LIGHTING
LUMINAIRE, ITT. 400

LUMINAIRE DATA:
   TYPE III MED SEMICUTOFF
   400W MERCURY VAPOR
   20000 LUMENS
   DATA FOR M.H. = 30FT

FIGURE 3.   Photometric report. Footcandle data for luminaire mounted at 30 ft (mounting height). Luminaire is ITT. 400 horizontal. Lamp is 400-W mercury vapor. Initial lumens 20,000.

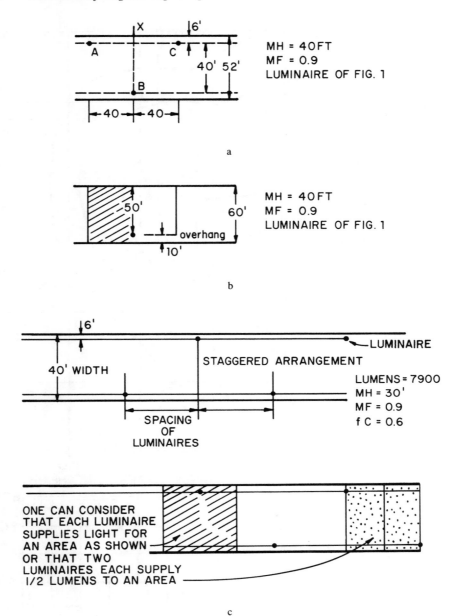

FIGURE 4.   Drawings and dimensions for illustrative examples.

4.   Accordingly, space poles 75 ft on center and 37.5 ft from the border edges, which provides a 225-ft length for the drying area. The spacing (and the half spaces) could be increased by 5 ft if desired.

DESIGN OF INTERIOR LIGHTING SYSTEMS FOR AGRICULTURAL STRUCTURES

The purpose of this section is to present acceptable procedures for determining wattage ratings of lamps for seeing tasks involving agricultural structures. The procedures are neither

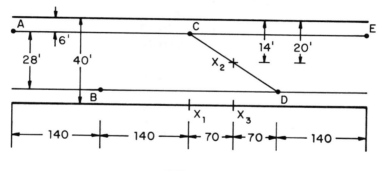

FIGURE 4d

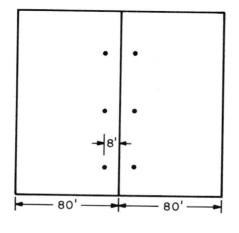

MH = 40'
MF = 0.8
LUMINAIRE DATA
  IN FIGURE 3
LUMENS = 20,600

FIGURE 4e

based in nor intended for physiological or biological effect purposes. Light intensity for visual tasks is the thrust of this presentation.

The requirements for visual tasks have been studied and evaluated by the IES. This society was founded in 1906 and has its headquarters in New York. Their handbooks and recommended levels of lighting are standards in the lighting industry. Many values presented in Table 4 were drawn from this source, extrapolating to agricultural situations as required. In addition, the research literature in agricultural engineering has established minimum values for well-established uses in agriculture. Some values in Table 4 came from this source.

The lux and footcandle values in Table 4 are the recommended minimum values of light for the task (lux [lx] = lm/m^2 and footcandles [fc] = lm/ft^2). As presented previously (IES Definitions), the term lumens expresses the total light flux on the task. It does not matter in the calculations whether one uses lx and m^2 or fc and ft^2, since 1 m^2 = 10 ft^2 and 10 lx = 1 fc. The light flux is produced by a light source (lamp plus luminaire), and the initial value of lumens from the source is specified by the manufacturer. See Table 1 for these data.

The more precise decisions about interior lighting system design involve room size (shape), color of walls and ceilings, MH of the light source, and photometric data for the luminaire. Agricultural structures do not usually fit into the residential and office mold since the surfaces are often unpainted. The spaces are large, perhaps there is no ceiling, and the MH for lighting outlets is greater than that of residential and office rooms. When known, these factors should be considered, because dark colors and long narrow spaces waste light. Dust is also an adverse factor. In consideration of the wide variation in agricultural structural

Table 4
RECOMMENDED
ILLUMINATION
VALUES FOR
AGRICULTURAL
STRUCTURES

Area and/or task	Minimum light on task	
	lx	fc
Broiler houses	50	5
Building en-trances, general	50	5
Calf pens	70	7
Egg handling and packing	500	50
Egg storage and shipping	200	20
Feed lots, outdoor	30	3
Feed processing areas	100	10
Haymow	30	3
Laying houses	30	3
Loading platforms	200	20
Machine storage	50	5
Milk handling	200	20
Milking parlor	200	20
Cow's udder	500	50
Pumphouse	100	10
Service areas	30	3
Shop, office	700	70
General	300	30
Storage area	100	10
Swine growing/ finishing	50	5
Nursery	200	20
Tobacco	200	20
Handling		
Grading	2000	200

conditions, it is suggested that a luminaire factor of 0.75 be used for reflector-type luminaires (or lamps) and that 0.50 be used for bare-bulb, no-reflector outlets. Also, a maintenance factor of 0.75 is recommended to account for the accumulation of dust and poorer reflecting conditions. The considerations have been accounted for in the following equations:

$$\text{fc} = \text{lm/ft}^2$$

$$\text{fc} = \frac{\text{(Initial lumens) (Luminaire factor) (Maint. factor)}}{\text{floor area per lighting outlet}}$$

$$\text{fc} = \frac{\text{Initial lumens} \times \text{LF} \times \text{MF}}{\text{Longitudinal spacing} \times \text{Across spacing}}$$

A procedure similar to that for outdoor lighting can now be followed to calculate the initial lumen requirement (and corresponding lamp wattages):

1.
$$\text{Initial lumens} = \frac{\text{fc} \times \text{Long space} \times \text{Across space}}{\text{LF} \times \text{MF}}$$

2. fc Selected from Table 4.
3. The longitudinal spacing for agricultural structures is adequately selected as factor times MH. The factors are 1, 1.5, 2.0, and 2.5. The maximum factor is 3.0, but this many cause objectional shadows in the space. Accordingly, if the MH is 10 ft, the longitudinal spacing should be from 10 to 15 ft for best uniformity and up to 30 ft if shadows are acceptable. A good compromise value is $2.0 \times 10 = 20$ ft.
4. The across spacing is influenced by the geometry and the structural framing of the building, and at the same time must also remain within the factor limitations of step 3. If the structure is 32 ft wide, one may choose two rows of lighting outlets, located 8 ft in from the walls and 16 ft apart. Thus $8 \cdot 16 \cdot 8$, and the across spacing is 16 ft and the factor $16/10 = 1.6$ is acceptable.

Example 8

Given a broiler house 40 ft wide and 200 ft long with rooftruss members 10 ft above the floor, select the lighting-outlet locations and wattage rating.

If across spacing is selected $10 \cdot 20 \cdot 10$ (assuming two rows of outlets 20 ft apart), the width per lighting outlet is 20 ft.

Longitudinal spacing may be $2 \times 10 = 20$ ft or $1.5 \times 10 = 15$ ft. The 20-ft space works out well for the 200-ft length of space, providing ten outlets in each of the two rows.

The footcandle value in Table 4 for a broiler house is 5.0. Calculate:

$$\frac{\text{Initial lumens}}{\text{per outlet}} = \frac{5 \times 20 \times 20}{0.75 \times 0.75} = \frac{2000}{0.525}$$

$$\text{Initial lumens} = 3810$$

Note in Table 2 that a 150-W lamp supplies only 2880 lm. Accordingly, a 200-W lamp would be needed, which is not the usual choice in agricultural structures (too large). Therefore change the longitudinal or across spacing or both.

Note that $6 \cdot 14 \cdot 6$ ft would serve for across spacing but requires three rows of outlets. Keep long space at 20 ft.

$$\text{Initial lumens} = \frac{5 \times 14 \times 20}{0.525} = 2667$$

Another choice might be $6 \cdot 14 \cdot 14 \cdot 6$ across spacing and reducing longitudinal spacing to 10 ft (this requires 20 outlets per row and 63 outlets overall). Then,

$$\text{Initial lumens} = \frac{5 \times 14 \times 20}{0.525} = 2667$$

and a 100-W lamp would provide this.

REFERENCES

1. **Kaufman, J. E. and Christensen, J. F.,** IES Lighting Handbook, 5th ed., Illuminating Engineering Society, New York, 1972.
2. **Brown, R. H.,** *Farm Electrification,* McGraw-Hill, New York, 1956.
3. **Moore, G. E.,** Photometric Data Reports, ITT Outdoor Lighting, Southhaven, Miss., 1983.
4. IES American National Standard Practice for Roadway Lighting, RP-8 Illuminating Engineering Society, New York, 1974.
5. Tech. Editor, Outdoor Lighting Design Manual, 3rd ed., Line Material Industries, McGraw Edison Co., Milwaukee, Wis., 1979.

EXAMPLES OF HEAT CALCULATIONS FOR AGRICULTURAL APPLICATIONS

Robert H. Brown

INTRODUCTION

Ever since man first learned to alter his environment he has made progress, and he has consistently strived to reach the ultimate in control over his surroundings. Electricity may be assigned a recognized top tole as a tool for man's use in this undertaking. The over 2 trillion kWh of electrical energy used annually in the U.S. represents a significant contribution to the upgrading of the capacity of man to deal more efficiently with his environment. One of the uses of this energy is to supply heat for many and varied uses, and electricity will be used as the illustration of the heat source in the examples presented in this chapter.

In the changing climate of modern farming operations, including the emphasis on efficient and mass production, man must continue to alter the natural environment whenever it makes his operations more competitive. Quite frequently, it is to his advantage to add artificial heat (liquid fuel, gas, wood, electric, etc.).

Artificial heat in agriculture is common practice today. It is used in greenhouses, hotbeds, crop dryers, types of production areas for live-stock and poultry, and work areas and also for thawing, preventing moisture accumulation, and in altering the temperature of water.

The decision to use electric heat is reached after having answered two questions: will it pay to add heat, and is electricity the best choice. Information needed for answering these questions comes from both empirical and theoretical sources. Design procedures for agricultural applications are presented in this chapter, with emphasis on the requirements of those areas and applications where beneficial results have been shown.

ELECTRIC HEAT

As a heat source, electricity is versatile, lends itself to automatic and accurate control, and is adaptable to a wide variety of situations. It merits special consideration because of its unique qualities — no flame, no products of combustion, and almost complete freedom of choice of shape and size of heating element. Its selection permits and sometimes dictates certain changes in the structural design of the heated area from that which would be employed with other heat sources.

Although dielectric heating, induction heating, and IR radiations utilize electrical energy as the input, the conventional electric heater is properly categorized as a resistance heater. The energy converter is pure resistance and is 100% efficient, producing 3412 Btu/kWh of input energy. Accordingly, when electric heat is referenced, resistance heating is assumed and the problem of quantity to install simplifies to that of calculating heat loss (in Btu/hr), dividing by 3412, and installing the resulting kilowatt heating capacity. If the decision is reached to employ another type of heat source, a similar consideration and analysis would need to be done.

HEAT TRANSFER

Heat is transferred from one part to another or from one region to another by any one of a combination of three modes: conduction, convection, and radiation. The heat transfer occurs whenever a temperature difference exists and transfers toward the lower temperature.

The *conduction* of heat is expressed by the following equation:

Table 1
THERMAL CONDUCTIVITY AND
EMISSIVITY OF SOME MATERIALS USED
IN AGRICULTURE

Material	Btu/hr ft, °F ft	Emissivity
Aluminum, commercial sheet	119.0	0.09
Copper, polished	220.0	0.023
Concrete, tiles		0.63
Concrete, stone[a]	0.54	0.63
Flat black lacquer		0.98
Glass, smooth, window	0.3—0.61	0.94
Iron, cast, freshly emeried	27.0	0.24
Iron, oxidized	27.0	0.74
Mineral wool	0.024	
Roofing paper		0.91
Sand, dry	0.19	
Sand, 10% moisture	0.60	
Soil, fresh clay, 30% moisture	1.40	
Steel, sheet	26.0	0.58
Water (32°F)	0.363	0.95—0.96
Water (100°F)	0.340	0.95—0.96
Wood, white pine	0.087	

[a] When calculating heat transfer through concrete floors and basement walls, use U in place of k and use U = 0.2 Btu/hr ft² °F for basement walls below ground level, U = 0.1 for concrete floors on dry soil, and U = 0.03 for concrete basement floors located at least 6 ft below ground level.

From McAdams, W. H., *Heat Transmission*, McGraw-Hill, New York, 1954. With permission.

$$\frac{dQ}{d\theta} = -kA\,\frac{dt}{dx}$$

where $dQ/d\theta$ is the instantaneous rate of heat flow, A is the area of the section in question, k is the thermal conductivity, and dt/dx is the temperature gradient (the latter is negative, thus the minus sign). If it is assumed that heat flow is steady and therefore independent of time (steady state), then the equation becomes:

$$q = -kA\,\frac{\Delta t}{x}$$

and the units are

$$\frac{Btu}{hr} = \left(\frac{Btu}{hr\ ft^2\ °F\ ft}\right)(ft^2)\left(\frac{°F\ temp\ diff}{ft\ thickness\ of\ mat'l}\right)$$

The conductivity, k, is a physical property of the substance involved, and its values are determined by experiment. The k-values for many agricultural products are yet to be evaluated. Those values that have been established are recorded in journals and in the various texts on heat transfer.[1,3] They range from approximately zero for a vacuum to several thousand for certain metals. Some typical values are given in Table 1.

Heat transfer by *convection* is described by the equation:

$$q_c = hA \; \Delta t$$

where q_c is the heat flow, h is the film coefficient of heat transfer, A is the area of a surface, and Δt is the temperature difference between the surface area and the fluid in contact with it. Physical properties of the fluid, velocity, product geometry, duct or pipe dimensions, and Δt influence the value of h. For certain conditions, h can be calculated by methods illustrated in the examples to follow. Some h-values are shown in Table 1.

Based upon the equation for the total thermal radiation from a black body and the equation for the net radiation between two surfaces (developed by Hottel), the following relation is recommended for calculating heat loss by *radiation*.

$$q_R = \alpha \; \epsilon \; A_1(T_1^4 - T_2^4)$$

where α is the Stefan-Boltzmann dimensional constant (0.174×10^{-8}, ϵ is the emissivity, A_1 the area of the heated surface, T_1 is the absolute temperature of A_1, and T_2 is the absolute temperature of the surrounding space. By definition, $\epsilon = 1$ for a perfectly black body. Other materials have smaller values of emissivity and again, as was the case with the previous heat transfer methods, a value for substituting into the equation must be secured from a table that lists the emissivity of a particular surface. Emissivity values do vary with temperature, but are essentially constant for temperatures below 600°F. Accordingly, the values in Table 1 are appropriate for most agricultural applications.

Example of heat transfer by radiation — Calculate the radiant heat transfer from a linear foot of $1\frac{1}{2}$-in. black, PVC water pipe containing 160°F water and located in 80°F still air.

From Table 1 $\qquad\qquad \epsilon = 0.92$

Calc. surface area $\qquad \pi R^2 L = 0.147 \; \text{ft}^2/\text{linear ft}$

Calc. q $\qquad\qquad\qquad q = A\epsilon\sigma(T_1^4 - T_2^4)$

$$T_1 = 160 + 460 \qquad T_2 = 80 + 460$$

$$q = 0.147 \times 0.92 \times 0.174\{(6.20)^4 - (5.40)^4\}$$

$$q = 14.75 \; \text{Btu/linear ft/hr}$$

HEAT TRANSFER CALCULATION

Watering Troughs

It is common knowledge that serious production declines are immediately experienced with layers if their water supply is cut off. This is sufficient reason for preventing the water from freezing. Yung and Brunig[4] report that little is to be gained by heating drinking water for livestock except to prevent freezing. The capacity of electric heating cable to install in watering troughs so as to prevent freezing could be determined experimentally or by calculation. For example, the heat loss for a 4 in. × 4 in. × 8 ft galvanized metal trough with 40°F water and 10°F air is computed as follows:

Heat Loss by Convection (per foot of trough — 29.7Btu/hr)
    For the sides:

$$q_c = h \ A \ \Delta t$$

$$h = 0.29\left(\frac{\Delta t}{L}\right)^{1/4} \text{ for a vertical plate}[1]$$

$$h = 0.29\left(\frac{30}{4/12}\right)^{1/4} = 0.89$$

$$q_c = 0.89 \times 1/3 \times 30 = 8.9 \text{ Btu/hr/ft of trough}$$

and there are two sides; therefore:

$$q_c = 17.8 \text{ Btu/hr/ft of trough}$$

For the water surface:

$$h = 0.27\left(\frac{\Delta t}{L}\right)^{1/4} \text{ for a horizontal plate facing upward}[1]$$

$$h = 0.27\left(\frac{30}{4/12}\right)^{1/4} = 0.82$$

$$q_c = 0.82 \times 1/3 \times 30 = 8.2 \text{ Btu/hr/ft of trough}$$

For the bottom of the trough:

$$h = 0.12\left(\frac{\Delta t}{L}\right)^{1/4} \text{ for a horizontal plate facing downward}[1]$$

$$h = 0.12\left(\frac{30}{4/12}\right)^{1/4} = 0.37$$

$$q_c = 0.37 \times 1/3 \times 30 - 3.7 \text{ Btu/hr/ft of trough}$$

$$q_R = \sigma \ \epsilon \ A_1(T_1^4 - T_2^4)$$

From table of emissivity it is found that $\epsilon = 0.96$ for water and $\epsilon = 0.25$ for the trough. This latter value was estimated by judging the surface to be about like that of iron freshly emeried.[1] Herein lies the chance for significant error since the more polished the surface, the lower is its emissivity. For the water:

$$q_R = 0.174 \times 10^{-8} \times 0.96 \times 1/3(500^4 - 470^4)$$

$$q_R = 7.63 \text{ Btu/hr/ft of trough}$$

For the trough surfaces:

$$q_R = 0.174 \times 10^{-8} \times 0.25 \times 1(500^4 - 470^4)$$

$$q_R = 5.95 \text{ Btu/hr/ft of trough}$$

Table 2
SOME MAJOR APPLICATIONS OF
ELECTRIC HEAT IN AGRICULTURE

Applications	Watts required
Plant propagation benches	8 W/ft²
Farrowing pens	25—40 W/ft²
Growing-finishing pens	15—40 W/ft²
Brooders, electric cover	2.5 W/chick
Water pipes 1 in., prevent freezing	6—12 W/ft
Sweet potato, curing house	4—10 W/bu
Hotbeds, outside	10 W/ft²
Drying, curing, and space heating	Varies with conditions

Heat Loss by Radiation (per foot of trough) = 13.6 Btu/hr

The sum of the heat losses above gives a total heat loss of 43.6 Btu/hr/ft of trough. Dividing the sum by 3.412, it is seen that at least 12.8 W is needed per foot of trough. The calculated heat loss and the corresponding wattage requirement do not include evaporation, which would contribute an additional loss of 1000 Btu/lb of water evaporated.

For frequently used materials, such as watering troughs, it is likely that evaporative losses have been determined experimentally. Such is the case with this trough, and the average loss in weight due to evaporation was 0.03 lb/hr/ft of trough, which is equivalent to 0.09 lb/hr evaporated per square foot of water surface. This amount of evaporation occurred with an average air velocity of 100 fpm at the surface of the water. Using a value of 1000 Btu/lb, it is seen that evaporation resulted in a heat loss of 30 Btu/hr/ft of 4-in. trough. Refer to Table 2 for watts required in other situations.

Accordingly, to provide for both the direct and the evaporative losses from the trough, it is recommended that 20 W be installed per foot of trough in poultry houses of open or loose construction and that 15 W be used for houses of tight construction.

Heating of Concrete Slabs

Heating cable embedded in concrete has been used successfully for brooding poultry and for warming floor areas for livestock. The heating capacity normally installed ranges from 5 to 40 W/ft² of surface area. The heating capacity required for a specific instance can be computed as shown in the following typical example.

Determine the watts per square foot required to maintain an 85°F slab temperature in 35°F still air. The 4-in., 4 ft × 20 ft slab is to be placed on a 55°F dry sandy floor (with water barrier film) and is to be constructued of 3-in. light-weight concrete with a 1-in. sand-cement topping. The heating cable would be installed on top of the light-weight concrete and a 6-in. concrete insulating border will be placed around the perimeter of the slab. The heat losses are as follows.

Conduction to Sand Floor

$$q = k \, A \, \frac{dt}{dx} \qquad k = 0.10 \text{ for limit weight concrete}$$

$$q = \frac{A \, \Delta t}{\dfrac{x}{k}} = \frac{80 \times (85 - 55)}{\dfrac{0.25}{0.10}} = 960 \text{ Btu/hr}$$

Convection

$$q_c = h \, A \, \Delta t$$

$$h = 0.38(\Delta t)^{1/4} \text{ for a heated horizontal surface facing up,}$$
$$\text{over 3 ft}^2 \text{ surface}^2$$

$$h = 0.38(50)^{1/4} = 1.01$$

$$q_c = 1.01 \times 80 \times 50 = 4040 \text{ Btu/hr}$$

Radiation

$$q_R = \sigma \, \epsilon \, A_1(T_1^4 - T_2^4)$$

$$q_R = 0.174 \times 0.63 \times 80(5.45^4 - 4.95^4)$$

$$q_R = 2450 \text{ Btu/hr}$$

The total heat loss is 7450 Btu/hr, the watts required 2183, and 27 W are needed per square foot. In these calculations, the effects of the edges and corners have been neglected owing to their small surface area and to the 6-in. insulating border. Kinard[6] used 20 W/ft² for electric underheat brooders constructed especially for laboratory trials. This capacity was sufficient to achieve a temperature of 95°F within the upper heated portion of the slab. Of course air temperatures greater than 35°F are frequently encountered in such areas as broiler housing, farrowing barns, and greenhouses. The same procedure as outlined above should be followed in those instances. For a 60°F ambient air and an 85°F slab, the heat loss is seen to be 3824 Btu/hr requiring 14 W/ft².

Hotbed-Type Spaces

Electric hotbeds have been used quite extensively for the production of sweetpotato slips and nursery crops and for growing plants in the greenhouse. From 8 to 10 W/ft² supplied by the proper lengths of electric heating cable or by electric heating mats is the watt-density value in common usage. A sample calculation is presented below for an electric hotbed situation, but the procedures followed should be helpful elsewhere, such as in solar collectors.

A 6 ft × 50 ft hotbed is constructed with 2 in. × 8 in. wooden sides, covered tightly with 6-mil polyethylene film, has the heating cable placed on top of 2 in. of 10% moisture sand and is covered with 1 in. of soil; the following calculations apply. (Assume 70°F soil, 35°F outside air, 50°F hotbed air, and 65°F ground temperature.)

Heat Loss Through the Cover by Conduction

$$q = \frac{A \, \Delta t}{\dfrac{1}{h_i} + \dfrac{x}{k} + \dfrac{1}{h_o}}$$

h_i = inside surface conductance, use 1.65; h_o = outside surface conductance (15 mi/hr wind), use 6.0. Thus the calculations become

$$q = \frac{300 \times (50 - 35)}{\dfrac{1}{1.65} + 0^* + \dfrac{1}{6.0}} = (1.30)(300)(15)$$

* neglect 6 mil thickness

$$q = 5850 \text{ Btu/hr}$$

Heat Loss Through the Walls by Conduction

$$q = \frac{A \, \Delta t}{\dfrac{1}{1.65} + \dfrac{2/12}{0.087} + \dfrac{1}{6}} = 0.373 \times 75 \times 15 = 420 \text{ Btu/hr}$$

Heat Loss Through the Sand by Conduction

$$q = \frac{A \, \Delta t}{\dfrac{2/12}{0.6}} = 3.6 \times 300 \times 5 - 5400 \text{ Btu/hr}$$

Note that no air film is present at the floor; accordingly $\dfrac{1}{h_i}$ and $\dfrac{1}{h_o}$ are not appropriate when determining the loss through the sand.

If the bed is tightly covered, the infiltration may be neglected and the heat loss totals 11,670 Btu/hr under the conditions assumed. This bed would require $\dfrac{11670}{3.412}$ or·3420 W, which is equivalent to 11.4 W/ft². It is seen from the above example that improved insulation over the ground and/or a better cover would greatly reduce the heat loss.

Example of Calculating the Supplemental Heat For a Swine Barn

The procedures when heat, moisture, and ventilation are involved differ from those previously illustrated, and a psychrometric chart is normally used in making some of the computations. Assume that inside air conditions of 50°F and 75% RH are satisfactory for high production in a swine barn. Also for this example, assume a cold humid night with outside air conditions of 10°F and 100% RH. The average weight of the pigs is taken as 125 lb.

1. With inside air of 50°F and 75% RH, the dew point temperature for the inside wall surface is determined from the psychrometric chart to be 43°F (t_w). The minimum overall U-value necessary, if condensation on the wall and ceiling is to be prevented, is calculated as follows:

$$U = \frac{t_i - t_w}{0.606(t_i - t_o)} + \frac{(50 - 43)}{0.606(50 - 10)}$$

$$U = 0.289 \text{ Btu/hr/ft}^2 \text{ °F}$$

2. A 125-lb pig produces 190 Btu/hr latent heat and 420 Btu/hr sensible heat, and approximatley 0.03 lb of water is added per pig per hour by manure and cleaning water.[9] Similar data are available in the literature for other animals and poultry. These values are needed to determine ventilation and moisture removal requirements. The latent heat value provides a means of determining respired moisture. Since 1035 Btu is required to evaporate 1 lb of water, it is calculated that $190 \div 1035 = 0.184$ lb

of respired moisture is given off per pig each hour. Making use of a psychrometric chart, it is determined that each pound of ventilating air entering at 10°F 100% RH and leaving at 50°F 75% RH picks up 0.0044 lb of water. Thus $\frac{0.184}{0.0044}$ = 41.8 lb of air per pig per hour is required to remove the respired moisture. An additional $\frac{0.030}{0.0044}$ = 0.682 lb of air is needed to remove the litter-cleaning moisture, making the total 42.5 lb of ventilating air per pig per hour.

3.   The total heat loss for the swine barn is obtained as follows. From the psychrometric chart, the heat contents of incoming and exhaust air are seen to be 3.6 and 18.2, respectively. For the 42.5 lb of ventilating air, and (18.2 − 3.8) = 14.4 Btu/lb difference in heat content of incoming and discharge air, the product is 14.4 × 42.5 = 612 Btu/pig/hr and is the heat loss due to ventilation requirements. To compute the heat loss through the walls and ceiling, assume 16 ft^2 of surface per pig, then: q = U A Δt = (0.289 from step 1) (16)(50 − 10) = 0.289 × 16 × 40 = 185 Btu/hr per pig.

4.   The total heat input supplied by the pig is

Sensible heat	420 Btu/hr
Latent heat	190Btu/hr
	610 Btu/hr per pig

5.   The additional heat required for balance is (797 − 610) = 187 Btu/hr per pig. This heat can be furnished by installing 5.5 W of electric heat per pig.

Example of Heat Transfer Calculations For a Greenhouse

Greenhouse structures present special design considerations for heating, ventilating, and cooling. A method for determination of heat loss will be presented here. Refer to the *ASHRAE Handbook and Product Directory*[11] and to fan manufacturer's literature for ventilating and cooling guidelines.

Most of the heat losses occur at night, and the maximum design heating requirement is based on this condition. Use 15°C (60°F) inside temperature (for most greenhouse plants) and select the outside temperature from charts or tables for a particular geographical location. Again, the ASHRAE Handbook[11] (chapter entitled Weather Data and Design Conditions) is an excellent reference. Use the design dry-bulb column for 99% of the time. For Athens, Ga, latitude 34°, the design temperature values are outside (t_o) = −8°C (18°F) and the inside (t_i) = 15°C (60°F).

The heat losses from a greenhouse are adequately represented with two calculations, one for heat loss by conduction (Q_e) and the second for heat loss by infiltration (Q_i).

$$Q_c = UA\ \Delta T = Ug\ A(t_i - t_o)$$

A is the total exposed surface area of the structure and Ug is best selected from Table 3, which presents approximations that have proven adequate for design purposes.

$$Q_i = 0.5\ V\ Ng(t_i - t_o)$$

V is the total volume of air inside the greenhouse, and N is the number of air changes per hour for greenhouse structures. These values are furnished in Table 4. The design heat loss is the sum of Q_c and Q_i.

As an example, a certain greenhouse is constructed of double-layer polyethylene and the

Table 3
APPROXIMATIONS OF HEAT TRANSFER COEFFICIENTS FOR GREENHOUSE GLAZING METHODS AND MATERIALS

	Value of Ug	
Greenhouse covering	**W/m² °C**	**Btu/hr °F ft²**
Single glass (sealed)	6.3	1.1
Single plastic	6.8	1.2
Single fiberglass	6.8	1.2
Double plastic, polyethylene	4.0	0.7
Double glass (sealed)	3.0	0.5

From *ASAE Yearbook,* EP406, American Society of Agricultural Engineers, St. Joseph, Mich., 1983—1984, 387. With permission.

Table 4
NATURAL AIR EXCHANGES FOR GREENHOUSES

Construction system	Value of Ng air exchanges/hr
New construction, glass or fiberglass	0.75—1.5
New construction, double layer plastic film (separation) 20—100 mm)	0.5—1.0
Old construction glass, good condition	1.0—2.0
Old construction glass, poor condition	1.0—4.0

From *ASAE Yearbook,* EP406, American Society of Agricultural Engineers, St. Joseph, Mich., 1983—1984, 387. With permission.

geometry yields the following calculated values: exposed surface area = 191 m² (2056 ft²) and total volume = 258 m³ (9120 ft³). Calculate the heat losses for 15°C inside and −8°C outside design temperatures.

$$Q_c = 4.0 \times 191 \times 23 = \quad 17,572$$

$$Q_i = 0.5 \times 1.0 \times 258 \times 23 = \underline{\quad 2,967}$$

$$\text{Total} \quad 20,539$$

It is seen that this structure would require the equivalent of 20,500 W of electric heating capacity in order to provide for the (−8°C) nighttime hours. The heat could be furnished via black pipes, air or water unit heaters (steam, air, water, electric), or by electric unit heaters. The 20,500-W value converts to 20,400 (3.412) = 69,946 Btu/hr.

REFERENCES

1. **McAdams, W. H.,** *Heat Transmission,* McGraw-Hill, New York, 1954.
2. **Henderson, S. M. and Perry, R. L.,** *Agricultural Process Engineering,* John Wiley & Sons, New York, 1955.

3. **Holman, J. P.,** *Heat Transfer,* McGraw-Hill, New York, 1981.
4. **Yung, F. D. and Brunig, M. P.,** Heating drinking water for livestock, *Agric. Eng.,* 37, 411, 1956.
5. **Junnila, W. A.,** Poultry Water Warming, paper presented at the Annu. Meet. American Society of Agricultural Engineers, Roanoke, Va., 1956.
6. **Kinard, D. T.,** Electric Underheat Brooder, College Exp. Stn. Bull. No. 3, University of Georgia, Athens, January 1953.
7. **Kreith and Black,** *Basic Heat Transfer,* Harper & Row, 1980.
8. **McFate, K. L.,** Progress Reports of the Farm Electric Utilization Project, University of Missouri, Columbus, April 1958.
9. **Hazen, T. E. and Mangold, D. W.,** Functional and basic requirements of swine housing, *Agric. Eng.,* 41, 585, 1960.
10. Psychrometric Chart, Normal Temperatures, Carrier Corporation, original by Dr. Willis H. Carrier, 1911.
11. *ASHRAE Handbook and Product Directory,* Fundamentals, American Society of Heating, Refrigerating and Air Conditioning Engineers, New York, 1977.

Feed and Crop Storages

WALL PRESSURES IN CROP STORAGES

Harvey B. Manbeck

INTRODUCTION

Crops stored in bins exert both lateral and vertical pressures upon the sidewalls. The magnitude of the pressures depends upon the type of material being stored and the geometry of the bin. The techniques for predicting the pressures exerted by fluids, grains, and silage on shallow and deep grain bin walls are discussed in this chapter.

SHALLOW AND DEEP BINS

A bin is defined as a shallow bin if the failure plane of the grain taken from the wall-floor intersection intersects the grain surface before the opposite sidewall. Conversely, if the failure plane intersects the opposite sidewall first, the bin is classified as being deep.

The failure plane of granular media is inclined at an angle ($45° + \phi/2$), where ϕ is the internal angle of friction of the stored material. The internal angle of friction may be estimated by the emptying angle of repose.

Deep and shallow bins are illustrated in Figure 1a and b. Note that a bin is shallow if $H \leq B \tan(45 + \phi/2)$ and deep if $H > B \tan(45 + \phi/2)$.

FLUID PRESSURES

The pressure exerted upon the walls of any fluid container equals

$$L = \rho g h \tag{1}$$

where ρ = fluid density (ML^{-3}), h = depth below the fluid surface (L), L = lateral pressure at a depth h below the fluid surface (Fl^{-2}), and g = gravitational constant (1.0 lb_f/lb_m or 9.807×10^{-3} kN/kg) (FM^{-1}). The fluid pressure always acts normal to the wall surface and, at a given point, is equal in all directions. No tangential wall pressures act on a fluid storage container.

GRANULAR NONCOHESIVE MATERIALS

Lateral pressures on shallow bin walls are predicted by Rankine's equation. The lateral wall pressure at a depth h below a horizontal grain surface is

$$L = (wgh) \tan^2(45 - \phi/2) \tag{2}$$

where L = lateral wall pressure (FL^{-2}), w = bulk density of stored material (ML^{-3}), h = depth below surface of stored grain (L), φ = emptying angle of repose of stored material (degrees) (see Table 1), and g = gravitational constant (1.0 lb_f/lb_m or 9.807×10^{-3} kN/kg) (FM^{-1}).

The lateral pressure distribution varies linearly with depth below the grain surface. The resultant wall load acting on a unit length of the wall is

$$L_T = \frac{LH^2}{2} \tag{3}$$

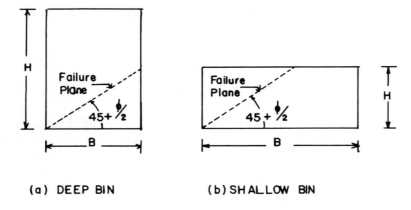

(a) DEEP BIN   (b) SHALLOW BIN

FIGURE 1.   Definition sketches for deep and shallow bins.

Table 1
BULK GRAIN PROPERTIES

Grain	Angle of[a] repose (degrees)	Bulk density (kg/m³)	Bulk density[a] (lb/ft³)
Oats			
Central U.S.	32	538	33.6
Pacific Northwest and Canada	32	564	35.2
Rough Rice, American Pearl	36	666	41.6
Barley			
Eastern	28	641	40.0
Western	28	692	43.2
Soybeans	29	743	46.4
Flaxseed	25	692	43.2
Corn, shelled	27	769	48.0
Rye	26	743	46.4
Wheat			
Soft red winter	27	782	48.8
Hard red spring	28	833	52.0
Hard red winter	27	820	51.2
Durum	26	833	52.0
Extra heavy	26	881	55.0

[a]   Taken from Brubaker, R. G. and Pos, J., Grain Bin Requirements, USDA
    Circular 850, U. S. Department of Agriculture, Washington, D.C., 1950.

where H = total depth of stored grain. In shallow bins, the vertical wall load is nearly zero. If the grain surface is not horizontal, the lateral pressure increases to

$$L = wgh\left[(\cos\delta) \left(\frac{\cos\delta - \sqrt{\cos^2\delta - \cos^2\phi}}{\cos\delta + \sqrt{\cos^2\delta - \cos^2\phi}} \right) \right] \qquad (4)$$

where δ = angle the grain surface makes with the horizontal (Figure 2), often the filling angle of repose.

In deep bins, both lateral and vertical wall pressures are significant. For static conditions,

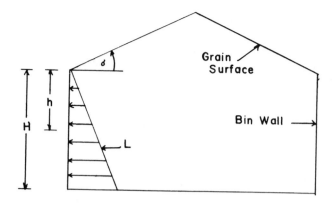

FIGURE 2.   Definition sketch for surcharge angle, δ.

Janssen's equation satisfactorily predicts lateral wall pressures. For a bin with a circular cross section, the lateral pressure at depth h equals

$$L = \frac{R(wg)}{\mu'} \left[1 - e^{-\frac{k\mu'}{R}h} \right] \tag{5}$$

where R = bin hydraulic radius = A/P (L), A = bin cross-sectional area (L²), P = bin circumference (L), μ' = coefficient of friction between the bin wall and the stored grain, k = (1 − sinφ)/(1 + sinφ), D = bin diameter (L), and other terms are as previously defined.

Equation 4 may be used to predict lateral pressures in deep bins of square or rectangular cross section by using an equivalent diameter. If the bin length is less than or equal to 1.5 times the bin width, then

$$D_{eq} = \frac{4 \text{ (cross-sectional area)}}{\text{Bin Perimeter}} \tag{6}$$

If the bin length is more than 1.5 times the bin width, then use the bin width for D_{eq}.

The vertical pressures exerted on the wall at depth h below the grain surface are

$$V = \mu'L \tag{7}$$

The total horizontal load per unit length of bin wall perimeter at a depth, H, below the grain surface is

$$F_H = \frac{R(wg)}{\mu'} \left[H - \frac{R}{k\mu'} \left(1 - e^{-\frac{k\mu'}{R}H} \right) \right] \tag{8}$$

where H = total bin wall height (L). The total vertical load acting on a unit length of bin wall footer is

$$F_v = \mu'F_H \tag{9}$$

The maximum load exerted on the floor of a deep bin is the difference between the total weight of the bin contents minus the total vertical friction load carried by the walls. For a circular bin the floor load, Q, is

$$Q = (wg)AH - F_v(\pi D) \tag{10}$$

The bin pressures and loads predicted by Equations 2 to 5 and 7 to 10 are for static conditions only. Storage conditions such as temperature changes, moisture content changes, and emptying alter the magnitude of these loads.

Thermal strains and stresses are induced in bin walls when ambient temperatures decrease. Bin wall thermal stresses can be conservatively predicted by assuming the granular mass to be semirigid and calculating the thermal hoop strain and stress by Equations 11 and 12:

$$\epsilon_T = \delta(\Delta T) \tag{11}$$

$$\sigma_T = E \, \epsilon_T \tag{12}$$

where ϵ_T = thermal strain, δ = coefficient of thermal expansion of the bin wall material $(1/°C \text{ or } 1/°F)$, ΔT = temperature change (°C or °F), E = bin wall modulus of elasticity (FL^{-2}), and σ_T = bin wall thermal stress (FL^{-2}). Recent research with model bins indicates that σ_T is the order of 10 to 15% of static lateral pressures.

When the moisture content of stored grain decreases, there is no change in wall pressures. However, significant increases in wall pressure do occur when the moisture content of a hygroscopic material increases in storage. Based upon limited data, the following guidelines are suggested for pressure increases.

1. Lateral wall pressures increase approximately sixfold for a 4% increase in moisture content.
2. Lateral wall pressures increase approximately tenfold for a 10% increase in moisture content.

The most advisable strategy for coping with these pressure increases is to plan, locate, and manage grain storage facilities so as to minimize increases in moisture content.

Lateral pressures in shallow bin walls during filling or emptying are the same as those predicted by Rankine's equation. In deep bins, wall pressures during filling are adequately predicted by Janssen's equation. However, during emptying of deep bins, bin wall pressures increase significantly, and these increases must be considered in wall design.

During concentric unloading the lateral pressures in deep bins may be predicted by applying appropriate overpressure factors, C_d, to the static pressures predicted by Janssen's equation (Equation 5)

$$L_d = C_d L \tag{13}$$

where L_d = dynamic lateral wall pressure at depth h below the grain surface during centric emptying (FL^{-2}), L = static pressure from Equation 5, and C_d = overpressure factor (Table 1). Overpressure factors range from 1.35 to 1.85 (ACI).

Typical values for the angle of repose and bulk density of various grains are given in Table 1. Typical coefficients of friction of grains on wall surfaces are given in Table 2.

SILAGE PRESSURES

The National Silo Association recommends that lateral silage pressures, L_s, be predicted by Rankine's equation

$$L_s = (wg)h \, \text{Tan}^2(45 - \phi/2) \tag{14}$$

Table 2
COEFFICIENTS OF FRICTION (μ') FOR GRAINS

Material	Moisture Content (%)	Concrete			Wood				Plastic		Metal	
		Plastic smooth finish	Steel trowel finish	Wood float finish	Oak		Douglas fir		Teflon®	Poly-ethylene	Mild steel C.R.	Galvanized sheet metal
					Grain par.	Grain perp.	Grain par.	Grain perp.				
Barley	10.7	0.23	0.56	0.50	0.23	0.29	0.27	0.32	0.17	0.23	0.20	0.20
	14.3	0.24	0.57	0.51	0.21	0.28	0.30	0.32	0.13	0.28	0.23	0.20
	16.4	0.33	0.62	0.55	0.30	0.33	0.37	0.41	0.11	0.35	0.21	0.34
Oats	10.6	0.28	0.40	0.43	0.20	0.23	0.27	0.29	0.13	0.20	0.20	0.22
	14.0	0.33	0.51	0.42	0.23	0.25	0.34	0.36	0.13	0.28	0.21	0.18
	17.3	0.50	0.65	0.64	0.46	0.48	0.48	0.50	0.14	0.50	0.44	0.32
Shelled corn	7.5	0.27	0.41	0.46	0.24	0.25	0.27	0.29	0.17	0.22	0.23	0.20
	9.9	0.25	0.59	0.62	0.28	0.31	0.31	0.31	0.18	0.27	0.20	0.24
	13.9	0.35	0.64	0.54	0.29	0.36	0.37	0.38	0.12	0.38	0.24	0.37
Soybeans	7.1 (7.0)	0.25	0.39	0.39	(0.29)	(0.35)	0.29	0.31	(0.23)	0.25	(0.32)	0.21
	9.8 (11.6)	0.31	0.47	0.37	(0.34)	(0.39)	0.33	0.31	(0.19)	0.29	(0.41)	0.18
	12.2 (15.4)	0.36	0.55	0.52	(0.42)	(0.45)	0.35	0.44	(0.17)	0.43	(0.45)	0.20
Wheat	11.2 (9.7)	0.36	0.52	0.51	(0.30)	(0.32)	0.31	0.35	(0.19)	0.27	(0.33)	0.10
	13.0 (11.9)	0.46	0.52	0.55	(0.28)	(0.32)	0.35	0.38	(0.17)	0.35	(0.33)	0.14
	15.7 (15.1)	0.56	0.68	0.69	(0.35)	(0.40)	0.48	0.50	(0.12)	0.45	(0.38)	0.33

For silos containing corn silage at 68 to 72% moisture content, the bulk density may be taken as 1041 kg/m^3 (65 lb/ft^3) and the ϕ may be taken as 32°.

The vertical pressure, V_s, exerted by silage on the wall at depth h below the silage surface is predicted by empirical Equation 15

$$V_s = 5.5 \, h^{1.08} \tag{15}$$

where V_s = vertical wall pressure in lb/ft^2 and h = depth below the silage surface in feet. By integrating Equation 15 over the entire height of a 1-ft-wide strip of silo wall, the total vertical wall load per foot of perimeter exerted by the silage on the wall footer, F_v, is predicted by Equation 16:

$$F_v = 2.64 \, h^{2.08} \tag{16}$$

REFERENCES

1. American Concrete Institute, Recommended practice for design and construction of concrete bins, silos and bunkers for storing grannular materials, (ACI 313-77) and commentary, A.C.I., Detroit, Mich., 1975.
2. **Brubaker, R. G. and Pos, J.,** Static coefficients of friction for some grains on various surfaces, ASAE paper no. 63-628, American Society of Agricultural Engineers, St. Joseph, Mich., 1963.
3. **Dale, A. C. and Robinson, R. N.,** Pressures in deep grain storage, *Agric. Eng.,* 35(8), 570, 1954.
4. **Janssen, H. A.,** Versuche uber Getreidruck in Silozellen, *Z. Ver. Dtsch. Ing.,* 39, 1045, 1895.
5. **Manbeck, H. B. and Muzzelo, L. M.,** Measurement of thermally induced pressures in a model grain bin, *Trans. ASAE,* 28(4), 1253, 1985.
6. Midwest Plan Service, *Structures and Environment Handbook* (MWPS-1), 11th ed., Midwest Plan Service, Ames, Iowa, 1983.
7. **Riembert, M. L. and Riembert, A. M.,** *Silos: Theory and Practice,* Trans Tech Publications, Clansthal, W. Germany, 1976.
8. United States Department of Agriculture, Gross bin requirements, USDA Circ. No. 835, USDA, Washington, D.C., 1950.

THERMAL PROPERTIES OF AGRICULTURAL MATERIALS

R. Paul Singh

INTRODUCTION

Thermal properties play an important role in the design and analysis of agricultural processes such as heating, cooling, storage, and drying of cereal grains. The most useful thermal properties in these analyses are specific heat, thermal conductivity, and thermal diffusivity. Several researchers in the past have conducted experimental studies to determine these properties for common cereal grains. Thermal properties vary with the moisture content and the stage of processing. Although published information on these properties is useful in engineering analysis, innovative processes and new varieties of cereal grains may require experimental determinations. In this section, methods used in the past to determine thermal properties are reviewed, followed by a discussion of experimental procedures that may be used for such determinations.

LITERATURE REVIEW

A method of mixture was used by Pfalzner[1] to measure the specific heat of wheat. A grain sample previously held at room temperature was enclosed in a small copper capsule and dropped into water in a calorimeter. The water in the calorimeter had been previously cooled to about 1.11°C. The specific heat of the grain sample was calculated from the measured temperature rise.

A method of mixture similar to Pfalzner's[1] method was used by Wratten et al.[2] and Kazarian and Hall[3] to measure the specific heat of rough rice, wheat, and corn. However, in their approach, the grain sample was not enclosed in a capsule. A calorimeter method was used by Disney[4] and Haswell[5] to determine the specific heat of wheat and barley. The method involved dropping a known amount of grain sample at room temperature into an ice calorimeter and measuring the amount of ice melted.

A variety of procedures have been used by researchers to measure the bulk thermal conductivity of grains. Bakke[6] measured thermal conductivity of oats using a steady-state apparatus with oats placed between two concentric cylinders. The temperature gradient was obtained by placing ice in the inner cylinder. Both cylinders were then placed in a constant-temperature hot water bath. The heat flow was determined by measuring the amount of ice melted during the test. Kazarian and Hall[3] and Wratten et al.[2] also used a cylindrical device to measure the bulk thermal conductivity of rough rice, wheat, and corn. Grain samples were placed in a cylinder with a line-heating source in the center. The temperature change at the center of the cylinder was measured to determine the solution of the one-dimensional, nonsteady-state heat-conduction equation in cylindrical coordinates.

Oxley[7] measured the bulk thermal conductivity of wheat, maize, and oats by using a spherical apparatus to solve the steady-state, one-dimensional heat-conduction equations in spherical coordinates. The grain samples were placed between two concentric spheres. The inner sphere was equipped with a heating element that provided the heat flow across the grain sample. The temperature difference across the grain sample was measured and used to calculate the bulk thermal conductivity.

Determination of the bulk thermal diffusivity of grains can be computed from the relationship: $\alpha = k/\rho C$ where k, ρ, and C are the bulk thermal conductivity, bulk density, and specific heat, respectively. Kazarian and Hall,[3] however, determined the bulk thermal diffusivity of wheat and corn by using transient heat flow in a slab, initially at a uniform temperature, with the faces suddenly lowered to and held at zero. The grain sample was enclosed in a rectangular box. The temperature of the grain sample and the box was held at room temperature. The box was then placed in the ice bath and the temperature change at its center was measured to solve the equations.

Thermal properties of grains very with moisture content. Therefore, moisture content should be accurately determined and reported along with the thermal property. The most widely used procedure involves use of an oven. Table 1 provides various conditions for operating an oven to measure moisture content of cereal grains.

Thermal properties of various grains obtained by several researchers are listed in Table 2. The property values are for bulk granular samples.

EXPERIMENTAL METHODS

Specific Heat

Several researchers have used the method of mixture with enclosing sample grains in a capsule. A disadvantage of this method is the time to reach equilibrium, 15 to 20 min, which is relatively long, and, consequently, the temperature correction term being used in the calculation of the specific heat is large compared to the measured temperature change in a calorimeter.

The specific heat of grains may be determined using the following procedure:

1. Take about 50 g of ice water in a calorimeter and allow it to warm up to approximately 7.2°C.
2. Add approximately 30 g of the grain sample, which had been held at room temperature, into the calorimeter.
3. Stir the grain-water mixture continuously by hand using a copper stirring bar.
4. Measure and record the temperature of the mixture with a thermocouple (such as 20-gage copper-constantan).
5. Equilibrium is usually reached in less than 30 sec for the majority of the measurements.
6. Record temperatures continually for 2 to 3 min after equilibrium has been reached.
7. Calculate the average specific heat of the sample from the following equation:

$$C = \frac{(\Delta T + \eta)W}{(t_a - t_e - \eta)m} \qquad (1)$$

where ΔT = measured rise in temperature, (°C), η = correction term for heat transfer into a calorimeter (°C), W = water equivalent of a calorimeter (kJ/°C), t_a = temperature of grain sample before dropping into a calorimeter (°C), t_e = final common temperature of grain sample and a calorimeter (°C), m = mass of grain sample (kg), and C = average specific heat of grain sample between the temperature t_a and t_e (kJ/kg°C).

8. The temperature correction term η is calculated from the Regnault-Pfaundler formula.[8] In this formula, the curve for the temperature against time is divided into three consecutive periods: the anterior period, the X-period, and the rating period (Figure 1). The term η accounts for correction for heat transfer into a calorimeter. This temperature correction term η is described by the following equation:

$$\eta = \Theta_1 - \Theta_2 + \frac{\Theta_3 - \Theta_4 - \Theta_1 + \Theta_2}{\Theta_r - \Theta_a} (\Theta_x - \Theta_a) \qquad (2)$$

where Θ_1 = temperature at the start of the anterior period (°C), Θ_2 = temperature at

Table 1
OVEN TEMPERATURE AND HEATING
PERIOD FOR MOISTURE CONTENT
DETERMINATIONS

Seed	Oven temperature (±1°C)	Heating time hr	Heating time min	Ref.
Alfalfa	130	2	30	2
Barley	130	20	0	2
Beans, edible	103	72	0	1
Bentgrass	130	1	0	2
Bluegrass	130	1	0	2
Bluestem	100	1	0	2
Bromegrass	130	0	50	2
Cabbage	130	4	0	2
Carrot	100	0	40	2
Clover	130	2	30	2
Collard	130	4	0	2
Corn	103	72	0	1
Fescue	130	3	0	2
Flax	103	4	0	1
Kale	130	4	0	2
Mustard	130	4	0	2
Oats	130	22	0	2
Onion	130	0	50	2
Orchardgrass	130	1	0	2
Parsley	100	2	0	2
Parsnip	100	1	0	2
Radish	130	1	10	2
Rape	130	4	0	2
Rye	130	16	0	2
Ryegrass	130	3	0	2
Safflower	130	1	0	2
Sorghum	130	18	0	2
Sunflower	130	1	0	2
Timothy	130	1	40	2
Turnip	130	4	0	2
Wheat	130	19	0	2

REFERENCES

1. Oven Methods for Determining Moisture Content of Grain and Related Agricultural Commodities, Equipment Manual, Grain Division, Consumer and Marketing Service, U.S. Department of Agriculture, Washington, D.C., 1971.
2. **Hart, J. R., Feinstein, L., and Golumbic, C.,** Oven Methods for Precise Measurement of Moisture in Seeds, Marketing Res. Rep. No. 304, U.S. Government Printing Office, Washington, D.C., 1959.

Courtesy American Society of Agricultural Engineers.

the end of the anterior period (°C), Θ_3 = temperature at the end of the X-period (°C), Θ_4 = temperature at the end of the rating period (°C), Θ_a = mean temperature of the anterior period (°C), Θ_x = mean temperature of the X-period (°C), Θ_r = mean temperature of the rating period (°C). The temperature correction term η is always negative since the heat flow is from outside to inside the calorimeter.

Table 2
THERMAL PROPERTIES OF GRAIN AND GRAIN PRODUCTS

Grain or grain product	Moisture content (% wet basis)	Temperature range (°F)	Mean temperature (°F)	Specific heat (Btu/lb°F)	Conductivity (Btu/hr,ft°F)	Diffusivity (ft²/hr)	Ref.[a]
Corn, yellow dent	0.9	54.0—83.8 for sp ht	68.9 for sp ht	0.366	0.0812	0.00395	5
	5.1	47.7—74.0 for diffusivity	56.8 for diffusivity	0.404	0.0847	0.00381	5
	9.8			0.438	0.0878	0.00361	5
	13.2	80—88	—	—	0.102	—	6
	14.7	54.0—83.8 for sp ht	68.9 for sp ht	0.484	0.0919	0.00351	5
	20.1	47.7—74.0 for diffusivity	58.6 for diffusivity	0.531	0.0945	0.00336	5
	24.7			0.567	0.0982	0.00344	5
	30.2			0.588	0.0996	0.00358	5
	—		68.9	0.350 + 0.00851 M[b]	0.0814 + 0.000646 M[b]		5
			95.3				
Flour, wheat	11.8—17.6	—	—	0.397 + 0.0119 M[b]	—	—	9
Groats		—	—	0.257	—	—	4
Oats	9.1	—	—	—	0.0370	—	2
	12.7	80—88	—	—	0.075	—	6
	27.7		—	—	0.0537	—	2
	11.7—17.8	—	—	0.305 + 0.0078 M[b]	—	—	4
Rice, rough	10.2—17.0	—	—	0.265 + 0.0107 M[b]	—	—	4
Rice, shelled	9.8—17.6	—	—	0.287 + 0.0091 M[b]	—	—	4
Rice, finished	10.8—17.4	—	—	0.282 + 0.009 M[b]	—	—	4
Rice, rough, medium	10—20	—	—	0.22008 + 0.01301 M[b]	0.0500135 + 0.000767 M[b]	0.00523 + 0.0000965 M[b]	10

Material	Moisture (%)	Temperature range	Temperature				Ref.
Soybeans	19.7	75.2—129.2	—	0.47	—	—	7
	24.5	73.4—190.4	—	0.49	—	—	7
Starch, from wheat, rice, and potato	—	—	—	0.41	—	—	9
Starch, wheat	8.6	71.6—122	—	0.32	—	—	8
	22.6	71.6—122	—	0.38	—	—	8
Wheat, soft white	Dry	—	—	0.334	0.0676	—	5
Wheat, hard	9.2	—	—	0.370	0.081	0.00446	1
Wheat, hard red	9.6	71.6—122	—	0.39	—	—	11
Wheat	11.7	80—88	—	—	—	—	6
Wheat, hard red	12.5	—	87.0	—	0.0872	—	11
	12.5	—	97.2	—	0.074	—	11
	14.0	—	77.7	—	0.079	—	11
	14.0	—	91.2	—	0.079	—	11
Wheat	17.8	80—88	—	—	0.082	—	6
	19.5	80—88	—	—	0.092	—	6
Wheat, hard red	21.3	71.6—123.8	—	0.51	0.0891	—	11
	23.0	—	79.4	—	0.0867	—	11
	23.0	—	89.6	—	0.0891	—	11
	23.0	—	99.7	—	0.0925	—	11
Wheat, Manitoba	1.3	—	—	0.310	—	—	3
	4.9	—	—	0.333	—	—	3
	10.1	—	—	0.367	—	—	3
	17.5	—	—	0.447	—	—	3
Wheat, Bersee	0.1	—	—	0.307	—	—	3
	4.2	—	—	0.322	—	—	3
	13.7	—	—	0.405	—	—	3
	19.9	—	—	0.476	—	—	3
	25.8	—	—	0.525	—	—	3
	33.6	—	—	0.582	—	—	3
Wheat, soft white	0.7	51.2—89.9 for sp ht and 48.3—73.8 for diffusivity	70.9 for sp ht and 57 for diffusivity	0.347	0.0679	0.00359	5
	5.5	—	—	0.375	0.0706	0.00347	5
	10.3	—	—	0.428	0.0747	0.00331	5
	14.4	—	—	0.500	0.0786	0.00318	5
	20.3	—	—	0.522	0.0798	0.00310	5
	—	—	—	0.334 +0.00977 M^b	0.0676 +0.000654 M^b.	—	5

Table 2 (continued)
THERMAL PROPERTIES OF GRAIN AND GRAIN PRODUCTS

Grain or grain product	Moisture content (% wet basis)	Temperature range (°F)	Mean temperature (°F)	Specific heat (Btu/lb°F)	Conductivity (Btu/hr.ft°F)	Diffusivity (ft²/hr)	Ref.ᵃ

ᵃ References furnished at end of table.

ᵇ M = moisture content percent, wet basis, regression equation.

From ASAE Data: ASAE 243.2. Approved by the ASAE Committee on Technical Data; adopted by ASAE 1948; revised 1954, 1962; revised by Electric Power and Processing Division Technical Committee, December, 1967; revised editorially March 1972; revised December 1973; reconfirmed December 1978; revised editorially April 1982; reconfirmed December, 1983.

REFERENCES

1. **Babbit, J. D.**, The thermal properties of wheat in bulk, *Can. J. Res.*, F23, 388, 1945.
2. **Bakke, A. L. and Stiles, H.**, Thermal conductivity of stored oats with different moisture content, *Plant Physiol.*, 10, 521, 1935.
3. **Disney, R. W.**, The specific heat of some cereal grains, *Cereal Chem.*, 31, 229, 1954.
4. **Haswell, G. A.**, A note on the specific heat of rice, oats, and their products, *Cereal Chem.*, 31, 341, 1954.
5. **Kazarian, E. A. and Hall, C. W.**, Thermal properties of grains, *Trans. ASAE*, 8(1)33, 1965.
6. **Oxley, T. A.**, The properties of grains in bulk, *Soc. Chem. Ind. J. Trans.*, 63, 53, 1944.
7. **Ramstad, P. E. and Geddes, W. F.**, The Respiration and Storage Behavior of Soybeans, Tech. Bull. 156, University of Minnesota, Minneapolis, 1942.
8. **Rodewall, H. and Kattein, A.**, The specific heat of wheat starch as a function of water content and temperature, *Z. Phys. Chem.*, 33, 540, 1900.
9. **Winkler, C. A. and Geddes, W. F.**, The heat of hydration of wheat flour and starches, *Cereal Chem.*, 8, 455, 1931.
10. **Wratten, F. T., Poole, W. D., Chesness, J. L., Bal, S., and Ramarao, V.**, Physical and thermal properties of rough rice, *Trans. ASAE*, 12(6), 801, 1969.
11. From unpublished report by U.S. Bureau of Standards on thermal conductivity and specific heat of hard red spring wheat.

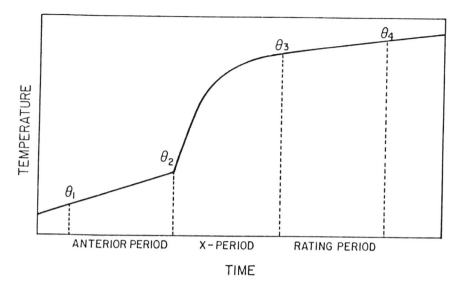

FIGURE 1.   Typical curve of temperature against time for measurement of a specific heat by using a calorimeter.

9.   The water equivalent of the calorimeter is determined experimentally. Copper, aluminum, and water may be used in the determination of the water equivalent of the calorimeter. The copper and aluminum are cut into small chips of diameter and length approximately 0.20 and 0.75 cm, respectively. The temperature rises in the calorimeter for the copper and aluminum chips can be measured in the same manner as for grain samples.

Bulk Thermal Conductivity

The majority of the reported values for the bulk thermal conductivity of grains have been determined by steady-state heat flow across the grain using spherical or cylindrical devices. The disadvantages of steady-state measurements are (1) the long time required to attain the steady-state conditions and (2) possibilities of moisture migration due to maintaining the temperature difference across the grain for long periods of time. Both of these difficulties can be minimized by the transient heat flow method.

The transient heat flow in an infinite mass, initially at a uniform temperature, heated by a line source of constant strength is described as follows. The basic equation for the heat flow from a line source is

$$\frac{\partial t}{\partial \Theta} = \alpha \left(\frac{\partial^2 t}{\partial r^2} + \frac{1}{r} \frac{\partial t}{\partial r} \right) \tag{3}$$

where t = temperature at radius r (°C), Θ = time (sec), r = radius from the heat source (m), and α = thermal diffusivity of the grain (m²/sec). The solution for the temperature is given by Hooper and Lepper[9] as:

$$t = \frac{Q}{2\pi k} I(rn) \tag{4}$$

where t = temperature (°C), Q = heat input (W/m), k = bulk thermal conductivity of the grain (W/m°C), r = distance from heat source (m), and $n = \frac{1}{2} (\alpha\Theta)^{-1/2}$ (m⁻¹). The function $I(rn)$ is

$$I(rn) = A - \ln(rn) + \frac{(rn)^2}{2} - \frac{(rn)^4}{8} + \dots \tag{5}$$

where A is a constant.

If (rn) is sufficiently small, all terms of the series except the first two may be dropped. Then $I(rn) = A - \ln(rn)$, and the temperature is given by

$$t = \frac{Q}{2\pi k} [A - \ln(rn)] \tag{6}$$

The temperature rise between time Θ_1 and Θ_2 is given by

$$t_2 - t_1 = \frac{Q}{4\pi k} \ln(\Theta_2/\Theta_1) \tag{7}$$

where t_1 = temperature Θ_1 (°C) and t_2 = temperature at time Θ_2 (°C), and finally

$$k = \frac{Q\ln(\Theta_1/\Theta_2)}{4\pi(t_2 - t_1)} \tag{8}$$

The schematic diagram of the apparatus that may be used for the determination of the bulk thermal conductivity of short-grain rough rice is shown in Figure 2.[10] The apparatus consists of an aluminum cylinder 30.4 cm high and 12.7 cm I.D. A 20-gage chromel resistance heating wire, 0.029 Ω/cm, is stretched between a 14-gage copper extension wire on the axis of the cylinder. The temperature of the heating wire is measured by a 20-gage copper-constantan thermocouple placed approximately 0.04 cm from the heating wire over a single layer of plastic electric tape at the middle of the heating wire. Power for the heater is supplied by a filtered DC power supply that is adjusted at 6 V of power and the current is controlled by a 50-Ω variable resistor. The current is measured by an ampere meter. The cylinder is insulated with 2.54-cm-thick glass-fiber, and both ends of the cylinder were insulated with 10-cm-thick styrofoam. The insulation of the top end is detachable for filtering samples.

The procedure used for the actual measurement of the bulk thermal conductivity is to fill the cylinder with grains by tapping it continuously during filling to obtain a uniform density. The recording potentiometer is then turned on to record the initial temperature at the center of the rice grains. When this temperature has stabilized, the current is turned on and recorded by reading the ampere meter visually. The temperature during the test run is recorded by the recording potentiometer, and the run is continued for about 5 min.

Temperature vs. time is plotted on semilog graph paper, and the straight-line portion is determined. Values in this portion of the curve are then utilized in the following equation expressed in English units to calculate the bulk thermal conductivity of the rice sample:

$$k = \frac{3.413 \ I^2R}{4\pi} \frac{\ln\Theta_2/\Theta_1}{t_2 - t_1} \tag{9}$$

where k = thermal conductivity (Btu/hr ft°F), I = current flowing through heating wire (ampere), R = resistance of heating wire (Ω/ft), t_1 and t_2 = temperatures of heating wire at times in the straight-line portion (°F), and Θ_1 and Θ_2 = times in the straight-line portion (hr). This procedure is repeated for different moisture levels.

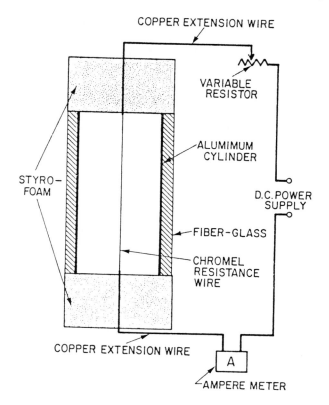

FIGURE 2.   Schematic diagram of the apparatus for measurements of the bulk thermal conductivity of short-grain rough rice.

Bulk Thermal Diffusivity

The bulk thermal diffusivity of short-grain rough rice is calculated by utilizing experimental values for the bulk thermal conductivity (k), specific heat (C), and bulk density in the equation

$$\alpha = \frac{k}{\rho C} \tag{10}$$

where α is the bulk thermal diffusivity.

Extensive discussion on determination of thermal properties of granular material is available in Mohsenin.[11]

REFERENCES

1. **Pfalzner, P. M.,** The specific heat of wheat, *Can. J. Technol.,* 29, 261, 1954.
2. **Wratten, F. T., Poole, W. D., Chesness, J. L., Bal, S., and Ramarao, V.,** Physical and thermal properties of rough rice, *Trans. ASAE,* 12(6), 801, 1969.
3. **Kazarian, E. A. and Hall, C. W.,** Thermal properties of grains, *Trans. ASAE,* 8(1), 33, 1965.
4. **Disney, R. W.,** The specific heat of some cereal grains, *Cereal Chem.,* 31, 229, 1954.
5. **Haswell, G. A.,** A note on the specific heat of rice, oats, and their products, *Cereal Chem.,* 31, 341, 1954.

6. **Bakke, A. L.,** Thermal conductivity of stored oats with different moisture content, *Plant Physiol.,* 10, 521, 1935.

7. **Oxley, T. A.,** The properties of grain in bulk, *Soc. Chem. Ind. Trans.,* 63, 53, 1944.

8. **White, W. P.,** The modern calorimeter, *Chem. Cat. Co.,* 42, 37, 1928.

9. **Hooper, F. C. and Lepper, F. R.,** Transient heat flow apparatus for determination of thermal conductivities, *ASHRAE Trans.,* 59, 463, 1953.

10. **Morita, T. and Singh, R. P.,** Physical and thermal properties of short-grain rough rice, *Trans. ASAE,* 22(3), 630, 1979.

11. **Mohsenin, N. N.,** *Thermal Properties,* Gordon & Breach, New York, 1980.

HYGROSCOPIC PROPERTIES OF AGRICULTURAL PRODUCTS

Do Sup Chung

Agricultural products are biological, heterogeneous, and hygroscopic substances. Each agricultural product exerts a characteristic partial pressure at a certain temperature and moisture content. If the partial pressure of the water held by an agricultural product is higher than that of water vapor in the surrounding air at a given temperature, then the product will lose moisture (desorption process). If the partial pressure of the water held by the product is lower than that of water vapor in the surrounding air, then the product will gain moisture (adsorption process). When the partial pressure of the water held by a product is equal to that of water vapor of the surrounding air, the moisture ceases to transfer from and to a product (equilibrium state).

The moisture content of a product at the equilibrium state is called the equilibrium moisture content (EMC) of a product at a given relative humidity and temperature of the surrounding air. The relative humidity of the air surrounding a product at a given temperature which is in equilibrium with the moisture content of a hygroscopic product is called the equilibrium relative humidity (ERH). It should be noted that food scientists or technologists prefer to call the ERH the water activity of a product (ERH/100).

Because the sorption relation between agricultural products and water vapor is important in product quality preservation, conditioning, storing, and drying, numerous investigations have been conducted in order to understand adsorption and desorption processes of water vapor on agricultural products, to determine hygroscopic properties of various agricultural products at many different temperatures (EMC vs. ERH or EMC vs. water activity), and to develop isotherm equations relating the equilibrium state of the product to environmental air conditions. Although many theories and isotherm equations for homogenous sorbents have been developed, sorption phenomena of heterogeneous materials such as agricultural products are still not completely understood.

This chapter was not intended to discuss the sorption theories or isotherm equations developed, but rather provide only hygroscopic properties of various agricultural products as published in the literature and presented in Table 1. Hygroscopic properties tabulated in Table 1 were basically determined experimentally by one of the following three methods. (1) A sample of known moisture content is allowed to come to equilibrium with a small headspace in a tight enclosure surrounded by a constant temperature, and the partial pressure of water vapor is measured monometrically or the relative humidity is measured. (2) A small sample is exposed to still air with a constant temperature and relative humidity. A constant relative humidity is obtained by using a sulfuric acid solution or a saturated salt solution. After equilibrium is reached, the moisture content is determined gravimetrically. (3) A sample is exposed to a constant temperature, and relative humidity of the air mechanically moved in a chamber. After equilibrium is reached, the moisture content of a sample is determined gravimetrically.

The variation in EMC values of a given product at a constant temperature and relative humidity determined by different investigators is given in Table 1. Such variation is mainly caused by a difference in the product variety, maturity and history (adsorption-desorption cycles experienced), drying practice and method, product conditions (soundness), EMC determination method, and the manner in which the EMC values were obtained (following the adsorption path or desorption path). Therefore, specific values of hygroscopic properties presented in Table 1 should be considered as approximate values, rather than exact values, when the reader refers to these values.

Table 1
HYGROSCOPIC PROPERTIES OF VARIOUS AGRICULTURAL PRODUCTS

Material	Temp (°C)	Relative humidity (%) Moisture Content (% Wet Basis)										Ref.
		10	20	30	40	50	60	70	80	90	100	
Shelled corn Y D	−6.7				10.4	11.8	13.3	15.0	16.6			1
	−1.1		6.7	9.2	10.7	12.7	14.1	17.7				2
	0				10.1	11.3	12.6	14.0	15.8			1
	0				11.0	12.5	14.0	15.8	18.0	21.8		3
	4.4	6.3	8.6	9.8	11.0	12.4	13.8	15.7	17.6	21.5		4
	10				9.2	10.7	12.1	13.6	15.5			1
	10	6.6	8.0	9.3	10.8	12.2	13.8	15.2	17.5	21.8		5
	10	6.7	8.5	9.9	11.2	12.5	13.9	15.4	17.3	20.3		6
	15.6	7.5	7.8	9.0	10.3	11.3	12.4	13.9	16.3	19.8		4
	15.6		4.8	6.8	8.6	10.7	12.1	13.9				2
	21.1			7.1	8.3	9.3	11.4	13.2				1
	25	5.1	7.0	8.3	9.8	11.2	12.9	14.0	15.6	19.6	23.8	1
	30			11.0	12.5	14.0	15.8	18.0	21.8			3
	30	4.4	7.4	8.2	9.0	10.2	11.4	12.9	14.8	17.4		4
	32.2	4.9	6.6	7.7	9.3	10.8	12.4	14.0	16.2	19.3		5
	32.2			5.3	6.5	8.3	10.2	12.1	13.9			2
	37.8	4.0	6.0	7.3	8.7	9.0	11.0	12.5	14.2	16.7		4
	48.9				5.3	6.5	10.2	12.1	13.9			2
	48.9				8.6	10.0	11.2	13.1	14.9			5
	50	3.1	5.0	6.5	7.9	9.3	10.7	12.4	14.5	17.6		6
	50	3.6	5.5	6.7	8.0	9.2	10.4	12.0	13.6	16.1		4
	60	3.0	5.0	6.0	7.0	7.9	8.8	10.3	12.1	14.6		4
	68.3				7.4	8.4	10.0	11.5	12.2			4
	71.1	3.9	6.2	7.6	9.1	10.4	11.9	13.9	15.2	17.9		2
W D	25	5.1	7.2	8.5	9.8	11.2	12.9	13.9	15.5	18.9	24.6	7
Shelled popcorn	25	5.6	7.4	8.5	9.8	11.0	12.2	13.1	14.2	18.4	23.0	7

Crop	Temp (°C)											Ref.
Barley	0		9.2	10.6	12.1	13.1	14.4	16.4	18.3	21.1		8
	20		8.2	9.5	10.9	12.0	13.4	15.2	17.5	20.5		8
	25		7.0	8.5	9.7	10.8	12.1	13.5	15.8	19.5		7
	30		7.6	9.1	10.4	12.2	12.2	14.3	16.6	19.0		8
Buckwheat	25	4.4	7.6	9.1	10.2	11.4	12.9	14.2	16.1	19.1	26.8	7
Cotton seed	25	5.0			6.9	7.8	9.1	10.1	12.9	19.6	24.5	9
Dry beans												
Michelite	4.4							14.4	17.0			10, 12
	10							15.3	18.0			10, 12
	25	5.6	7.4	8.6	9.8	11.2	12.9	14.9	17.5[a]			11, 12
	37.8							14.2	18.5			10, 12
	54.4							14.3	18.5			10, 12
Red Mexican	25	6.0	7.5	8.6	9.8	11.0	12.8	15.2	18.6[a]			11, 12
Great Northern	25	5.9	7.4	8.5	9.6	10.6	12.6	15.0	18.0[a]			11, 12
Light red kidney	25	6.1	7.5	8.7	9.9	11.1	12.9	15.0	18.0[a]			11, 12
Dark red kidney	25	5.4	7.2	8.4	9.9	10.7	12.5	15.0	18.6[a]			11, 12
Flat, small white	25	6.0	7.1	8.3	9.6	11.0	12.6	15.0	18.1			11, 12
Pinto	25	6.1	7.4	8.5	9.8	11.0	12.6	15.2	18.2	20.7		11, 14
Soybeans	25		5.5	6.5	7.1	8.0	9.3	11.5	14.8	18.1		12
	25				7.0	8.0	10.1	12.2	16.0	20.7		13, 12
Edible	10	5.1	7.2	8.9	10.4	11.8	13.4	15.2	17.4	20.6		6
	30	4.3	6.4	8.1	9.6	11.1	12.7	14.5	16.7	20.0		6
	50	3.5	5.7	7.3	8.9	10.4	12.0	13.8	16.1	19.5		6
Sugar beet seeds	4.4			10.0	11.5	12.7	13.9	15.3	17.6	22.6		14
	15.6			9.0	10.0	11.5	12.5	14.1	16.2	19.9		14
	26.7			8.0	9.1	10.4	11.6	12.9	14.7	18.0		14
	57.8			7.0	8.3	9.2	10.4	11.5	13.2	15.8		14
Wheat												
Soft, red winter	-6.7				11.3	12.8	14.1	15.3	17.0			1
	0				11.0	12.2	13.5	14.7	16.2			1
	10				10.2	11.7	13.1	14.4	16.0			1
	21.1				9.7	11.0	12.4	14.0	15.7			1
Soft	25	4.3	7.2	8.6	9.7	10.9	11.9	13.6	15.7	19.7	25.6	7, 12
	50	5.2	6.6	7.6	8.6	9.6	10.6	11.8	13.4	15.7		6
	30	6.2	7.6	8.6	9.6	10.5	11.5	12.7	14.2	16.5		6
	10	7.6	8.6	9.9	10.8	11.7	12.7	13.9	15.3	17.5		6
Durum	25	5.1	7.4	8.5	9.4	10.5	11.5	13.1	15.4	19.3	26.7	7, 12
	50	4.8	6.5	7.9	9.2	10.4	11.8	13.3	15.2	18.1		6
	30	5.4	7.2	8.2	9.8	11.0	12.3	13.8	15.7	18.6		6
	10	6.2	7.9	9.2	10.4	11.7	13.0	14.4	16.3	19.1		6

Table 1 (continued)
HYGROSCOPIC PROPERTIES OF VARIOUS AGRICULTURAL PRODUCTS

Material	Temp (°C)	Relative humidity (%)										Ref.
		Moisture Content (% Wet Basis)										
		10	20	30	40	50	60	70	80	90	100	
White	25	5.2	7.5	8.6	9.4	10.5	11.8	13.7	16.0	19.7	26.3	7, 12
Hard, red winter	25	4.4	7.2	8.5	9.7	10.9	12.5	13.9	15.8	19.7	25.5	7, 12
	25	4.4	7.2	8.5	9.7	11.1	12.5	13.9	15.9	19.7	25.0	7, 12
Wheat	−1.1		6.9	9.2	10.4	11.8	13.2	14.5	16.3			6
	15.6	6.1		7.8	9.6	10.7	12.7	13.8	15.3			2
	25	5.8	7.6	9.1	10.7	11.6	13.0	14.5	16.8	20.6		15
	32.2		5.3	7.0	8.6	10.3	11.5	12.9	14.3			2
	48.9			6.2	7.4	9.6	10.4	11.9	13.6			2
	50	4.0	5.8	6.7	8.1	10.0	10.8	12.6	15.1	19.4		15
	0					13.0	14.2	16.0	18.0	21.1		3, 12
	20					12.2	13.2	15.1	17.1	20.3		3, 12
	40						12.6	14.1	16.3	19.8		17
	10		7.9	9.2	10.6	12.7	14.2	15.0	17.3			16
	20	5.6	7.1	8.3	9.6	10.9	12.2	13.5	15.0			17
	40	4.6	6.2	7.4	8.6	10.0	11.3	12.3	14.2			17
	80	2.5	3.7	4.8	5.7	6.7	8.0	9.8	11.5			17
Flax seed	25	3.3	4.9	5.6	6.1	6.8	7.9	9.3	11.4	15.2	21.4	7, 12
	25				6.1	6.8	7.9	9.3	11.4	15.7		13
Oats	25	4.1	6.6	8.1	9.1	10.3	11.8	13.0	14.9	18.5	24.1	7, 12
	21.1—26.7							13.1	15.4	18.5	31.4	18
	30		7.2	7.9	8.7	9.8	11.6	13.8	16.2	19.0		8
	20		6.7	8.2	9.4	10.8	12.0	14.4	16.8	19.9		8
	0		7.8	9.1	10.5	11.8	12.9	15.2	19.9	20.9		8
Peanuts, pod	10			5.5	6.1	7.1	8.6	9.8	11.9			19
	50	3.4	4.2	4.9	5.5	6.2	6.9	7.9	8.7	10.4		6
	30	4.2	5.1	5.8	6.4	7.1	7.8	8.6	9.6	11.2		6
	10	5.8	6.6	7.3	7.9	8.5	9.2	9.9	10.9	12.5		6

Material												Ref.
Kernels	10			4.8	5.5	6.0	6.6	7.3	9.0		10.3	19
	50	0.1	2.1	3.1	3.9	4.8	5.7	6.8	8.2		11.0	6
	30	1.8	3.0	3.9	4.8	5.6	6.6	7.6	9.0		12.0	6
Rice, milled	10	3.1	4.2	5.1	5.9	6.8	7.7	8.7	10.0		18.1	6
Whole grain	25	5.1	7.6	9.0	10.3	11.5	12.6	12.8	15.4	23.6	19.1	7, 12
Rough	25	5.9	8.0	9.5	10.9	12.2	13.3	14.1	15.2		19.1	20
	37.8	4.9	7.0	8.4	9.8	11.1	12.3	13.3	14.8		17.1	20
	26.7				9.2	10.4	11.7	13.2	15.0		16.5	21
	43.9						10.3	12.3	14.3			21
	0	4.6	8.2	9.9	11.1	12.3	13.3	14.5	16.6		19.2	22
	20	4.9	7.5	9.1	10.4	11.1	12.5	13.7	15.2		17.6	22
	30	5.3	7.1	8.5	10.0	10.9	11.9	13.1	14.7		19.1	22
	25	5.9	6.5	7.9	9.4	10.8	12.2	13.4	14.8		16.7	23
	22.8	7.4	7.3	8.7	9.8	10.9	12.4	13.5	15.9		19.0	24
	50	5.2	6.3	7.4	8.5	9.6	10.7	12.0	13.6		16.1	6
	30	7.8	7.4	8.5	9.5	10.6	11.7	12.9	14.5		17.0	6
	10	8.3	8.8	9.9	10.9	11.9	12.9	14.2	15.8		18.1	6
Rye	22.8	8.7	7.6	8.7	9.9	10.9	12.2	13.5	15.7		02.6	7
	30	9.0	9.0	10.4	11.5	12.8	14.3	16.5	20.3			8
	20	9.5	9.5	10.9	12.2	11.5	15.2	17.4	20.8			8
	0	10.4	10.4	11.6	12.7	13.9	15.8	18.3	21.9			8
Sorghum	−1.1	8.2	8.2	10.1	11.2	12.3	13.5	14.5	15.8			2, 12
	15.6	7.5	7.5	9.5	10.7	11.8	12.9	14.0	15.5			2, 12
	25	4.4	7.3	8.6	9.8	11.0	12.0	13.8	15.8	21.9	18.8	7, 12
	32.2	7.0	7.0	10.2	11.8	12.2	13.1	14.8				2, 12
	48.9	6.6	6.6	8.0	9.4	10.7	11.6	12.7	14.3			2, 12
Sorghum, kafer	4.4	6.8	8.5	9.7	11.0	12.2	13.8	15.3	17.3		19.0[a]	25
	21.1	6.0	7.7	9.1	11.3	11.5	12.8	14.2	16.0		17.0[a]	25
	32.2	5.0	7.0	8.4	9.6	10.8	12.0	13.2	14.7		17.7	25
Sorghum	50	5.4	7.0	8.3	9.5	10.6	11.8	13.2	15.0		18.2	6
	30	6.1	7.7	8.9	10.0	11.2	12.4	13.8	15.5		18.7	6
	10	6.8	8.4	9.6	10.7	11.8	13.0	14.4	16.1			6
Rape seed	25—35	3.9	3.9	4.9	5.7	6.6	7.4	8.8	11.8		17.4	26
Linseed	25—35			5.1	6.3	7.3	8.4	10.0	12.4		18.1	26
Caraway seed	25—35			6.7	7.8	9.0	10.3	12.0	14.5		19.8	26
Hoodder beet seed	25—35				7.1	8.6	10.3	13.0	16.1		22.2	26
Opium poppy seed	25—35			4.9	5.9	6.9	8.0	9.5	11.7		17.0	26

Table 1 (continued)
HYGROSCOPIC PROPERTIES OF VARIOUS AGRICULTURAL PRODUCTS

Material	Temp (°C)	Relative humidity (%)										Ref.
		Moisture Content (% Wet Basis)										
		10	20	30	40	50	60	70	80	90	100	
Onion seed	25–35			6.9	8.2	9.6	10.8	12.6	16.2	23.4		26
Perennial rye grass seed	25–35		5.9	7.6	9.1	10.6	11.9	14.0	17.3	24.0		26
Roughed stalked meadow grass seed	25–35			7.5	9.0	10.5	12.0	13.5	16.1	21.3		26
Red fescue seed	25–35			7.0	8.8	10.3	11.6	13.8	17.3	23.1		26
Creeping bentgrass seed	25–35			6.7	7.9	9.2	10.5	12.1	14.4	19.2		26
Brown beans	25–35			7.1	9.1	11.1	13.2	15.8	20.5	27.8		26
Dwarf French beans	25–35				9.1	11.2	13.4	15.7	20.2			26
Runner beans	25–35				8.8	11.2	12.9	15.9	20.8	29.7		26
Field beans	25–35				9.5	11.2	13.1	15.7	21.0	29.1		26
Green peas	25–35			8.1	9.7	11.3	13.1	15.3	19.3	27.2		26
Peas, whole	25–35	6.6		9.0		11.2		14.1	17.1			27
Bran	21.1–26.7							14.0	18.0	22.7	38.0	13, 14
Bread	25	0.9	1.8	3.1	4.4	6.1	7.7	10.3	12.3	16.0		28
Bone meal	21.1–26.7							14.1	10.8	12.4	22.0	18
Cotton cloth	25	2.4	3.6	4.7	4.8	6.3	7.8	9.1	10.4	12.0		28
Flour	25	2.0	3.6	5.2	5.7	7.5	9.6	11.2	13.7	16.0		29
Alfalfa hay												
No. 1	20.9	6.6	7.4	7.8	8.7	10.0	11.5	13.8	17.4			30
No. 2	25.6	3.9	4.8	5.7	6.6	8.3	10.0	13.0	14.5			30
No. 3	25.6	3.9	5.7	6.6	7.4	9.1	11.1	14.2	16.0			30
Alfalfa hay	15.6				13.5	20.0	26.0	38.0				14, 31
	37.8	7.6	8.0	9.1	10.5	13.3	17.5	23.0	36.0			14, 31
	65.6	5.1	5.6	6.3	7.1							14, 31
Alfalfa hay, various fractions												
Ordinary				7.1		9.6	11.1	12.7	16.5	25.2[a]		32
Shaded				6.8		9.9	11.0	12.9	18.0	23.6[a]		32
Leached				5.3		7.5	9.2	11.5	15.3	21.5[a]		32

Material	°C	1	2	3	4	5	6	7	8	9	Ref.
Top of stem				8.7		10.0	11.4	13.6	17.3	23.4[a]	32
Bottom of stem				6.8		8.9	10.7	12.8	15.1	22.1[a]	32
Bromegrass						8.4		15.3	20.5	29.8	32
Clover	37.8		7.4	8.1	9.1	10.1	11.2	12.4	15.0		33
Immature ladino						12.4		17.2	24.5[a]	29.9[a]	32
Oats straw	28.9		5.7	6.8	7.6	8.5	10.0	11.5	14.5		30
Prairie hay				7.0		7.4			21.9[a]		32
Red clover	28.9		5.7	7.7	7.8	8.7	10.0	10.8	12.7		30
No. 1	26.7		7.0	7.4	8.4	9.5	11.2	13.5	16.8		
No. 2	28.9		6.8		8.2	9.3	11.2	13.4	17.0		30
Rye straw	30				7.5	8.1	10.4	12.0	12.7	18.8	30
Blanched potato dice	30	0.9	2.4	4.2		9.4		10.9	13.5	20.8	32
Blanched carrot dice	30	1.1	1.6	4.8	10.9	15.3	16.9	17.8	21.0	40.0	34
Onion slices	30	1.1	2.1	5.4	10.0	15.2	20.2	23.3	28.6	41.7	34
Garlic, slices	30	0.7	1.4	3.4	6.7	10.8	14.6	15.9	18.1	27.8	34
Powder	30	1.5	3.3	5.5	8.1	11.0	14.8	17.5	18.8	33.3	34
Apple slices	30	0.7	2.8	6.0	11.4	17.6	21.1	22.6	25.5	45.1	34
Apricots	30	0.8	1.2	3.2	7.0	11.1	18.5	20.5	25.0	38.9	34
Seedless grapes	30	0.7	1.0	2.9	8.3	16.7	20.9	22.1	28.1	44.4	34
Orange powder	30	1.1	2.1	5.4	10.3	11.2	21.6	24.3	28.9	47.8	34
Potato	20		7.5		10.5	13.2			19.2		34
Starch	25		6.8		10.1	13.0			16.3		35
Beef	10		6.7		8.9	11.8					35
Dried peaches	26.7	0.3	1.0	2.3	4.2	7.0	11.0	12.7	25.1	38.8	35
	37.8	0.2	0.9	2.1	3.9	6.6	10.4	15.8	23.9	37.3	36
	48.9	0.2	0.9	2.0	3.7	6.2	9.8	15.0	22.8	35.8	36
	82.2	0.2	0.7	1.7	3.2	5.3	8.4	12.9	19.9	31.9	36
Dried prunes	26.7	1.6	3.8	6.4	9.4	12.9	16.9	21.9	28.1	37.2	36
	37.8	1.6	3.7	6.2	9.1	12.4	16.4	21.2	27.3	36.2	36
	48.9	1.5	3.5	5.9	8.7	12.0	15.8	20.5	26.5	35.3	36
	82.2	1.4	3.2	5.3	7.9	10.8	14.4	18.7	24.3	32.7	36
Egg albumin (spray-dried)	25	2.8	5.1			7.3	8.7	10.7	15.0		35
	10	2.3	4.3			.65	8.3	10.0			37
	37.2	1.2	2.7			4.5	5.8	6.7			37
	80										37

Table 1 (continued)

HYGROSCOPIC PROPERTIES OF VARIOUS AGRICULTURAL PRODUCTS

Material	Temp (°C)	Relative humidity (%)										Ref.
		Moisture Content (% Wet Basis)										
		10	20	30	40	50	60	70	80	90	100	
Cotton	26.7	2.2	2.4	4.4	5.4	6.4	7.4	8.6	10.1	12.1		36
	37.8	2.2	3.3	4.3	5.3	6.2	7.3	8.5	9.9	11.9		36
	48.9	2.1	3.2	4.2	5.2	6.1	7.2	8.3	9.7	11.7		36
Cabbage												
Air-dried	0	3.1		4.1		7.3		15.5	22.1			37
Scalded	25	1.3		4.7		9.1		16.6	23.3			37
Savoy	37.2	1.7		4.7		8.6		16.5	22.7			37
Carrot												
Scalded	10	3.2	3.7	4.5	6.3	8.8	12.5	17.4	24.7			37
	25	2.1		4.4		9.5		18.2	26.6			37
Air-dried	37.2	1.2	2.3	4.4	6.4	9.4	13.0	19.1	29.2			37
Milk, powder	10	2.7	3.0	3.4	4.8	7.0	6.5	7.6				37
Full-cream	37.2	2.6	3.3	4.1	4.0	4.5	6.5	7.9				37
Spray-dried	80	1.2	1.8	1.6	2.3	2.4	4.3	7.7				37
Peas, whole	23.9	6.6		9.0		11.2		14.1	17.1			18
Scotch beans	21.1—26.7							13.8	17.0	22.0	33.9	18
Tobacco	26.7	1.6	2.9	4.0	5.2	7.1	10.0	13.8	17.6			38
Bright	37.8	1.4	2.7	3.7	4.9	6.6	9.2	12.9	16.6			38
Strip	48.9	1.2	2.4	3.5	4.6	6.0	8.4	12.0	15.5			38
	60	1.1	2.1	3.1	4.2	5.6	7.7	11.1				38
N.C. leaf	25	6.7	9.9	12.2	14.2	16.5	18.7	21.3	24.9			39
Tobacco	26.7	2.5	3.5	4.6	6.2	8.3	11.0	15.1				38
Turkish	37.8	2.1	3.1	4.4	5.4	7.4	10.1	13.7				38
	48.9	1.8	2.7	3.2	4.5	6.4	8.9	12.3	17.8			38
	60	1.5	2.2	2.8	3.7	5.4	7.9	11.0	15.8			38
Tobacco	26.7	3.4	5.0	5.6	6.3	7.4	9.5	13.3				38
Burley	37.8	2.6	4.3	5.2	6.0	6.6	8.3	11.4	16.5			38
Strip	48.9	1.8	3.5	4.7	5.5	6.2	7.3	9.7	13.8			38
	60	1.2	2.7	4.1	5.1	6.0	6.8	8.5	11.9			38

a   Unreliable because mold growth on sample.

REFERENCES

1. **Thompson, T. L.,** Predicted Performances and Optimal Designs of Convection Grain Dryers, Ph.D. thesis, Purdue University, West Lafayette, Ind., 1967.
2. **Haynes, B. C.,** Vapor Pressure Determination of Seed Hygroscopicity, Tech. Bull. 1219, ARS, U.S. Department of Agriculture, Washington, D.C., 1961.
3. **Pixton and Warbutton,** Moisture content-relative humidity equilibrium of some cereal grains at different temperatures, *J. Stored Products Res.,* 1, 283, 1971.
4. **Rodriguez-Arias, J. H.,** Desoprtion Isotherms and Drying Rates of Shelled Corn in the Temperature Range of 40 to 140°F, Ph.D. thesis, Michigan State University, East Lansing, 1956.
5. **Gustafson, R. J.,** Equilibrium Moisture Content of Shelled Corn from 50 to 155°F, M.S. thesis, University of Illinois, Urbana, 1972.
6. **Chung, D. S., Pfost, H. B., Maurer, S. G., and Milliken, G. A.,** Summarising and Reporting Equilibrium Moisture Data for Grains, Winter Meeting, American Society of Agricultural Engineers, Chicago, 1976.
7. **Coleman, D. A. and Fellows, H. C.,** Hygroscopic moisture of cereal grains and flaxseed exposed to different relative humidities, *Cereal Chem.,* 2, 275, 1955.
8. **Bakharev, I. Y.,** Ventilirovanie Serna, Zagotizdat (ventilation of grain), Grain Supply Press, 1948.
9. **Karon, M. L.,** Hygroscopic equilibrium of cotton seed, *J. Am. Oil Chem. Soc.,* 40, 1361, 1947.
10. **Dexter, S. T., Anderson, A. L., Pjahler, P. L., and Beene, G. J.,** Response of white pea beans to various humidities and temperatures of storage, *Agron. J.,* 47, 246, 1955.
11. **Waston, W. J. and Morries, H. J.,** Hygroscopic equilibria of dry beans, *Food Technol.,* 8, 353, 1954.
12. **Brooker, D. B., Bakker-Arkema, F. W., and Hall, C. W.,** *Drying Cereal Grains,* AVI Publishing, Westport, Conn., 1974, 70.
13. **Larnmour, R. K., Sallens, H. R., and Craig, R. M.,** Hygroscopic equilibrium of sunflower seed, flax seed, and soybeans, *Con. J. Res.,* 22, 1, 1944.
14. **Hall, C. W.,** *Drying Farm Crops,* Edwards Brothers, Ann Arbor, Mich., 1957, 21.
15. **Becker, H. A. and Sallans, H. R.,** A study of desorption isotherms of wheat at 25 and 50°C, *Cereal Chem.,* 33, 79, 1956
16. **Young, J. H. and Nelson, G. L.,** Research of hysteresis between sorption and desorption isotherms of wheat, *Trans. ASAE,* 10, 756, 1967.
17. **Pichler, H. J.,** Sorption isotherms for grain and rape, *J. Agric. Eng. Res.,* 2, 159, 1957.
18. **Snow, D., Crichton, M. H., and Wright, N. C.,** Mold deterioration of feedstuffs in relation to humidity of storage, *Ann. Appl. Biol.,* 31, 102, 1944.
19. **Beasley, E. O.,** Moisture Equilibrium of Vergina Bunch Peanuts, thesis, North Carolina State University, Raleigh, 1962.
20. **Houston, D. H. and Kester, E. B.,** Hygroscopic equilibrium of whole-grain edible forms of rice, *Feed Technol.,* 8, 302, 1954.
21. **Hogan, J. T. and Karm, M.,** Hygroscopic Equilibrium of Rough Rice at Elevated Temperatures, Special Report, American Chemical Society Meeting, Cincinnati, March 29, 1955.
22. **Bakharev, I. Y.,** Moisture equilibrium of peanuts, in *Grain Drying and Grain Dryers,* Gerzhoi, A. P. and Samochetov, V. F., Eds., Israel Program for Scientific Translations, Haifa, 1960.
23. **Breese, M. H.,** Hysteresis in the hygroscopic equilibrium of rough rice at 25°C, *Cereal Chem.,* 33, 481, 1955.
24. **Henderson, S. M.,** Equilibrium moisture content of small grain hysteresis, *Trans. ASAE,* 13, 762, 1970.
25. **Fenton, F. C.,** Storage of grain sorghums, *Agric. Eng.,* 22, 185, 1941.
26. **Sijbring, P. H.,** Results of some experiments on the moisture relationship of seeds, *Proc. Int. Seed Test Assoc.,* 28(4), 837, 1963.
27. **Morris, T. N.,** *The Dehydration of Food,* Van Nostrand, New York, 1947.
28. **Washburn, G. W.,** *Internation Critical Tables,* Vol. 2, McGraw-Hill, New York, 1927, 324.
29. **Bailey, C. H.,** The hygroscopic moisture of flour exposed to atmospheres of different relative humidity, *Ind. Eng. Chem.,* 12, 1104, 1920.
30. **Zink, F. J.,** Equilibrium moistures of some hays, *Agric. Eng.,* 16, 451, 1935.
31. **Davis, R. B., Jr., Garlow, G. E., Jr., and Brown, D. B.,** Supplemental heat in mow drying of hay. III. *Agric. Eng.,* 31, 223, May 1950.
32. **Dexter, S. T., Sheldon, W. H., and Huffman, C. F.,** Better quality hay, *Agric. Eng.,* 28, 291, 1947.
33. **Milne, C. A.,** Latent Heat of Hay, M.S. thesis, Department of Agricultural Engineering, Purdue University, West Lafayette, Ind., 1954.
34. **Charm, S. E.,** *The Fundamentals of Food Engineering,* AVI Publishing, Westport, Conn., 1971, 613.
35. **Earle, R. L.,** *Unit Operations in Food Processing,* Pergamon Press, Oxford, 1966, 153.
36. **Henderson, S. M. and Parry, R. L.,** *Agricultural Process Engineering,* AVI Publishing, Westport, Conn., 1976, 309.
37. **Morris, T. N.,** *The Dehydration of Food,* Van Nostrand, New York, 1947.

38. **Locklain, E. E., Veasey, L. G., and Samfield, M.,** Equilibrium desorption of water vapor in tobacco, *Agric. Food Chem.,* 5, 294, 1957.
39. **Washburn, E. W.,** *Internation Critical Tables,* Vol. 2, McGraw-Hill, New York, 1927, 324.

ON-FARM GRAIN DRYING

Donald B. Brooker

HISTORY

Widespread use of grain drying has been a development of relatively recent years. Immediately after World War II, farmers began heavy use of fertilizers, and this, coupled with the adoption of hybrid corn, gave a tremendous increase in corn yields. A serious shortage of corn storage space developed. The corn combine was, at the same time, taking its place in agriculture. The round metal bin was a logical storage structure for the shelled corn from the combines, and such bins soon replaced the slotted-side ear corn cribs.

The metal bins were easily rodent and bird proofed and held twice as many bushels of corn for a given volume than did the cribs. However, the cribs had an advantage over the bins in one respect; the moisture content of the corn going into storage could be as high as 20 or even 22% wet basis, depending on location, without the ears spoiling. Shelled corn stored in the tight bins had to be much dryer than 20% if spoilage was to be averted. Harvest and safe storage moistures for corn and other grains[1-5] are given in Table 1.

Unfortunately, corn seldom dries in the field to a safe storage moisture content. In order for the shelled corn to keep in storage, it had to be further dried. The era of forced air drying of corn began. The drying of other grains developed simultaneously.

SYSTEM CHARACTERISTICS

Airflow

Grain is dried by forcing air through the mass with fans. The resistance of the grain mass to airflow was studied by Shedd[6] and his curves (Figure 1) are classical.

Shedd's curves are principally for clean grain. Fines, broken kernels, trash, and moisture content all affect airflow resistance. Haque et al.[7] presented curves (Figures 2 and 3) for shelled corn containing fines.

Airflow resistance is also a function of grain moisture content. Data of Haque et al.[8] are plotted in Figures 4 to 6 for shelled corn, grain sorghum, and wheat, respectively. As moisture content increases, airflow resistance decreases.

Stephens and Foster[9,10] found that the method of filling the bin had a marked effect on airflow resistance. The results of their studies are shown in Figures 7 to 9 for shelled corn, grain sorghum, and wheat, respectively.

Fans

The fans used in connection with grain dryers are commonly either vaneaxial or backward curved centrifugal. Fan characteristics are given as output (ft³/min) vs. static pressure. Typical dryer fan performance data are given in Tables 2 and 3.

Equilibrium Moisture

If grain is left in air at a given temperature and relative humidity, it will eventually reach a moisture content called the equilibrium moisture for that temperature and relative humidity. The 1982-83 Yearbook of the American Society of Agricultural Engineers in ASAE Data ASAE D245.4, Moisture Relationship of Grains, shows a table and several graphs of the equilibrium moistures of various grains.

Under the same data heading, the ASAE Yearbook also gives the constants for two equilibrium moisture equations, the modified Henderson Equation[11] and the Chung Equa-

Table 1
MOISTURE CONTENT DURING HARVEST AND FOR SAFE STORAGE (PERCENT, w. b.)

Cereal	Maximum during harvest	Optimum at harvest for minimum loss	Usual when harvested	Required for safe storage	
				For 1 year	For 5 years
Barley	30	18—20	10—18	13	11
Corn	35	28—32	14—30	13	10—11
Oats	32	15—20	10—18	14	11
Rice	30	25—27	16—25	12—14	10—12
Rye	25	16—20	12—18	13	11
Sorghum	35	30—35	10—20	12—13	10—11
Wheat	38	18—20	9—17	13—14	11—12

Note: Above applies to climates where cereal grain is usually grown. For example, oats in a warm climate may need to be at 10 to 11% for safe-keeping; rye in the Soviet Union is stored at 13 to 14%.

From Brooker, D. B., Bakker-Arkema, F. W., and Hall, C. W., *Drying Cereal Grains*, AVI Publishing, Westport, Conn., 1974. With permission.

tion.[12] The constants for the two equations and the equations are presented in Tables 4 and 5.

ON-FARM DRYING SYSTEMS

In-Bin Drying

Grain is dried in a storage bin by several methods. The methods described are layer drying, batch-in-bin, and in-bin counterflow.

Layer Drying

When grain is added to the bin periodically in batches, the process is called layer drying. How much grain is added is a function of grain moisture, the airflow rate, and drying air conditions. Drying is begun as soon as the first layer is in place. A special case of layer drying is called full-bin drying, and in this case conditions are such that the entire bin is filled as a single layer.

Layer drying can be accomplished with unheated air. Needed airflow rates are a function of grain moisture and geographical location of the bin. Airflow rates are expressed as cubic feet per minute per bushel (ft³/min/bu) of undried grain. Pierce and Thompson[13] simulated drying corn with unheated air and with air that was heated to raise its temperature 3°F. In each case, it was assumed that the fan added heat that raised the air temperature 2°F. The results of Pierce and Thompson's simulation for 24% moisture corn at several locations in the midwestern U.S. are shown in Table 6.

In most cases, layer drying is carried out by adding enough heat to lower the relative humidity to 55%. A humidistat is placed in the plenum, and a heater is used to maintain the desired relative humidity. Layer adding schedules for various bin-fan-heater combinations have been published by dryer companies, and Table 7 is one such table for shelled corn; Table 8 is for other grains.

Filling rates in bushels per day are given in Tables 7 and 8. However, a further restriction in filling rates is that undried grain depth is limited by grain moisture. These restrictions are given in Table 9.

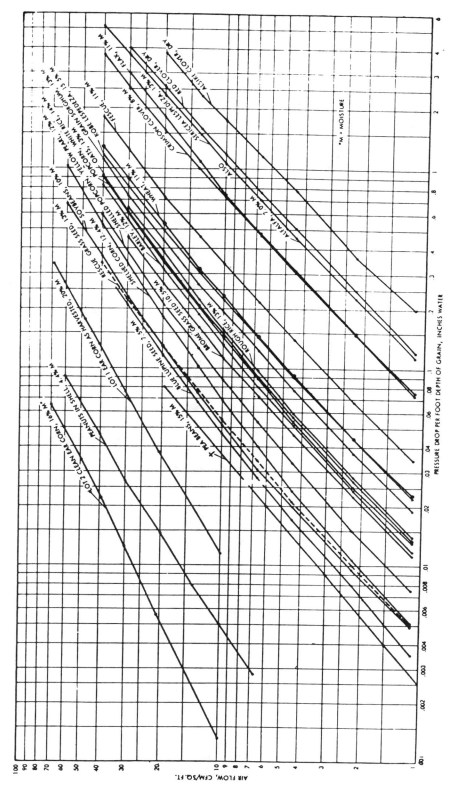

FIGURE 1. Resistance of grains and seeds to airflow. (1) Gives values for a loose fill (not packed) of clean, relatively dry grain. For a loose fill of clean grain having high moisture content (in equilibrium with relative humidities exceeding 85%), use only 80% of the indicated pressure drop for a given rate of airflow. (2) Packing of the grain in a bin may cause 50% higher resistance to airflow than the values shown. (From Shedd, C. K., *Agric. Eng.*, 34, 616, 1953. With permission.)

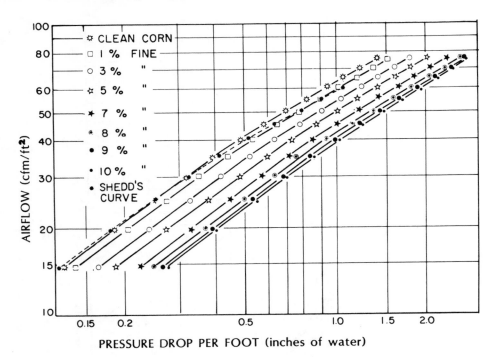

FIGURE 2.   Resistance of a mixture of corn and its fines to airflow: 0 to 10%. (From Haque, E., Foster, G. H., Chung, D. S., and Lai, F. S., Static Pressure Drop Across a Corn Bed Mixed with Fines, ASAE Paper No. 76-3528, American Society of Agricultural Engineers, St. Joseph, Mich., 1976. With permission.)

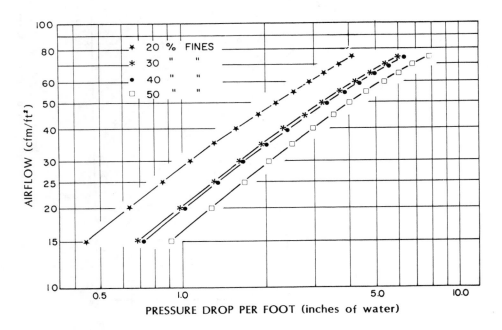

FIGURE 3.   Resistance of a mixture of corn and its fines to airflow: 20 to 50% fines. (From Haque, E., Foster, G. H., Chung, D. S., and Lai, F. S., Static Pressure Drop Across a Corn Bed Mixed with Fines, ASAE Paper No. 76-3528, American Society of Agricultural Engineers, St. Joseph, Mich., 1976. With permission.)

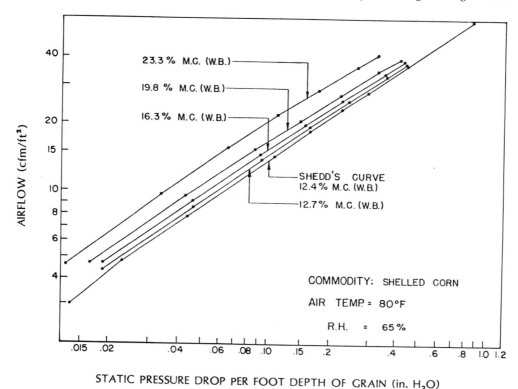

FIGURE 4.   Airflow resistance of shelled corn at various moisture contents. (From Haque, E., Ahmed, Y. N., and Deyoe, C. N., Static Pressure Drop in a Fixed Bed of Grain as Affected by Grain Moisture Content, ASAE Paper No. MCR-81-301, American Society of Agricultural Engineers, St. Joseph, Mich., 1981. With permission.)

Batch-in-Bin Drying

When a few feet of grain are placed in a bin, dried, and transferred to another bin for storage, the process is called batch-in-bin drying. Normally the depth of grain is selected so that the entire process is completed in 1 day. A 24-hr cycle may consist of filling, drying for 20 hr, cooling with unheated air, and transferring the grain to storage. Plenum air temperature is controlled with a thermostat and drying air temperatures of 120 to 160°F are common. The batch sizes for various bin, fan-heater, and grain moisture combinations when the plenum air is heated to 120°F are given in Table 10.

A unique type of batch-in-bin dryer is shown in schematic in Figure 10. When the batch is dry, it is dumped to the lower portion of the bin and cooled there. Features of this dryer are it unloads the drying chamber quickly, no time is lost by cooling while the grain is still in the drying chamber, and the heat removed from the grain during cooling is reclaimed by mixing the exhaust cooling air with the air coming from the fan-heater.

Stirring Devices

When grain is dried in deep layers, as is the case with layer or batch-in-bin drying, the grain is often overdried. In layer drying, the fan-heater must be operated until the drying zone passes through the entire grain mass. The operator cannot stop drying when the average moisture of the mass is ideal because the upper layers then have a moisture content too high for safe storage. The higher temperatures of batch-in-bin drying cause severe overdrying of the lower layers and underdrying of the upper layers. Overdrying is expensive in two ways: removing too much moisture means less grain weight at the market place and energy is wasted in removing that moisture.

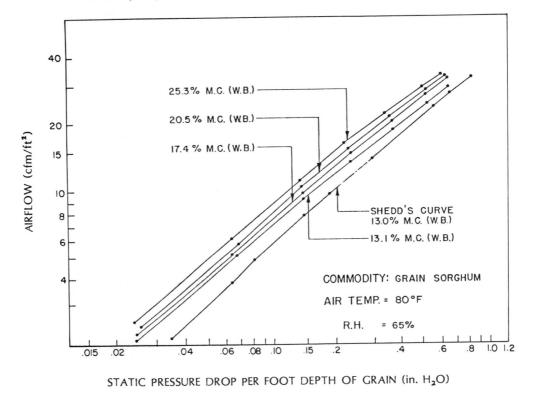

FIGURE 5. Airflow resistance of grain sorghum at various moisture contents. (From Haque, E., Ahmed, Y. N., and Deyoe, C. N., Static Pressure Drop in a Fixed Bed of Grain as Affected by Grain Moisture Content, ASAE Paper No. MCR-81-301, American Society of Agricultural Engineers, St. Joseph, Mich., 1981. With permission.)

Machines that mix grain while it is drying are called stirring devices. They consist of vertical augers or "drops" that are suspended from a radial arm at the top of the bin. The outer end of the radial arm is carried on a track along the bin wall with the arm pivoted at the bin center as shown in Figure 11. As the vertical augers travel around the bin center with the radial arm, they may or may not move back and forth along the arm.

The vertical augers remove grain from the lower portions of the bin and raise it to the surface. At the same time, grain from upper layers moves downward and a mixing action is obtained. The augers loosen the grain to allow more air to flow. They also mix spots of high-moisture grain or trash into the grain mass to lessen the chance of "hot spots" developing during storage. When stirring devices are used, the drying can be stopped when a desirable average moisture is reached.

Counterflow Drying

A third method of in-bin drying is called counterflow. With these systems, an auger is permanently mounted to sweep grain from the perforated floor to a sump at the bin center. From the sump, the grain is removed from the bin by either an underfloor auger (Figure 12) or by a vertical auger that discharges into a transfer auger (Figure 13). The on-floor auger is of a special design so that grain is removed evenly from all portions of the floor; that is, the entire grain mass moves downward vertically at a somewhat constant rate.

The drying air moves upward into the downward moving grain; hence the term counterflow is used. The plenum temperature is controlled at some temperature between 120 and 180°F. Controls regulate the rate at which grain is removed to correspond with the drying capacity of the unit.

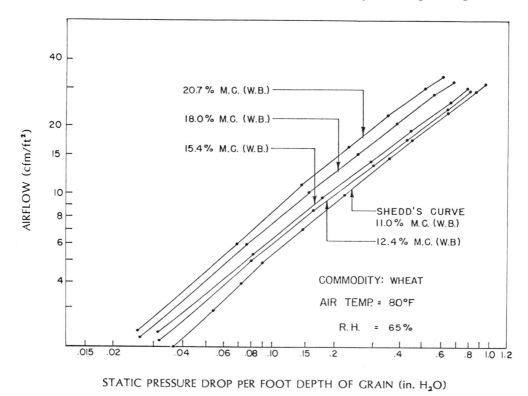

FIGURE 6. Airflow resistance of wheat at various moisture contents (From Haque, E., Ahmed, Y. N., and Deyoe, C. N., Static Pressure Drop in a Fixed Bed of Grain as Affected by Grain Moisture Content, ASAE Paper No. MCR-81-301, American Society of Agricultural Engineers, St. Joseph, Mich., 1981. With permission.)

All grain is removed at the same moisture content, and overdrying is not a problem. The grain is removed while warm, and cooling is carried out in the storage bin with aeration fans. Drying rates depend on bin size and number and size of the fan-heater units used. Rates can vary from 1450 bu/day (5-hp fan and 18-ft diameter bin) to 6000 bu/day (two 20-hp fans and 36-ft diameter bin).

One manufacturer of a counterflow in-bin dryer has raised the false floor to the upper portion of the bin (see Figure 14). The dried grain that is removed from the floor is spread into the space below where it is cooled. Heat removed during cooling is reclaimed and used in the drying process.

Column Dryers

Column dryers, as compared to in-bin dryers, use large amounts of air and high drying air temperatures. The air passes through the grain while it is held in a relatively narrow column. Column dryers are classified as batch or continuous flow.

Column Batch

A typical column batch dryer is shown in Figure 15. The column of grain is on either side of a central plenum. Column widths are in the order of 12 to 14 in. Airflow rates above 100 ft³/min/bu are common. Drying temperature is usually in the 180 to 200°F range.

The column batch dryer can be equipped with controls so that it is filled, the batch dried and cooled, and the dryer emptied, all automatically. If the dryer is to be operated at maximum capacity, an adequate grain handling system must be used in conjunction with it.

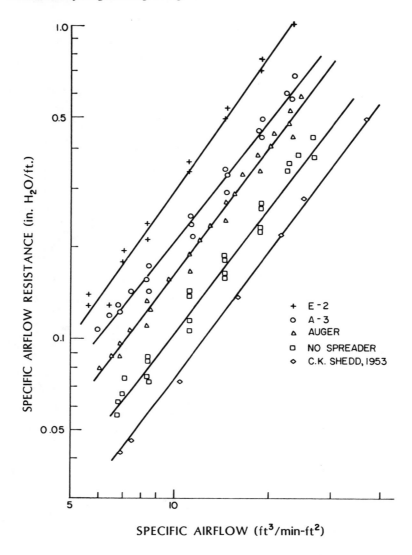

FIGURE 7.   Airflow resistance of shelled corn produced by filling with three types of spreaders, filling with no spreader, and by loose fill as given by Shedd.[6] (From Stephens, L. E. and Foster, G. H., *Trans. ASAE*, 19, 354, 1976. With permission.)

The dryer is filled from a wet-holding bin. The wet-holding bin can be installed above the dryer or can be located adjacent to it at ground level. Overhead bins fill the dryer by gravity and are preferred over ground storages that fill the dryer by means of an inclined auger. An auger handling high-moisture corn has high horsepower requirements, and the capacity is about two thirds of that which it has handling dry grain. The discharge rate from the wet holding bin should be equal to that of the leveling auger at the top of the dryer.

Also, the conveyor taking the grain from the dryer should have a capacity equal to the unloading auger in the bottom of the dryer. The greater the time spent in loading and unloading, the less the overall drying rate of the dryer.

Typical data for several sizes of on-farm batch dryers are provided in Table 11. The drying rates are given in two ways: dry and cool and full heat. The expression full heat means that the cooling cycle of the dryer is not used, and therefore the grain is discharged

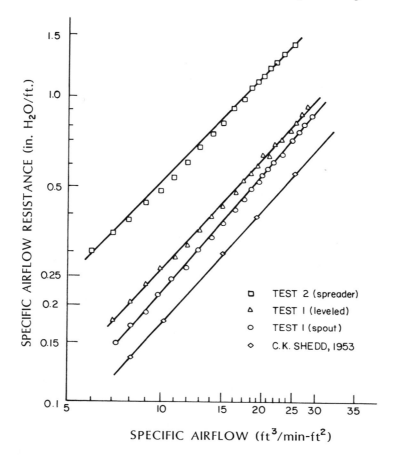

FIGURE 8.   Airflow resistance of grain sorghum produced by various filling methods. (From Stephens, L. E. and Foster, G. H., *Trans. ASAE*, 21, 1217, 1978. With permission.)

while hot and must be cooled in storage external to the dryers. The dryers can be "stacked", one unit above another, for additional capacity.

Continuous Flow
Cross Flow

Farm-size column continuous flow dryers are similar in appearance to the column batch and are classified as cross flow. The grain flows vertically down the column, and the air is forced across the column. The plenum is divided so that hot air enters the grain from the upper portion and cool air enters the grain from the lower portion.

Automatic controls regulate the discharge rate of the dryer to correspond to the drying rate. Airflow rates and air temperatures are in the same range as those used for the column batch.

Dryers can be "stacked" so that as many as five plenums can deliver heated air with the sixth (the lowest plenum) providing the cooling air. Output of a unit depends on the number of fan-heaters used. A continuous flow dryer with one 10-hp fan-heater that discharges cool grain may deliver 115 dried bushels of shelled corn per hour when drying shelled corn from 25 to 35% moisture. A dryer with five 20-hp fan-heaters supplying drying air and another 20-hp fan used for cooling may deliver over 800 dried bushels of shelled corn per hour.

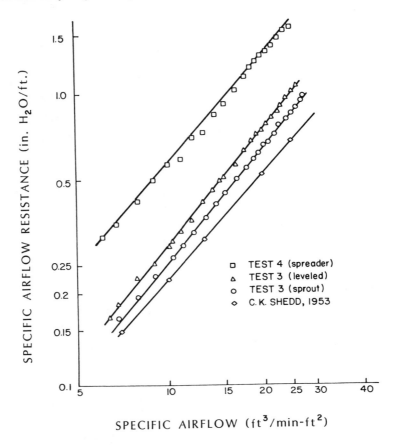

FIGURE 9.   Airflow resistance of wheat produced by various filling methods. (From Stephens, L. E. and Foster, G. H., *Trans. ASAE*, 21, 1217, 1978. With permission.)

Table 2
FAN PERFORMANCE: AXIAL FLOW FANS

			Air volume (ft³/min) at indicated static pressure							
HP	Diam. (in.)	rpm	1 in.	1 ½ in.	2 in.	2 ½ in.	3 in.	3 ½ in.	4 in.	5 in.
5—7	24	3,450	11,050	10,400	9,600	8,650	7,500	6,300	5,200	3,100
7—9	24	3,450	13,200	12,400	11,650	10,800	9,800	8,750	7,200	4,500
10—13	28	3,450	19,400	18,400	17,300	16,300	15,100	14,000	12,800	9,500
10—16	36	1,750	26,000	24,400	22,600	20,600	18,100	12,600	10,700	6,700

Data from Farm Fans, Inc.

The grain handling system that brings grain to and takes grain from a continuous-flow dryer need only have a capacity equal to the drying rate of the dryer. Therefore, the handling system capacity can be less than that required by batch dryers. Since less handling capacity is needed, the wet holding bin can logically be at ground level.

Concurrent Flow
When grain and air travel in the same direction through the dryer, the dryer is classified

Table 3
FAN PERFORMANCE: BACKWARD CURVED CENTRIFUGAL FANS

HP	Air volume (ft³/min) at indicated static pressure								
	0 in.	1 in.	2 in.	3 in.	4 in.	5 in.	6 in.	7 in.	8 in.
3	5,300	4,500	3,800	3,400	3,000	2,600			
5	9,500	8,900	7,650	6,700	6,000	5,000			
7½	12,000	11,300	10,400	9,400	8,500	7,700	6,700		
10	16,000	15,150	14,200	13,200	12,200	11,200	10,100	7,800	3,300
15	19,000	18,000	17,100	16,200	15,200	14,200	13,200	11,600	8,200
20	23,600	22,800	21,750	20,650	19,600	18,500	17,200	15,800	14,500

Data from Sukup Manufacturing Company.

Table 4
EQUILIBRIUM MOISTURE CONTENT: MODIFIED HENDERSON EQUATIONS AND CONSTANTS[12]

$$M = \left[\frac{\ln(1 - RH)}{-K(T + C)} \right] \frac{1}{N}, \% \text{ dry basis}$$

$$RH = 1 - EXP[-K \cdot (T + C) \cdot M^N]$$

	K	N	C	Standard error moisture
Barley	0.000012733	2.0123	319.481	0.0080
Beans, edible	0.000011611	1.8812	425.614	0.0138
Corn, yellow dent	0.000048078	1.8634	57.658	0.0127
Peanut, kernel	0.000361341	1.4984	59.010	0.0126
Peanut, pod	0.000036993	2.5362	9.972	0.0303
Rice, rough	0.000010659	2.4451	60.090	0.0097
Sorghum	0.000004740	2.4757	172.705	0.0087
Soybean	0.000169626	1.2164	209.445	0.0173
Wheat, durum	0.000014299	2.2110	94.527	0.0068
Wheat, hard	0.000012782	2.2857	68.467	0.0071
Wheat, soft	0.000006833	2.5558	83.823	0.0122

Note: M = grain moisture, decimal dry basis; RH = relative humidity, decimal; T = temperature, °F.

as concurrent flow. Figure 16 is a schematic of such a dryer. Since the hottest air is introduced into the wettest grain, the plenum air temperature can be higher than that used with the cross-flow dryers. The air temperature and grain moistures shown in Figure 17 are typical for this type of dryer. Maximum grain temperature is about 130°F.

The cooling section of the dryer is counter flow. Since the coolest air is introduced into the coolest grain, there is less stress-cracking of the discharged grain. Drying rates for typical farm-sized concurrent-flow dryers vary from about 250 to 650 bu/hr of shelled corn when the corn is dried from 25 to 15% moisture.

Table 5
EQUILIBRIUM MOISTURE CONTENT: CHUNG EQUATIONS AND CONSTANTS[12]

$$M = E - F \ln\{-(T + C) \cdot \ln(RH)\}$$

$$RH = \exp\left\{\frac{-A}{(T + C)} \exp(-B \cdot M)\right\}$$

Grain	A	B	C	E	F	Standard error moisture
Barley	1370.988	19.889	132.381	0.36318	0.050279	0.0055
Beans, edible	1732.644	15.975	257.132	0.46680	0.062596	0.0136
Corn, yellow dent	562.140	16.958	22.369	0.37338	0.058970	0.0121
Peanut, kernel	458.820	29.243	29.006	0.20951	0.034196	0.0133
Peanut, pod	939.618	37.903	−9.763	0.18061	0.026383	0.0322
Rice, rough	1070.298	21.732	32.265	0.32099	0.046015	0.0096
Sorghum	1979.406	19.644	153.128	0.38641	0.050907	0.0086
Soybean	590.940	13.917	148.518	0.45854	0.071853	0.0191
Wheat, durum	1658.970	18.077	170.230	0.41013	0.055318	0.0057
Wheat, hard	952.974	17.609	59.796	0.38954	0.056788	0.0061
Wheat, soft	1307.682	23.607	32.192	0.30398	0.042360	0.0147

Note: M = grain moisture, decimal dry basis; RH = relative humidity, decimal; T = temperature, °F.

Table 6
MINIMUM AIRFLOW REQUIREMENTS (ft³/min/bu) FOR ONE LOCATION IN EACH OF THE NORTH CENTRAL REGION STATES, ASSUMING AN INITIAL MOISTURE CONTENT OF 24% WET BASIS AND A HARVEST DATE OF OCTOBER 15 (VALUES ARE BASED UPON SIMULATED DRYING RESULTS FOR A 10 YEAR PERIOD)

Location	Natural air[a]		Electrical heat[b]	
	Next to worst year	Worst year	Next to worst year	Worst year
Chicago, Ill.	2.70	3.03	2.23	2.79
Indianapolis, Ind.	4.10	4.29	1.87	1.98
Des Moines, Iowa	2.18	4.42	1.96	3.70
Dodge City, Kan.	2.22	3.32	2.12	3.11
Lansing, Mich.	2.75	3.58	2.34	2.58
St. Cloud, Minn.	1.91	3.07	1.74	3.35
Columbia, Mo.	2.48	2.99	2.19	2.79
Lincoln, Neb.	2.07	4.42	1.79	3.82
Bismarck, N.D.	0.57	0.75	0.54	0.89
Mansfield, Ohio	2.55	3.92	1.85	3.48
Huron, S.D.	1.40	2.10	1.77	2.43
Madison, Wis.	2.24	2.44	1.77	2.22

[a] A 2 °F temperature rise due to drawing the air over the fan motor was assumed.
[b] Enough heat added to raise air temperature 5°F ~ 2°F temperature rise from fan and fan motor, 3° rise from heater.

From Pierce, R. O. and Thompson, T. L., *Trans. ASAE*, 22, 182, 1979. With permission.

Table 7
FILLING AND DRYING RATES FOR LAYER DRYING SHELLED CORN

Average Filling Rate in Bushels per Day (Upper Number) and
Average Drying Rate in Bushels per Day (Lower Number) Shown in Parentheses

Fill bin to 16-ft depth

Fan heater	1 unit 24 in. - 5 hp 700,000 Btu/hr			1 unit 24 in. - 7½ hp 700,000 Btu/hr				2 units 24 in. - 5 hp 700,000 Btu/hr			2 units 24 in. - 7½ hp 700,000 Btu/hr			3 units 24 in. - 5 hp 700,000 Btu/hr			4 units 24 in. - 7½ hp 700,000 Btu/hr
Bin diameter	18 ft	21 ft	24 ft	21 ft	24 ft	27 ft	30 ft	30 ft	33 ft	36 ft	30 ft	33 ft	36 ft	36 ft	42 ft	48 ft	48 ft
Initial moisture content 20%																	
25%	501 (179)	590 (231)	540 (259)	697 (246)	689 (289)	685 (328)	694 (367)	1160 (452)	1080 (518)	1100 (558)	1375 (480)	1378 (578)	1370 (620)	2062 (720)	2060 (879)	2055 (1014)	2750 (1156)
30%	213 (153)	266 (204)	266 (204)	293 (210)	308 (234)	334 (263)	362 (292)	522 (384)	532 (418)	561 (448)	575 (412)	616 (468)	645 (498)	862 (618)	1015 (811)	1118 (813)	1232 (936)

Fill bin to 16-ft depth

Fan heater	1 unit 28 in. - 10 hp 1,500,000 Btu/hr					2 units 28 in. - 10 hp 1,500,000 Btu/hr				3 units 28 in. - 10 hp 1,500,000 Btu/hr			4 units 28 in. - 10 hp 1,500,000 Btu/hr	
Bin diameter	24 ft	27 ft	30 ft	33 ft	36 ft	30 ft	33 ft	36 ft	42 ft	36 ft	42 ft	48 ft	42 ft	48 ft
Initial moisture content 20%														
25%	1280 (358)	1100 (400)	1103 (456)	1125 (512)	948 (512)	2150 (550)	2560 (716)	2210 (748)	2200 (900)	3220 (825)	3330 (1014)	3290 (1245)	4370 (1128)	5120 (1432)
30%	466 (315)	452 (330)	495 (373)	538 (395)	485 (397)	736 (488)	932 (598)	865 (622)	984 (738)	1100 (732)	1240 (879)	1390 (1026)	1500 (1000)	1864 (1260)

Table 7 (continued)

FILLING AND DRYING RATES FOR LAYER DRYING SHELLED CORN

Fan heater		1 unit 36 in. - 10 hp 1,800,000 Btu/hr				2 units 36 in. - 10 hp 1,800,000 Btu/hr				3 units 36 in. - 10 hp 1,800,000 Btu/hr		
Bin diameter		**30 ft**	**33 ft**	**36 ft**	**42 ft**	**36 ft**	**42 ft**	**48 ft**		**42 ft**	**48 ft**
				Fill bin to 16-ft depth							
Initial	20%										
moisture	25%	1150	1243	1340	1225	2060	2280	2530		2320	3260
content		(464)	(535)	(609)	(679)	(730)	(914)	(1108)		(1050)	(1242)
	30%	522	588	656	640	860	1030	1213		1160	1422
		(388)	(445)	(504)	(527)	(618)	(766)	(922)		(885)	(966)

Note: Bushels shown are wet bushels as determined by volume. Table is based on ultimate grain depth of 16 ft, rates must be decreased for greater depths. Heat controlled by humidistat set at 55%. Assumed outside air — 45°F and 100% RH. Final average moisture contents — 13%.

From Bulletin U-03-1, Farm Fans, Inc.

Table 8
FILLING AND DRYING RATES FOR LAYER DRYING: WHEAT, OATS, SOYBEANS, ROUGH RICE, AND GRAIN SORGHUMS

Average Filling Rate in Bushels per Day (Upper Number) and Average Drying Rate in Bushels per Day (Lower Number) Shown in Parentheses

Fan heater	24 in. - 5 hp 705,000 Btu/hr			24 in. - 7 1/2 hp 705,000 Btu/hr			28 in. - 10 hp 1,000,000 Btu/hr		
Bin diameter	18 ft	21 ft	24 ft	24 ft	27 ft	30 ft	30 ft	33 ft	36 ft
Grain Sorghum									
20%									
25%	208 (119)	253 (152)	301 (187)	344 (203)	389 (244)	454 (286)	622 (345)	716 (406)	665 (417)
30%	118 (97)	147 (123)	178 (150)	197 (164)	232 (195)	269 (229)	348 (283)	406 (333)	384 (327)
Wheat									
20%	467 (198)	552 (254)	643 (314)	759 (339)	860 (408)	965 (481)	1395 (568)	1590 (670)	1485 (715)
25%	184 (150)	230 (190)	278 (233)	308 (255)	363 (304)	421 (356)	543 (439)	632 (514)	608 (513)
Rough Rice									
20%	1243 (251)	1317 (324)	1426 (402)	1872 (431)	1970 (521)	2090 (617)	4000 (718)	3620 (819)	3280 (910)
25%	265 (194)	328 (246)	395 (302)	441 (328)	517 (392)	597 (459)	794 (568)	780 (590)	836 (648)
Soybeans									
				560 (263)	510 (282)	525 (313)	797 (393)	827 (440)	733 (443)
Oats									
	570 (337)	695 (429)	825 (528)	935 (517)	1085 (684)	1240 (804)	1490 (902)	1570 (1002)	1660 (1103)

(Note: "Fill bin to 16-ft depth" indicated for Grain Sorghum, Soybeans, and Oats.)

Note: Bushels shown are wet bushels as determined by volume. Tables are based on ultimate grain depth of 16 ft, rates must be decreased for greater depths. Assumed outside air — 45°F and 100% RH. Final average moisture contents — 13%. Heater controlled by humidistat set at 55%.

From Bulletin U-03-1, Farm Fans, Inc.

Table 9
MAXIMUM ALLOWABLE DEPTH OF WET SHELLED CORN DURING LAYER DRYING

Initial grain moisture content	Maximum allowable wet grain depth[a]
20%	16 ft
25%	6 ft
30%	3 ft

[a]   With heater controlled with humidistat set at 55%. These depths are average and may be increased slightly during the initial stages of filling, but should be reduced during the final stages when total grain depth approaches 16 ft.

From Bull. U-03-1, Farm Fans, Inc. With permission.

Table 10A

BATCH-IN-BIN DRYING SHELLED CORN WITH AIR HEATED TO 120°F

Fan heater	1 unit 24 in. - 5 hp 1,070,000 Btu/hr			1 unit 24 in. - 7½ hp 1,070,000 Btu/hr			2 units 24 in. - 7½ hp 1,070,000 Btu/hr				3 units 24 in. - 7½ hp 1,070,000 Btu/hr			4 units 24 in. - 7½ hp 1,070,000 Btu/hr
Bin diameter	18 ft	21 ft	24 ft	21 ft	24 ft	27 ft	27 ft	30 ft	33 ft	36 ft	33 ft	36 ft	42 ft	48 ft
Initial moisture content 20%	7.9 ft	6.9 ft	6.0 ft	7.6 ft	6.7 ft	5.9 ft	8.4 ft	7.7 ft	7.0 ft	6.4 ft	8.5 ft	7.7 ft	6.7 ft	6.7 ft
	1600	1850	2200	2100	2400	2750	3850	4400	4800	5200	5800	6300	7500	9700
25%	5.6 ft	3.9 ft	4.1 ft	5.4 ft	4.6 ft	4.0 ft	6.0 ft	5.4 ft	3.9 ft	4.3 ft	6.1 ft	5.5 ft	4.6 ft	4.6 ft
	1150	1350	1500	1500	1650	1850	2750	3000	3350	3450	4200	4450	5100	6700
30%	4.7 ft	4.0 ft	3.3 ft	4.4 ft	3.7 ft	3.3 ft	5.0 ft	4.4 ft	3.9 ft	3.3 ft	5.1 ft	4.5 ft	3.6 ft	3.7 ft
	970	1120	1220	1240	1360	1550	2300	2520	2700	2750	3500	3750	4000	5450

Note: Total bushels (by volume) dried in 20 hr, and equivalent depth. Where depth exceeds 4 ft, the grain must be dried in two separate batches to avoid excessive overdrying in lower portion. Bushels shown are wet bushels as determined by volume. Table is based on ultimate grain depth of 16 ft; rates must be decreased for greater depths. Assumed outside air — 45°F and 100% RH. Final average moisture contents — 13%. Recommended depth for grain sorghum is 75% of the depth shown above for shelled corn. If adjusted depth is over 3 ft, dry in two separate batches. The volume dried in a 20-hr period will be about 75% of the shelled corn volume.

From Bulletin U-03-1, Farm Fans, Inc. With permission.

Table 10B
BATCH-IN-BIN DRYING SHELLED CORN WITH AIR HEATED TO 120°F

Fan heater		1 unit 28 in. - 10 hp 1,530,000 Btu/hr			2 units 28 in. - 10 hp 1,530,000 Btu/hr				3 units 28 in. - 10 hp 1,530,000 Btu/hr			4 units 28 in. - 10 hp 1,530,000 Btu/hr	
Bin diameter		24 ft	27 ft	30 ft	30 ft	33 ft	36 ft	42 ft	36 ft	42 ft	48 ft	42 ft	48 ft
Initial	20%	8.4 ft	7.4 ft	6.6 ft	9.1 ft	8.6 ft	7.9 ft	6.7 ft	9.3 ft	8.0 ft	7.0 ft	9.3 ft	8.3 ft
moisture		3,000	3,400	3,800	5,200	5,900	6,500	7,500	7,600	8,900	10,000	10,500	12,200
content	25%	5.7 ft	5.1 ft	4.4 ft	6.5 ft	5.9 ft	5.4 ft	4.5 ft	6.6 ft	5.6 ft	4.9 ft	6.6 ft	5.7 ft
		2,070	2,350	2,500	3,650	4,050	4,350	5,000	5,350	6,200	7,000	7,200	8,300
	30%	4.4 ft	3.9 ft	3.3 ft	5.1 ft	4.5 ft	4.1 ft	3.5 ft	5.1 ft	4.4 ft	3.7 ft	5.1 ft	4.4 ft
		1,600	1,800	1,900	2,900	3,150	3,400	3,900	4,200	4,950	5,450	5,700	6,400

Note: Total bushels (by volume) dried in 20 hr, and equivalent depth. Where depth exceeds 4 ft, the grain must be dried in two separate batches to avoid excessive overdrying in lower portion. Bushels shown are wet bushels are determined by volume. Table is based on ultimate grain depth of 16 ft; rates must be decreased for greater depths. Assumed outside air — 45°F and 100% RH. Final average moisture contents — 13%. Recommended depth for grain sorghum is 75% of the depth shown above for shelled corn. If adjusted depth is over 3 ft, dry in two separate batches. The volume dried in a 20-hr period will be about 75% of the shelled corn volume.

From Bulletin U-03-1, Farm Fans, Inc. With permission.

Table 10C
BATCH-IN-BIN DRYING SHELLED CORN WITH AIR HEATED TO 120°F

Fan heater	1 unit 36 in. - 10 hp 3,000,000 Btu/hr					2 units 36 in. - 10 hp 3,000,000 Btu/hr			3 units 36 in. - 10 hp 3,000,000 Btu/hr	
Bin diameter	**24 ft**	**27 ft**	**30 ft**	**33 ft**	**36 ft**	**36 ft**	**42 ft**	**48 ft**	**42 ft**	**48 ft**
Initial moisture content 20%	8.1 ft	7.9 ft	7.5 ft	6.9 ft	6.6 ft	8.1 ft	7.5 ft	6.7 ft	8.2 ft	7.9 ft
	2,900	3,650	4,300	4,750	5,400	6,600	8,400	9,700	9,200	11,500
25%	6.3 ft	6.0 ft	5.5 ft	5.0 ft	4.5 ft	6.0 ft	5.4 ft	4.6 ft	6.2 ft	5.6 ft
	2,300	2,750	3,100	3,450	3,650	4,850	5,900	6,700	6,900	8,100
30%	5.2 ft	4.8 ft	4.3 ft	3.7 ft	3.3 ft	3.5 ft	4.3 ft	3.5 ft	4.9 ft	4.4 ft
	1,900	2,250	2,450	2,600	2,750	4,000	4,800	5,100	5,500	6,400

Note: Total bushels (by volume) dried in 20 hr, and equivalent depth. Where depth exceeds 4 ft, the grain must be dried in two separate batches to avoid excessive overdrying in lower portion. Bushels shown are wet bushels as determined by volume. Table is based on ultimate grain depth of 16 ft; rates must be decreased for greater depths. Assumed outside air — 45°F and 100% RH. Final average moisture contents — 13%. Recommended depth for grain sorghum is 75% of the depth shown above for shelled corn. If adjusted depth is over 3 ft, dry in two separate batches. The volume dried in a 20-hr period will be about 75% of the shelled corn volume.

From Bulletin U-03-1, Farm Fans, Inc. With permission.

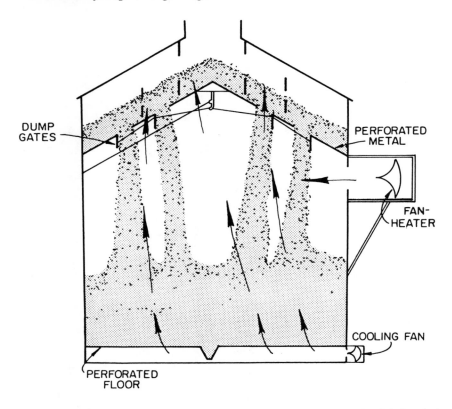

FIGURE 10.   A batch-in-bin dryer suspended in the upper portion of the bin. Dried grain is transferred through dump gates to the lower portion for cooling. (Redrawn from brochure of Stormor, Inc. With permission.)

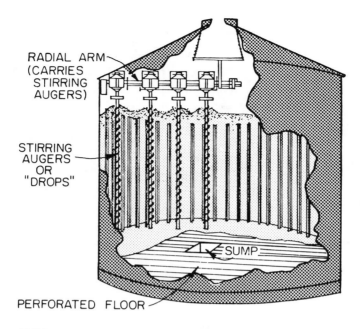

FIGURE 11.   Stirring device. The vertical augers mix the grain as they rotate about their own axes and are carried by the radial arm as it pivots about the bin center. (Redrawn from brochure of Sukup Manufacturing Company. With permission.)

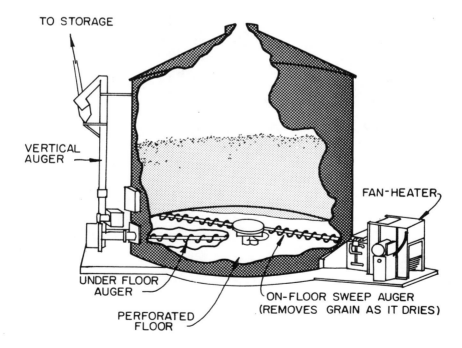

FIGURE 12.   This counterflow dryer discharges to a sump at the bin center. The warm grain is taken to storage via an underfloor auger and external augers. (Redrawn from brochure of Sukup Manufacturing Company. With permission.)

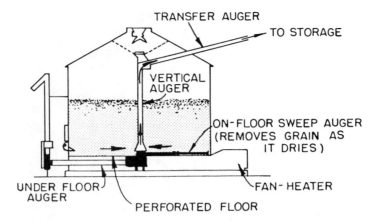

FIGURE 13.   A counterflow dryer that discharges via an auger at the bin center that empties into a transfer auger. (Redrawn from brochure of Nebraska Engineering Company. With permission.)

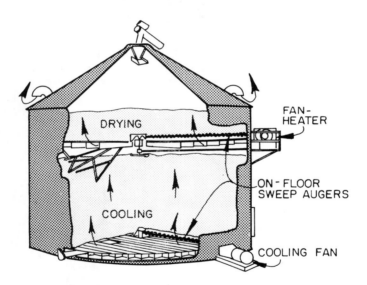

FIGURE 14.   A counterflow dryer with the drying floor and sweep auger in the upper portion of the bin. (Redrawn from brochure of Mathews Company. With permission.)

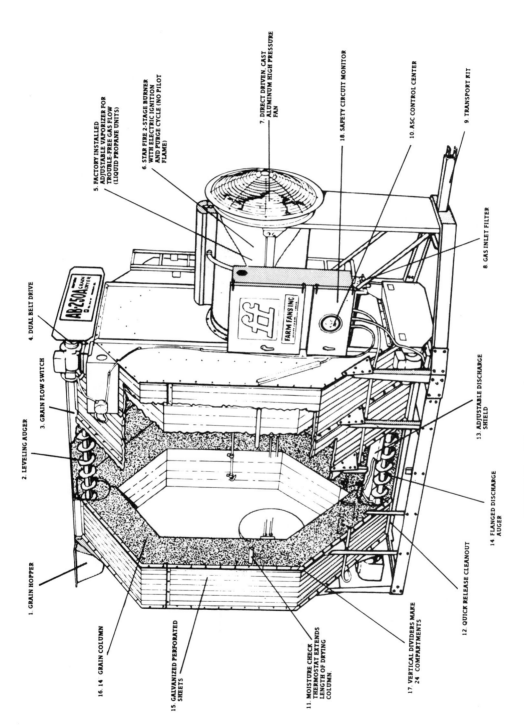

FIGURE 15.   Column batch dryer. (Courtesy of Farm Fans, Inc.)

Table 11
TYPICAL DATA FOR FARM-SIZE COLUMN BATCH DRYERS

	8	12	14	20
Grain column length (ft)	8	12	14	20
Total holding capacity (bu)	120	180	350	500
Less transport				
Length	13 ft 3 in.	18 ft 6 in.	21 ft 0 in.	27ft 0 in.
Width	6 ft 0 in.	6 ft 0 in.	7 ft 11 $^1/_2$ in.	7 ft 11 $^1/_2$ in.
Height	8 ft 8 in.	8 ft 8 in.	12 ft 8 in.	12 ft 8 in.
With transport				
Length	16 ft 2 in.	20 ft 2 in.	23 ft 1 $^1/_2$ in.	29 ft 1 $^1/_2$ in.
Width	7 ft 8 in.	7 ft 8 in.	7 ft 11 $^1/_2$ in.	7 ft 11 $^1/_2$ in.
Height	10 ft 0 in.	10 ft 0 in.	13 ft 4 $^1/_2$ in.	13 ft 4 $^1/_2$ in.
Fan horsepower	10—13	10—16	(2) 10—13	(2) 10—16
Fan diameter (in.)	28	36	28	36
Total heater capacity (Btu/hr)	3,000,000	4,500,000	6,000,000	9,200,000
Top auger (hp)	1	2	5	7 $^1/_2$
Top auger capacity (Btu/hr)	1,500	1,500	2,090	2,500
Bottom auger (hp)	1	2	5	7 $^1/_2$
Bottom auger capacity (Btu/hr)	900	1,100	2,090	2,500
Max. running amps, 1 phase, 230 V (with load and unload conv.)	90	110	160	200
Max. running amps, 3 phase, 220 V (with load and unload conv.)	50	70	100	125
Drying capacity,[a] wet bu shelled corn				
Dry and cool, 25—15%	115 bu/hr	165 bu/hr	270 bu/hr	390 bu/hr
Dry and cool, 20—15%	155 bu/hr	230 bu/hr	360 bu/hr	515 bu/hr
Full heat, 25—15%	150 bu/hr	225 bu/hr	370 bu/hr	530 bu/hr
Full heat, 25—15%	210 bu/hr	310 bu/hr	520 bu/hr	750 bu/hr

[a]   Excluding load and unload time.

Data from Farm Fans, Inc.

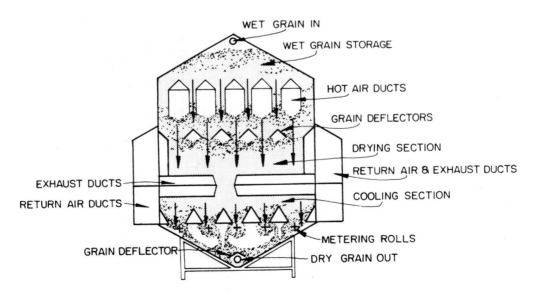

FIGURE 16.   Concurrent-flow dryer. Path of grain flow shown by arrows. (Redrawn from brochure of M and W Gear. With permission.)

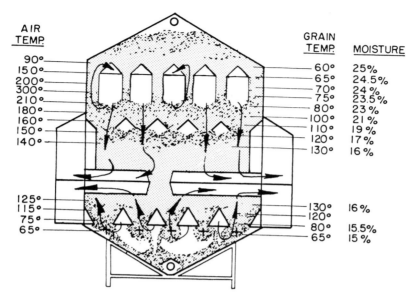

FIGURE 17.   Concurrent-flow dryer. Heated air enters grain from upper ducts and cooling air from the lower ducts. (Redrawn from brochure of M and W Gear. With permission.)

REFERENCES

1. **Brooker, D. B., Bakker-Arkema, F. W., and Hall, C. W.,** *Drying Cereal Grains,* AVI Publishing, Westport, Conn., 1974, 12.
2. **Hall, C. W.,** *Drying Farm Crops,* AVI Publishing, Westport, Conn., 1957, 10.
3. **Hall, D. W.,** Handling and Storing of Food Grains in Tropical and Subtropical Areas, FAO Development Paper 90, Food and Agriculture Organization, Rome,
4. **Matz, S. A.,** *Cereal Science,* AVI Publishing, Westport, Conn., 1969.
5. **Stewart, B. R. and Britton, M. G.,** Design of farm grain storage, in *Grain Storage: Part of a System,* Sinha, R. N. and Muir, W. E., Eds., AVI Publishing, Westport, Conn., 1973, chap. 12.
6. **Shedd, C. K.,** Resistance of grain and seeds to air flow, *Agric. Eng.,* 34, 616, 1953.
7. **Haque, E., Foster, G. H., Chung, D. S., and Lai, F. S.,** Static Pressure Drop Across a Corn Bed Mixed with Fines, ASAE Paper No. 76-3528, American Society of Agricultural Engineers, St. Joseph, Mich., 1976.
8. **Haque, E., Ahmed, Y. N., and Deyoe, C. N.,** Static Pressure Drop in a Fixed Bed of Grain as Affected by Grain Moisture Content, ASAE Paper No. MCR-81-301, American Society of Agricultural Engineers, St. Joseph, Mich., 1981.
9. **Stephens, L. E. and Foster, G. H.,** Grain properties as affected by mechanical grain spreaders, *Trans. ASAE,* 19, 354, 1976.
10. **Stephens, L. E. and Foster, G. H.,** Bulk properties of wheat and grain sorghum as affected by a mechanical spreader, *Trans. ASAE,* 21, 1217, 1978.
11. **Henderson, S. M.,** A basic concept of equilibrium moisture, *Agric. Eng.,* 33, 29, 1952.
12. **Pfost, H. B., Maurer, S. G., Chung, D. S., and Milliken, G. A.,** Summarizing and reporting equilibrium moisture data for grains, ASAE paper no. 76-3520, American Society of Agricultural Engineers, St. Joseph, Mich., 1976.
13. **Pierce, R. O. and Thompson, T. L.,** Solar grain drying in the north central region — simulation results, *Trans. ASAE,* 22, 182, 1979.

CROP DRYING USING SOLAR ENERGY

J. W. Goodrum and J. L. Butler

INTRODUCTION

Traditionally many crops — notably hay — are dried by timed exposure to solar heat in the open field. Tests have shown that solar heat may also be successfully applied to dry harvested crops if conventional crop dryer designs are modified for solar-derived heat inputs. In terms of energy, approximately 10^{14} kW of solar energy continuously reaches the surface of the earth. This great quantity of power, spread over half the surface of the earth, is available for drying purposes during the daylight hours. If heat is stored during the day, drying may continue at night. Over much of the U.S., the solar intensity averages about 250 Btu/ft²/hr. Therefore a surface area of 400 ft² is capable of collecting energy equivalent to the output of a 50,000-Btu/hr propane-fueled dryer (assuming 50% collector efficiency) for 5 or 6 hr/day.

Careful management of the drying process is essential if solar drying is to be successful. Good energy conservation measures, particularly in the use of insulation, must be incorporated into the drying system design. Since in general solar drying will be slower than conventional crop dryers, the minimal safe drying rates for each crop should be known. This permits the dryer operator to match the throughput rate for a specific crop to the capabilities of the solar dryer.

In order to utilize the available solar energy at a site, a collector system must be designed to (1) absorb the broad spectrum of wavelengths making up solar energy and to (2) transfer this heat energy to the product being dried. Much of solar energy is composed of visible wavelengths, but there are invisible UV (ca. 10%) and IR components (ca. 30%) which if collected greatly increase the output of a given collector area. One of the flat plate-type collector designs may be used for crop drying. The flat plate type is generally simplest to construct and the lowest in cost per square foot of energy collection area. These collectors can produce maximum temperatures of about 100°C, which is adequate for many types of crop drying. The heat in a collector is transferred to a moving stream of air (or water), which is transported to the crop.

The use of solar drying will be illustrated here by describing a collector that faces south and has a slope (compared to level ground) that gives maximum energy collection per square foot during the summer.

The quantity of solar energy available at a given geographical site will depend on:

1. The general location (see Figure 1)
2. The local climate (cloudiness, rain)
3. Physical features that block the sun from the collector site for some part of the day

These topics are discussed in the following sections.

PREDICTING HOURS OF SUN EXPOSURE ON A COLLECTOR SURFACE

Solar Nomographs

Solar nomographs have been developed for convenient calculation of the daily exposure hours of a given collector to solar radiation. If these exposure hours are multiplied by the corresponding solar intensity during those hours, one obtains the total energy on a daily basis. If one stands on a proposed collector site and faces south, the sun will move in a

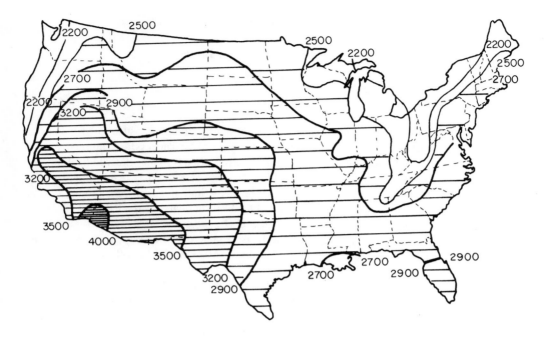

FIGURE 1.   Normal annual hours of sunshine.

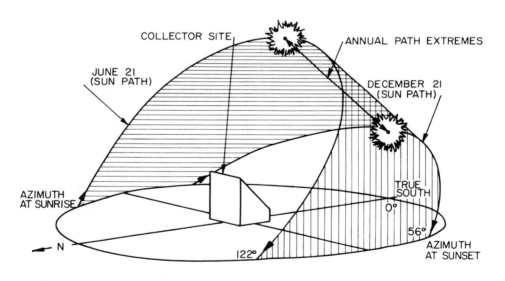

FIGURE 2.   Annual range of sun paths for Indianapolis.

circular arc across the sky in the course of a day (see Figure 2). This path of the sun may be plotted on a chart by using solar azimuth and solar altitude coordinates measured at hourly intervals during the day. The result is a solar nomograph.

The solar nomograph is a convenient means of showing the pattern of daily and annual movements of the sun above a given site. It may be regarded as the collector's view of the sun for each hour of the day and for each month of the year. The solar azimuth is the direction or angle in degrees that the sun is east or west of south. The solar altitude is the direction or angle (in degrees) of the sun in relation to the horizontal. Note that at noontime, the position of the sun slowly increases in altitude from December to June (Figure 2).

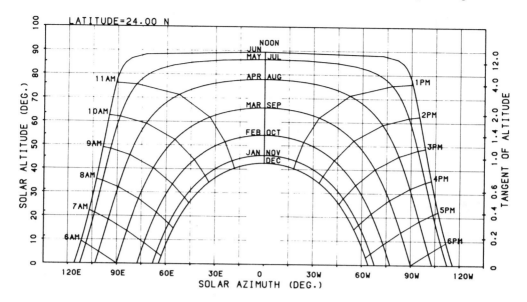

FIGURE 3.   Solar nomograph for 24°N latitude.

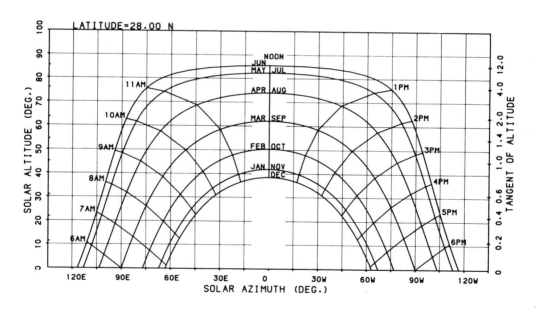

FIGURE 4.   Solar nomograph for 28°N latitude.

Thereafter, the noon altitude slowly decreases until the annual cyclic movement of the sun is completed in December.

The annual pattern of solar paths is a function of latitude. The solar nomographs for 4° latitude intervals for the U.S. and Canada given in Figures 3 to 13 are sufficiently accurate for solar crop dryer design.

Nomographs can help in determining the following:

1.   The number of hours the sun can be effectively used on a given day
2.   The best angle to tilt a south-facing solar collector at a selected time of the year

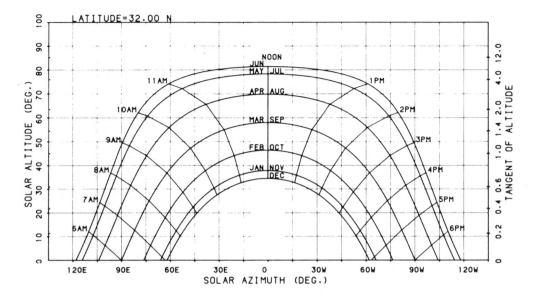

FIGURE 5. Solar nomograph for 32°N latitude.

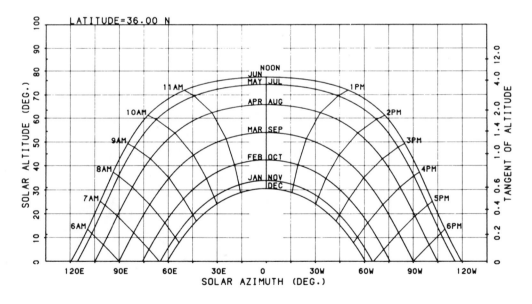

FIGURE 6. Solar nomograph for 36°N latitude.

3. How far and in which direction a silo, a building, or a tree casts a shadow at various times of the year and at different times of the day (a solar collector will not be effective if in the shade when the sun's energy is needed)

The nomographs are designed so that they are viewed facing south. The sun follows the curved line, rising to the left (east) and reaching a high point at solar noon (directly south) and setting to the right (west). On December 21, the sun would follow the lowest curve, while on June 21 it would follow the highest curve. The diagonal lines cutting across the curved lines are for the time of day from 6:00 a.m. on the left to 6:00 p.m. on the right.

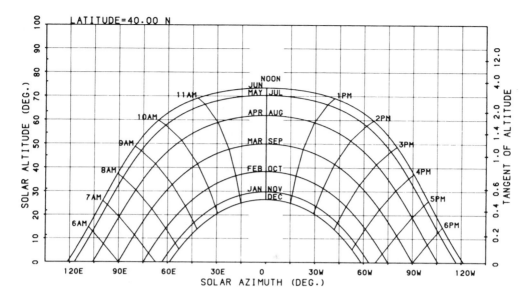

FIGURE 7.   Solar nomograph for 40°N latitude.

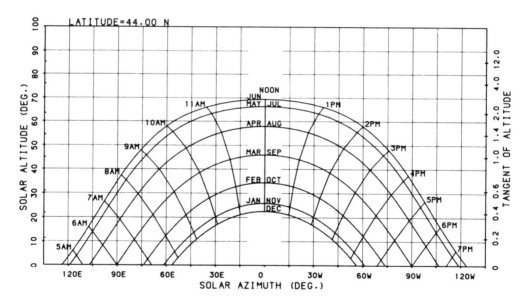

FIGURE 8.   Solar nomograph for 44°N latitude.

Solar altitude in degrees is shown on the vertical scale to the left. The tangent of the solar altitudes is tabulated along the vertical scale on the right. Along the bottom is the solar azimuth (the direction of the sun in degrees with respect to south). In the morning the angle would be to the left, or east, of south and in the afternoon to the right, or west, of south.

All nomographs are plotted for the 21st of the month. Other dates may be estimated by moving the appropriate distance between the curves. The hourly time shown is solar time, which may differ from local time. To convert to local time, add four times the difference between local longitude and the longitude on which local time is based. For example, Tifton, Ga. is at longitude 83.5°, located in the eastern time zone based at 75°. Local time is solar

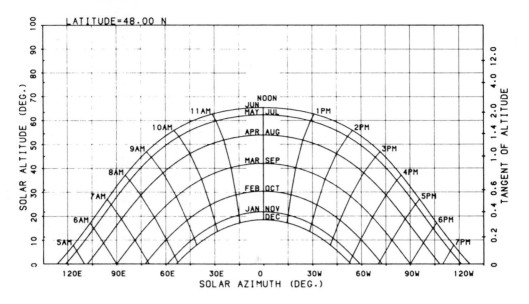

FIGURE 9.   Solar nomograph for 48°N latitude.

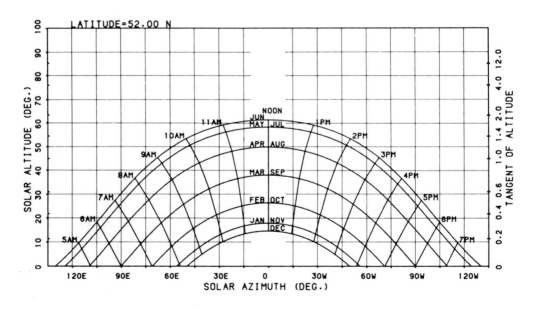

FIGURE 10.   Solar nomograph for 52°N latitude.

time plus 4 (83.5 − 75) or 34 min ahead of solar time. At noon, solar time, a clock would read 12:34 p.m. Further discussion of the nomograph may be found in References 2 and 3.

Using the Solar Nomographs

To use the nomographs (Figures 3 to 13), first determine the approximate latitude of the desired location. This can be done by referring to a map which shows the latitudes and contains the specific location. After determining the nearest latitude, locate the nomograph for that latitude. From the nomograph, the needed numbers can be obtained to calculate the

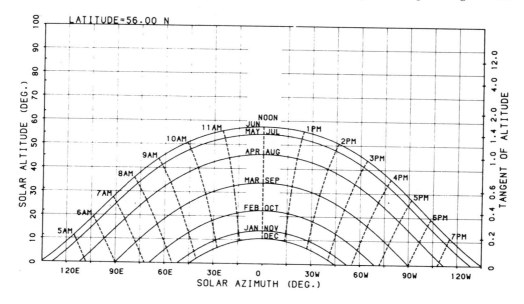

FIGURE 11.   Solar nomograph for 56°N latitude.

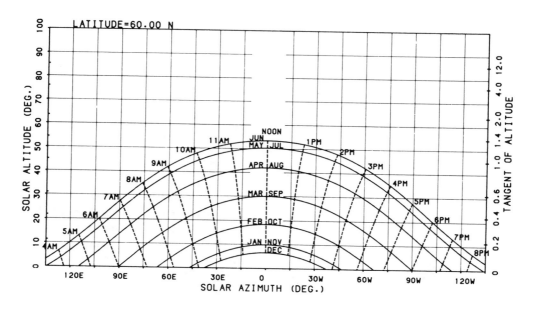

FIGURE 12.   Solar nomograph for 60°N latitude.

information desired. The following examples use the nomograph for 40°N latitude (Figure 7).

The approximate time of sunrise and its azimuth can be read along the horizontal base of the nomograph. On December 21, the sun will rise shortly after 7:00 a.m. at an azimuth of about 58°. On June 21 it will be near 120° at about 5:00 a.m.

Shading of Solar Collectors

Obstructions in front of a solar collector (trees, buildings, hills, etc.) can seriously reduce the amount of energy collected. A "shadow map" may be constructed on a solar nomograph

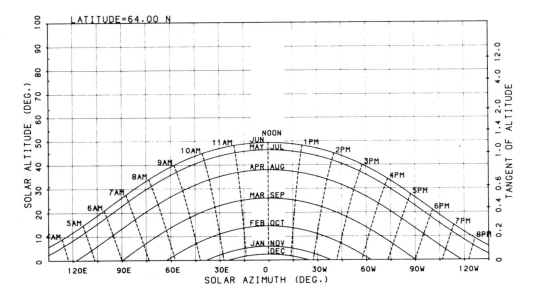

FIGURE 13.   Solar nomograph for 64°N latitude.

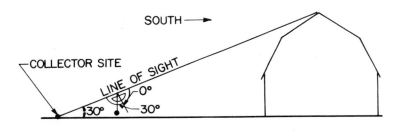

FIGURE 14.   Determining the angle of a building roof line.

so that the solar energy loss due to an obstruction may be calculated. If, for example, it were decided to locate a collector north of a long high building, it may be near enough to the building to be shaded at times. By standing in the proposed location and determining the angle between a line extending to the top of the building and the horizontal, the needed information will be available when the area will be shaded. Assume that an angle of 30° was determined by the use of a protractor and weighted string as shown in Figure 14. By following the 30° altitude line across the nomograph, it can be seen that from November 21 through January 21, the area will always be shaded because curved month lines for December and January to November are entirely below the 30° line. On February 21 and October 21, the sun will reach the area just before 10:00 a.m. and shine on the area until shortly after 2:00 p.m. The February to October line is above the 30° horizontal line between those two times. This approach can also be useful in determining when a row of trees shade a given area.

In general, the outline drawing or silhouette of a tree or other object may be measured (as pairs of azimuth and altitude coordinates around the silhouette) and plotted directly onto the nomograph. One may then read off the hours when the object shades the collector. To compute the daily energy available, multiply the actual hours of solar exposure by the intensity at each hour and then sum these products for each day (see the following section on Intensity).

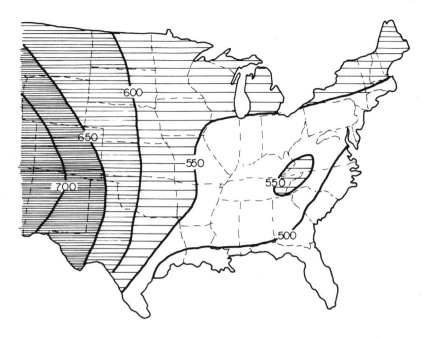

FIGURE 15. Example of solar intensity. Normal daily solar radiation of Langleys on horizontal surface received on June 21. (From Daniels, F., *Direct Use of the Sun's Energy,* Yale University Press, New Haven, Conn., 1964. With permission.)

PREDICTING INTENSITY AND TOTAL ENERGY AVAILABLE AT A COLLECTOR SURFACE

Solar radiation striking a collector has a beam fraction — traveling directly from the sun — and a scattered or diffuse radiation fraction. The latter strikes the collector from all directions. This means that collector orientation has little impact on the amount of diffuse radiation intercepted by a surface. Therefore, collectors are primarily oriented for beam collection. The clearness index, Kt, is defined[4] as

$$Kt = \frac{\overline{H}}{H_o}$$

where $\overline{H}$ = daily insulation at site and H_o = daily insulation above earth's atmosphere. This index is related to the average cloud cover of a site.

In order to compute either the optimal orientation for a flat plate collector or the total energy from a partially shaded collector, knowledge of daily and annual solar intensity variations is necessary. The intensity ($kJ/m^2/hr$ or $Btu/ft^2/hr$) is a function of time of day and time of year at a given site (see Figure 15).

Solar Incidence Angle vs. Solar Intensity

The solar incidence angle i is defined as the angle between (1) the normal to the collector surface and (2) a line from the collector to the sun. The incidence angle changes as the sun moves over a fixed collector surface during the day. It also changes slowly, for a given time of day, with the month of the year. The hourly intensity I of solar radiation is related to incidence angle i by

$$I = I_n \cos i \qquad (1)$$

where I_n = solar intensity normal to collector surface.

One may calculate the incidence angle for a given time, site, and collector orientation. For a south-facing collector:

$$COS \ i = SIN(\phi - S) \ SIN \ \delta + COS(\phi - S) \ COS \ \delta \ COS \ W \qquad (2)$$

where ϕ = the site latitude (degrees), S = the slope of collector surface (degrees), δ = the declination (see Table 1), and W = the solar hour angle = hours from solar noon × 15°.[5]

It may be seen from Equation 1 above that the maximum solar beam intensity corresponds to a zero solar incidence angle (i.e., the sun direction is normal to the collector surface). The collector ideally should be oriented so that the beam radiation is normal to the collector surface. Since large flat plate collectors are normally fixed in position, a single collector orientation must be selected for maximum thermal yield during the crop drying months. Solar intensity is greatest at solar noon and decreases toward zero in the morning and evening hours (see Table 2). A collector orientation for zero incidence angle near solar noon provides the greatest daily radiation input per square foot of collector area. From the solar nomographs it may be seen that the solar altitude at noon (on June 21, for example) varies with site latitude. During summer in the southern half of the U.S., the noon solar altitude ranges between 85° (at 28°N latitude) and 73° (at 40° N latitude). For the northern half of the U.S., solar altitude angles range between 73° (at 40°N latitude) and 65° (at 48°N latitude). From analysis of the above geometrical relationships and also from site tests, a general rule of thumb has emerged. For an s-facing collector, the optimal collector slope for summer months is equal to the site latitude plus 10°.[3]

Meteorological Data

The National Weather Service and other governmental agencies have published data on solar energy at a number of sites, particularly the major U.S. cities.[6-8] The average daily intensity for each month of the year is normally given, along with the clearness index, Kt.

Most of the site data is for a horizontal surface and must be corrected for analyzing performance of tilted collectors. Efforts are underway to provide data for tilted surfaces, and some data are currently available.[4]

The energy density for horizontal surfaces may be converted to sloped surface intensity, according to Anderson, if the diffuse radiation component is known for a site:[2]

$$H_T = \overline{R_b}\overline{H_b} + R_s\overline{H_d} + 0.2(1 - R_s) \overline{H}$$

where there is no snow ground cover, $\overline{H_T}$ = daily total solar energy or flux, $\overline{R_b}$ = geometric factor for beam or direct solar component, $\overline{H_b}$ = horizontal beam fraction of solar flux, $\overline{R_s}$ = geometric factor for diffuse solar flux, $\overline{H_d}$ = horizontal diffuse or scattered type solar flux, and $\overline{H}$ = daily total solar flux on horizontal surface.

Models Which Predict the Daily Pattern of Solar Intensity

Since there is a limited availability of site-specific hourly intensity data, it is useful to have models that approximately predict the hourly intensity for any collector site. From the hourly projections, one may evaluate the effect of an object that partially shades a collector. Currently, elaborate computer models have an accuracy as high as 95%[20] but require detailed weather information for the site.

The empirical model is one of several simple intensity models which may be used by the designer of a solar crop dryer for making quick design estimates. This model[9,10] approximates a body of experimental data with the equation:

Table 1
SOLAR DECLINATION

Date	Declination Deg	Min	Date	Declination Deg	Min
Jan. 1	− 23	4	Feb. 1	− 17	19
5	22	42	5	16	10
9	22	13	9	14	55
13	21	37	13	13	37
17	20	54	17	12	15
21	20	5	21	10	50
25	19	9	25	9	23
29	− 18	8			
Mar. 1	− 7	53	Apr. 1	+ 4	14
5	6	21	5	5	46
9	4	48	9	7	17
13	3	14	13	8	46
17	1	39	17	10	12
21	− 0	5	21	11	35
25	+ 1	30	25	12	56
29	3	4	29	14	13
May 1	+ 14	50	June 1	+ 21	57
5	16	2	5	22	28
9	17	9	9	22	52
13	18	11	13	23	10
17	19	9	17	23	22
21	20	2	21	23	27
25	20	49	25	23	25
29	21	30	29	23	17
July 1	+ 23	10	Aug. 1	+ 18	14
5	22	52	5	17	12
9	22	28	9	16	6
13	21	57	13	14	55
17	21	21	17	13	41
21	20	38	21	12	23
25	19	50	25	11	2
29	18	57	29	9	39
Sep. 1	+ 8	35	Oct. 1	− 2	53
5	7	7	5	4	26
9	5	37	9	5	58
13	4	6	13	7	29
17	2	34	17	8	58
21	+ 1	1	21	10	25
25	− 0	32	25	11	50
29	2	6	29	13	12
Nov. 1	− 14	11	Dec. 1	− 21	41
5	15	27	5	22	16
9	16	38	9	22	45
13	17	45	13	23	6
17	18	48	17	23	20
21	19	45	21	23	26
25	20	36	25	− 23	25
29	21	21	29	23	17

From The American Ephemeris and Nautical Almanac published each
year by the U.S. Government Printing Office, Washington, D.C.

Table 2

**THE AVERAGE HOURLY BEAM, DIFFUSE,
AND GLOBAL SOLAR INTENSITY FOR A
TYPICAL MARCH DAY IN KANSAS CITY,
CALCULATED BY THE EMPIRICAL METHOD**

Time	Total hourly intensity (MJ/m² · hr)	Diffuse hourly intensity (MJ/ m² · hr)	Beam hourly intensity (MJ/m² · hr)
7—8 a.m.	0.53	0.28	0.25
8—9	0.99	0.46	0.53
9—10	1.42	0.61	0.81
10—11	1.76	0.71	1.05
11—12	1.95	0.77	1.18

$$\gamma = \bar{I}/\bar{H} = (a + b\,COS\ w)\,r_d \tag{3}$$

where a = 0.409 + 0.5016 SIN $(w_s - 60)$, b = 0.6609 − 0.4767 SIN $(w_s - 60)$, r_d = $\pi/24$ (COS w − COS w_s)/(SIN w_s − $\dfrac{\pi\ w_s}{180}$ COS w_s), w = solar hour angle (degrees), and ws = sunrise hour angle (degrees) (see section on Solar Nomographs, paragraph 5). Note that $\bar{I}/\bar{H}$ is the hourly fraction of total daily energy. Equation 3 is used to prepare Figure 16 for an average day in the month of March.

FLAT PLATE COLLECTOR DESIGN

There are two general designs of solar collectors: the flat plate and the focusing design. To date, only the flat plate appears to have strong potential for economic use in crop drying. Focusing collectors concentrate solar energy so that higher temperatures are available. However, focusing units have greater structural complexity and greater cost power energy unit delivered. Flat plate collectors may be classified as air-type or water-type, depending on the heat transfer medium used to move the solar energy from collector to point of application. Air has many advantages for crop drying and the air-type flat plate collector has been predominately used in solar dryers.

Basic Collector Elements

The flat plate collector has four basic elements: the absorber, the cover plate, thermal insulation, and a heat transfer medium. These are shown schematically in Figure 17. The absorber converts solar radiation to thermal energy (IR radiation). A copper or aluminum sheet with flat black surface coating is often used. The black coating converts the solar radiation to IR energy, and the metal sheet conducts the heat to the transfer medium. In the case of air-type collectors, metal fins have been used to increase the heat transfer surface in contact with air.

The cover plate transmits some 90% of solar radiation striking a collector while its low IR transmission reduces the outflow of thermal radiation from the absorber (see Figure 17). In Figure 18, the approximate wavelength spectrum of solar radiation (0.3 to 2.0 μm) and thermal radiation (3.0 to 100 μm) emitted from a black absorber plate are shown. The transmittance of glass is shown as a dotted line on this figure. The high transmittance of glass across the solar spectrum and its low thermal radiation transmittance make glass an excellent collector cover material. A number of commercial polymeric materials are also

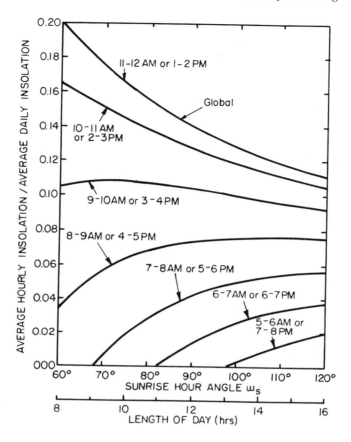

FIGURE 16.   Hourly intensity via the empirical model. (From Daniels, F., *Direct Use of the Sun's Energy*, Yale University Press, New Haven, Conn., 1964. With permission.)

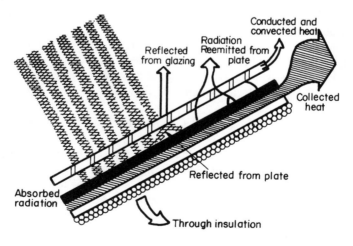

FIGURE 17.   Flat black absorber plate with glazing cover. (Courtesy of Copper Development Association, Inc., Stamford, Conn.)

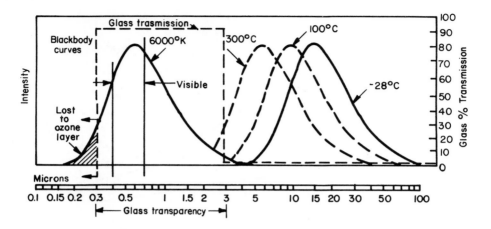

FIGURE 18.   Solar radiation compared with reradiation from thermal sources.

very effective cover materials. These include, for example, Tedlar® (Dupont Co.) and Sun-Lite® (Kalwal Co.).

The cover may be deleted in the interest of lowering costs and simplifying collector design. These savings must be carefully balanced against the reduced performance of the collector.

The product $(\tau\alpha)$, where τ is the transmittance of the cover and α is the absorbance of the absorber plate, provides a coefficient for comparing the efficiencies of cover/absorber pairs. This coefficient is valid where α is approximately unity.

Thermal insulation is essential in an efficient, cost-effective flat plate collector. The back and sides of the collector are capable of serious thermal losses by radiation and convection to the environment if not insulated. Construction grades of glass fiber insulation may be suitable for the maximum temperatures encountered in these collectors.

A heat transfer medium is usually necessary to transport the thermal energy in a collector absorber plate to the material to be dried. As stated earlier, air has many advantages in a solar crop drying system. However, air has relatively low heat transfer capacity, and efficient air-type collectors must have a large heat transfer surface between absorber and air. Also, the collector dimensions must be sized for large air flows. The heated air is forced through the porous bed material to be dried.

Although a water-type flat plate collector may be physically more compact for a given solar energy output, a liquid/air heat exchanger is necessary for crop drying. The additional complexities of circulating two fluids, freezing, and liquid leakage generally offset any advantages of the water-type collector for crop drying purposes.

EXAMPLES OF SOLAR CROP DRYING DESIGNS

The designs of solar crop drying systems have been predominantly of the flat plate type, using air for heat transfer. Somewhat different designs have been developed for drying of specific crops. Peanuts may be successfully dried at relatively low temperatures.[11-14] Grains require a collector system that can deliver higher temperature air for drying.[15-17]

Economics of solar drying are based on fossil fuel savings; the capital costs of a solar system must be offest by the fuel savings within a time acceptable to the user. The typical ''payback condition'' — time to recoup the extra capital investment of a solar system — is 1 to 10 years.[14]

Economic analysis often shows that a hybrid system of solar plus fossil fuel sources is most desirable from a financial return viewpoint. Systems with a projected payback condition on the order of 10 years or more are highly dependent on future fossil fuel prices and interest rates.

Bledsoe et al. have conducted tests of air-type collectors incorporated in the southern wall and south-facing roof sections of a utility building.[16-18] The design was developed at the University of Tennessee. Assuming use of propane gas (with 80% usage on hay drying and 20% on corn drying), the simple payback time for the added cost of the solar collector is about 2 years. The solar-dried hay was baled at approximately 30% moisture content and was worth from $29 to 48/ton more than conventionally harvested hay, which was baled at 20% moisture content. The increase in value was primarily due to the retention of the leaves when baled at the higher moisture content.

The Figure shows the incorporation of the Tennessee plan with an optional machine shed. This addition includes a rock storage area, which may be used to store energy. The stored energy may be used for nighttime drying or for supplementing the solar energy received on partially cloudy days.

The unit can dry 32 large round bales simultaneously or can provide heat to dry corn in storage. The ring-grates may be reversed when hay is not being dried to provide a level surface. The fans are turned 180° from the hay drying position when drying corn so that the air is directed through the grain instead of through the hay.

Troeger and Butler conducted extensive tests of solar peanut drying.[11-13] Their tests included parallel operation of air- and water-type flat plate collectors for comparative evaluation. They found that solar heat could readily supply the energy required for drying peanuts.

Lambert[14] has conducted an evaluation of ten on-farm systems in Florida. Results include cost per square foot of collector and time to reach the payback condition.

REFERENCES

1. **Daniels, F.,** *Direct Use of the Sun's Energy,* Yale University Press, New Haven, Conn., 1964.
2. **Anderson, E. E.,** *Fundamentals of Solar Energy Conversion,* Addison-Wesley, Reading, Maine, 1983.
3. **Plante, R. H.,** *Solar Domestic Hot Water,* John Wiley & Sons, New York, 1983.
4. **Lunde, P. J.,** *Solar Thermal Engineering,* John Wiley & Sons, New York, 1980.
5. **Kreith, F. and Kreider, J. F.,** *Principles of Solar Engineering,* McGraw-Hill, New York, 1978.
6. Climate Atlas of the United States, U.S. Department of Commerce, Washington, D.C., 1983.
7. Comparative Climatic Data for the United States, National Climatic Weather Center, Ashville, N.C., 1984.
8. **Knapp, Stoffel, and Whitaker,** Insolation Data Manual, S.E.R.I., U.S. Government Printing Office, Washington, D.C., 1980.
9. **Liu, B. Y. H. and Jordan, R. C.,** *Sol. Energy,* 22, 81, 1979.
10. **Collares-Pereira, M. and Rabl, A.,** The average distribution of solar radiation, *Sol. Energy,* 22, 451, 1979.
11. **Troeger, J. M.,** Proc. 3rd Solar and Biomass Conf., U.S. Department of Agriculture, Atlanta, 1983, 1.
12. **Troeger, J. M.,** Design of a solar peanut drying system, *Trans. ASAE,* 23(5), 763, 1982.
13. **Troeger, J. M. and Butler, J. L.,** Peanut drying with solar energy, *Trans. ASAE,* 23(5), 1250, 1980.
14. **Lambert, A. J.,** Proc. 3rd Solar and Biomass Conf., U.S. Department of Agriculture, Atlanta, 1983, 13.
15. **Foster, G. H.,** Proc. 3rd Solar and Biomass Conf., U.S. Department of Agriculture, Atlanta, April 1983, 242.
16. **Bledsoe, et al.,** Proc. 3rd Solar and Biomass Conf., U.S. Department of Agriculture, Atlanta, 1983, 5.
17. **McGraw, B. A., et al.,** Design and Testing of a Solar Collector Unit, Proc. ASME 6th Annu. Solar Conf., Orlando, April 1983.
18. **Bledsoe, et al.,** A Multi-Use Dryer for Large Round Bales Using Solar Heated Air, ASAE Paper No. 81-4553, American Society of Agricultural Engineers, St. Joseph, Mich., December 1981.
19. **Atwater, M. A. and Ball, J. T.,** A numerical radiation model based on meteorological observations, *Sol. Energy,* 21, 163, 1978.

CHEMICAL CONTROL OF STORED GRAIN INSECT PESTS

Phillip K. Harein

INTRODUCTION

Food supplies for millions of people depend almost entirely on the harvest each year. Efforts to overcome this situation usually involve consultations and recommendations by the major grain-producing countries to maintain minimum reserves of food grains. Food surpluses may occur in future years, but are likely to be temporary, undependable, and scattered.

Providing an adequate food bank is possible in two ways: production and conservation. We need not rely solely on production, contrary to the philosophies of many agricultural leaders, especially in regard to grain and cereal products. Conserving the quality of harvested crops will provide significant increases in available food resources. At least 10% of harvested foods are destroyed by insect pests in storage. Losses of 30% are too common throughout large areas of the world.[1] In fact, losses to stored grain from insects are often equal to that sustained by cereal grains in the field. Replanting or the partial recovery of damaged plants provides a means of regaining field losses. No such avenue exists following damage to stored grain.

The economic losses of stored grain in temperate regions of the world are usually lower than in the tropics or subtropics. Losses in quantity are of primary concern in tropical countries where insects are offered a near optimum environment most of the year. Joffe[2] illustrates the additional problems of protecting stored grain from insects under tropical conditions. Loss in quality takes priority in temperate areas.

Stored-grain insects have little concern for boundaries in regard to their distribution or proliferation, unless certain areas provide environmental conditions contrary to their needs. Their needs usually center on temperature, moisture, food, and space. The opportunity for insects to locate such environments depends on many variables including climate, culture, and technology. As pointed out by Freeman,[3] no country, whatever its geographical situation, can afford to dispense with efforts to manage stored-grain insects.

Unfortunately, the presence of insects in stored grain and cereal products is often accepted. Many agricultural producers and grain handlers believe that grain in storage "spontaneously generates" insect life and that infested grain is satisfactorily and desirably "matured" only when infested.

As alluded to earlier, outbreaks of storage insect pests depend primarily on an adequate supply of food in an area providing both optimum temperature and moisture.[4] Sinha[5] had previously reviewed the distribution and frequency of occurrence of stored-grain insects within various regions of Canada. He substantiated that regional differences in climate and in agricultural and storage practices are important factors in determining the success of specific insect species.

Pesticides remain essential to maintain the abundance and quality of stored grain in the world. In fact, insecticides are generally the most effective management method and, in many instances, provide the only feasible method available to reduce or hold pest insect populations to acceptable levels. Admittedly, insecticides should be used only when no other more effective or safer method is available and then only when their use will not produce undue hazards to nontarget organisms or the environment.

A continual review of all pesticides and their uses in relation to human and environmental safety is mandatory. Such reviews should be carried out objectively by qualified scientists, regulatory officials, and other informed individuals. When new reliable evidence which requires a change in pesticide usage becomes available, appropriate action should be taken.

The increased limitations of most insecticide applications have prompted investigation of integrated insect control systems. In essence, efforts must concentrate on developing compatible systems of insect management using various chemical and nonchemical methods in proper sequence and timing to achieve the desired results. This growing philosophy of using integrated methods to prevent or control stored-grain insects applies to small farm storages as well as to large terminal facilities.

It seems ironic that pesticides have fallen into public distrust, although we are still dependent on them. Condemning chemical controls is unrealistic unless consumers are either prepared to accept foods with increased concentrations of stored-grain insects or have some alternate insect-control method established.

The management of grain in storage to prevent its deterioration from all factors, including insects, is not an easy task, especially with ever-increasing food standards and supply requirements and increased restrictions on insecticide use. An acceptable level of insect infestation and insecticide residue in food supplies must be attained.

PREVENTING INSECT INFESTATIONS

The natural habitats of stored-grain insects are important sources of infestation for newly harvested grain.[6-9] Insects that are natural infesters of seed from wild plants, especially those gathered by other seed-harvesting animals, have little trouble adjusting to grain harvested for human consumption.

Sanitation

Thorough sanitation is the most effective means of preventing insect infestations. Storage areas should be clean and constructed tight enough to keep out insects and to keep in fumigant gases if such treatment becomes necessary. Loading and unloading docks should also be clean and constructed so that grain cannot accumulate under them.

All grain, dust, and chaff should be removed from harvesting machinery, transport equipment, and storage areas before using such facilities for the new crop. The surface of equipment and storage areas that will be in contact with the grain should be sprayed at least 2 weeks before harvest with one of the following insecticide formulations:

Insecticide formulation	Amount of insecticide per 7.56 ℓ of water
Methoxychlor 50% W.P. (wettable powder)	353 mℓ
Methoxychlor 25% E.C. (emulsifiable concentrate)	940 mℓ
Pyrethrins 6% E.C. combined with piperonyl butoxide 60%	646 mℓ
Malathion — premium grade 50—57% E.C.	235 mℓ
Chlorpyrifos-methyl (Reldan 2E)	302 mℓ

Spray one of the above to the point of "run-off" using 3.7 ℓ of total formulation per 46.5 m² of surface. Also spray the outside walls of the bins to a height of 1.8 m and the ground adjacent to the bin for a distance of 1.8 m out from the foundation. Applying these residual insecticides to storage areas not only reduces existing insect populations, but also provides an additional barrier against reinfestation.

Encapsulated formulations of malathion were more persistent than emulsions when applied to plywood surfaces as determined by bioassay with four species of stored-product insects.[10] Significantly higher insect mortalities can be expected if malathion emulsions are applied to wood vs. concrete surfaces.

A constant surveillance program of incoming grain is necessary to establish and maintain an effective insect prevention program. Stored grain should be inspected at 7-day intervals, especially during the summer and autumn months, to determine if some type of treatment is needed. Expensive equipment is not needed to detect the insects. Sifting samples of grain over a screen with 10 to 12 wires per inch to separate the kernels and most stages of insects is adequate. Damaged kernels, abnormal odors, webbing, or heating grain are also important clues to detecting insect activity.

Current Grain Protectants

Grain protectants have several advantages: they persist for extended periods at concentrations lethal to the target insects, and they are generally safer to apply compared to fumigants and require little, if any, special equipment for application. Protectants are applied to kill most indigenous insects but, of more importance, to prevent insects from establishing an infestation. One protectant application is usually sufficient during a single storage season, especially if applied in early summer. Protectants can also be used effectively in loosely constructed storage facilities, facilities that could not be fumigated successfully without extensive and expensive sealing efforts. Even the application of malathion, bromophos, or iodofenphos to the storage structure proved of value in translocating quantities to grain that were lethal to stored-grain insects.[11]

Inert dusts have some value for protecting stored grain as they may be both toxic and repellent to some insects. A few of these materials that have been evaluated in the U.S. include diatomaceous earths, silica aerogels, magnesium oxide, aluminum oxide, and activated clays.[12-15] Similar research has been conducted on inert dusts in India[16-18] and Kenya.[19] These materials lose much of their effectiveness on moist grain as they kill insects principally by abrasive action or by absorption of lipids. Richards[20] indicated that the adsorptive action of dusts on insects may involve both physical and physiological phenomena as his results were not explained as simply a product of adsorption. Dusts with high surface area per gram of substance and with particles exceeding 20 Å with low mineral sorptivity for water appear to be the most effective.[21]

There has been relatively little use of inert dusts for protecting grain in the U.S. Commercial interest in using inert dusts is greater in countries with the least stringent grading standards, especially if alternate insecticides are not available, are too costly, or are too dangerous to handle. The application of inert dusts can produce disadvantages, i.e., dust contaminating the air, damaging the machinery by abrasive action, increasing the risk of fire or explosion, causing lung damage (silicoses) to workers, and reducing the grain test weight or commercial grade.

Protectant dusts are diluted with inert ingredients such as clay, diatomaceous earth, grain flour, or talc and are either dispensed through applicators to the surface of the grain mass in the bin or are prorated onto the grain stream as it is being transferred to the storage site.

The relative effectiveness of inert dusts as protectants for corn, sorghum, and wheat compared to malathion, diazinon, and synergized pyrethrins was evaluated by LaHue in a series of publications.[22-24] In 1967, LaHue[25] compared diatomaceous earth dust impregnated with malathion with other protectants on seed corn and reported it to be very effective, but reemphasized its abrasive characteristics on all types of machinery. Diatomaceous earth dusts impregnated with pirimiphos-methyl and malathion also tested more effective than sprays containing equivalent insecticide dosages.

Liquid formulations of grain protectants are usually applied by pressure through a series of nozzles to obtain a uniform application onto grain as it is conveyed into storage. However, Minett and Williams[26] reported that treatment of a small proportion of a wheat bulk with high concentrations of malathion may prove a more effective method of application than attempting to treat all kernels uniformly. Results of bioassay and residue analysis at the U.S. Department of Agriculture Grain Inspection Laboratory in Manhattan, Kansas have substantiated their findings.

The only liquid formulations of chemical grain protectants registered for use in the U.S. today are malathion, pyrethrum, chlorpyrifos-methyl, and pirimiphos-methyl. Their characteristics and usefulness are cited below.

Malathion (*O,O*)-dimethyl phosphorodithioate of diethyl mercaptosuccinate) is an organophosphorus compound with relatively high toxicity to insects and low toxicity to man. Its chemistry, methods of analysis, and storage stability are adequately described by Miles et al.[27]

Malathion is effective against many species of stored-grain insects, although there is evidence of increased insect resistance.[28-31] Malathion applied, whether as dust or sprays at 8 ppm, gave excellent control of *Sitophilus oryzae* for 5 months in stored polished rice;[32] 4 ppm also provided complete protection from *Cryptolestes turcicus*, but was less effective against *Tribolium castaneum*. Malathion also provided the required protection at 10 ppm for stored seeds from potential infestations of *Oryzaephilus* spp., *Tribolium* spp., and *Sitophilus* spp.[33] Miles failed, however, to get adequate control of moths at 10 ppm, although such was obtained by Moore and Decker.[34] Malathion at 2 ppm or on 10% moisture wheat effectively controlled *Sitophilus* spp. and *Rhyzopertha dominica*.[35] This effectiveness is lost on grain of greater moisture content, as determined by Kadoum and LaHue.[36] Both the residual stability and the subsequent effectiveness of malathion are generally reduced by increases in grain moisture as well as temperature. Watters[37] obtained 99% mortality of *Laemophleous ferrugineus* on 13.5% moisture wheat with 2 ppm malathion, but 16 ppm was required to achieve comparable control with 15.5% moisture. Control was reduced when the moisture content was increased to 18%. Malathion residues that degradated from 19.7 to 3.4 ppm on grain sorghum during a 1-year storage period still provided excellent protection against mixed populations of stored-grain insects.[38] The effectiveness of malathion may also vary depending on the kind of grain treated.[39]

Malathion is frequently used as a reference standard to determine the relative effectiveness of candidate protectants. King et al.[40] reported that malathion, ronnel, and silica gel provided effective control of stored-grain beetles. Heavy infestations of *O. surinamensis* were controlled by 10 ppm malathion, 2 ppm fenitrothion, or 4 ppm dichlorvos.[41] Both malathion and fenitrothion were effective for 8 months. The relative toxicity and residual effectiveness for malathion and diazinon were compared against 17 species of stored-product insects by Strong et al.[42] Bromophos was persistent on concrete, fenitrothion moderately so, and malathion residues degraded rapidly.[43] Tyler and Green[44] reported that the persistence of fenitrothion at 2 ppm was similar to malathion at 10 ppm on warm, damp grain.

Pyrethrins (pyrethrolone esters of chrysanthemum carboxylic acid) have insecticidal characteristics and safety qualities that have been established for many years, although the exact date marking their use as insecticides is not known. Pyrethrum powder was introduced into the U.S. in the mid 19th century.

Oil-base formulations of pyrethrins were used against many species of insects prior to World War II.[45] They "knock-down" insects rapidly. However, death may not occur immediately; in fact, insects may recover from their paralyses to escape. Generally they are more effective against the adult stage of an insect, especially the most active or free-flying species.[46] The synergist piperonyl butoxide [2-(2 butoxyethoxy)ethoxy-4,5-(methylene-dioxy)-2-propyltoluene] greatly increases its pyrethrin toxicity, especially if the ratio of piperonyl butoxide to pyrethrins and the diluent are optimum for combating specific pest insects. Pyrethrins synergized with piperonyl butoxide in a 1:10 ratio are most widely accepted.

Several chemicals have been tested in different proportions as a potential alternate synergist for pyrethrins. The synthetic pyrethroid SBP-1382 [(±)-*cis,trans-s*-benzyl-3-furyl)methyl-2,2-dimethyl-3-(2-methyl propenyl)cycloproparrecarboxylate appears to be effective against a wide range of insects.[47] A related compound [(±)-*trans*-chrysanthemic acid ester of (±)-

allethrolone (bioallethrin)] also appears to be effective against four species of stored-product insects.[48]

Pyrethrins are generally used as space treatments, bulk grain applications, or as surface treatments on the walls and floors of the storage areas to control stored-grain insects. As space treatments, they are usually applied by thermal or mechanical generators to produce aerosols. Such applications are aimed to contact flying or crawling insects exposed directly to the aerosol particles. Thermal aerosol formulations usually consist of 0.2% pyrethrins and 2.0% piperonyl butoxide (percent by weight) plus the diluent. The thermal generators reportedly decrease the effectiveness of pyrethrins as a result of degradation while exposed to high temperatures. Formulations for mechanical aerosols frequently have 0.5% pyrethrins and 5.0% piperonyl butoxide. Whatever the formulation, the total dosage of pyrethrins applied for stored-product insects should be about 0.3 g/100 ft^3 of space.

Ultra low volume (ULV) application of synergized pyrethrins, as a space treatment, is a relatively new application technique. The advantage is more effective distribution of particles with less range in particle size, resulting in greater insecticidal efficacy. Some insect resistance to pyrethrins has been noted, but this fails to be of major concern currently.

The repellent properties of synergized pyrethrins are important in protecting stored commodities from insects. A high level of repellency to *S. oryzae, T. confusum, T. castaneum, C. pusillus*, and *O. surinamensis* throughout a 12-month storage period was observed by LaHue.[49] Brooke[50] reported that both the toxicity and repellency of 1.17 ppm pyrethrin with 27 ppm piperonyl butoxide protected wheat from infestations of *S. granarius, S. oryzae*, and *R. dominica*. Most researchers agree, however, that the toxic effect of pyrethrins is lost rapidly when applied to stored grain, while the repellent action remains the primary factor for protection. Subsequently, insect-free commodities can be protected partially against stored-grain insects by applying pyrethrins to their surface only. This should be applied evenly over the surface of the grain immediately after grain is leveled off upon reaching its storage site.

As with other insecticides, each species and stage of stored-grain insect differs in its susceptibility to synergized pyrethrins. Since most insect infestations of stored products involve more than one insect species or stage, the dosage should be effective against the least susceptible individuals.

As of this writing, the supply of pyrethrins for insecticidal formulation into the U.S. is inadequate. Attempts are in progress to substitute formulations containing a mixture of pyrethrins and synthetic pyrethroids.

Chlorpyrifos-methyl and pirimiphos-methyl are two organophosphate insecticides that were recently approved by the Environmental Protection Agency (EPA) in the U.S. for use as protectants on stored grain. Both are broad spectrum insecticides with moderate toxicity to warm-blooded animals.

Formulations of chlorpyrifos-methyl registered currently by EPA are Reldan 4E and Reldan 3% dust. In U.S. field experiments, 6 ppm of chlorpyrifos on grain (wheat, sorghum, rice, barley, and oats) protected the grain for 12 months under adverse conditions of high grain temperature and moisture. When applied at 10 ppm, chlorpyrifos-methyl has no adverse effect on seed germination. Neither is it known to be carcinogenic or teratogenic, does not effect reproduction or growth of laboratory test animals. Chlorpyrifos methyl [*O, O*-dimethyl-*O*-(3, 5, 6-trichloro-2-pyridinyl) phosphorothioate] at 3 ppm on wheat controlled *S. oryzae, S. granarius,* and *S. zeamaiz* and was stable for a longer period compared to malathion on high-moisture red winter wheat.[84]

Pirimiphos-methyl is currently registered in the U.S. as Actellic with an emulsified concentrate formulation containing 5 lbs of active ingredient per U.S. gallon. This organophosphate insecticide is registered for use as a grain protectant in 75 grain-producing countries. As with chlorpyrifos-methyl, pirimiphos methyl is significantly more persistent and effective than malathion in protecting grain stored under high-moisture conditions. Satisfactory control of a wide variety of stored-grain insects is obtained with pirimiphos-methyl at 6 to 8 ppm on shelled corn and grain sorghum, the only two grains approved for such treatment in the U.S. as of this writing. Chlorpyrifos-methyl is also registered as a stored-grain protectant in France, the United Kingdom, Turkey, Australia, and Argentina. Pirimiphos methyl [*O*-2(-(diethylamino)-6-methyl-4-pyrimidinyl)-*O*,*O*-dimethyl phosphorothioate] also shows promise as a protectant for rice.[83]

Potential Grain Protectants

Diazinon [*O*,*O*-diethyl-*O*-(2-isopropyl-4-methyl)-6 pyrimidinyl] has been reported by Strong et al.[51] to be more effective than malathion against 13 species of stored-grain insects. These results obtained by LaHue[52] showed diazinon to be effective for 3 months against *R. dominica*, but it did not inhibit their reproduction.

Telford et al.[53] noted considerable loss of diazinon residue following treatment. However, Roan and Srivastava[54] reported that diazinon penetrated the interior of a wheat kernel and maintained a stable level for 45 days, after which diazinon residue dissipation was extremely slow. About 70% of the diazinon applied adhered to the grain. The cleaning, tempering, and milling process appeared to dissipate about one half of the residue.

Fenitrothion [*O*,*O*-dimethyl-*O*-(3-methyl-4-nitrophenyl)phosphorothionate], out of 16 organophosphorous insecticides, was one of the most effective against *T. confusum* and *T. castaneum*[55] and it proved, in subsequent studies,[56] to be the most toxic compound to *S. granarius* in wheat. Even heavy infestations of *O. surinamensis* were killed in farm-stored barley with 2 ppm fenitrothion.[57] Fenitrothion was also highly effective in protecting bagged wheat and barley from infestation by *O. surinamensis*, *S. granarius*, and *T. confusum*.[58] Dosages of 2 and 4 ppm were effective for 6 and 10 months, respectively. The residues of fenitrothion disappeared gradually. Fenitrothion residual persistency was similar to malathion on damp, warm grain, but was superior on dry, cool grain.[59]

Wheat treated in a laboratory to provide a fenchlorphos (*O*,*O*-dimethyl-*O*-2,4,5-trichlorophenyl phosphorothionate) residue of 4 ppm resulted in a LD_{95} when the weevils were exposed for 14 days.[60] Although a dosage of 1 ppm killed only 7% of the adult weevils, it decreased their reproductive potential 94%.

Bromodan [5-(bromomethyl)-1,2,3,4,7,7-hexachloro-2-norbornene], because of its relatively low mammalian toxicity of 12,900 LD_{50}, was tested as a protectant for wheat against infestations of three species of stored-grain insect.[61] It was found that 20 ppm killed all the *S. oryzae* and *Lasioderma serricorne* and prevented their reproduction. At 8, 16, or 24 ppm on barley and wheat, bromodan killed the insects in farm bulk grain.

Bromophos (*O*,*O* dimethyl-*O*-2,5-dichloro-4 bromophenyl phosphorothionate) is also another candidate protectant for stored grain,[62] especially if malathion resistance continues to increase. It has a relatively low mammalian toxicity, and its residual life on high-moisture grain appears longer than that of malathion.

Phoxim (phenylglyoxylonitrile oxime *O*,*O*-diethyl phosphorothionate) appears promising as a protective treatment of bulk grain and other bulk stored commodities.[63] Strong[64] suggested that it be researched further regarding its usefulness as a space application and as a residual treatment on storage facilities. Phoxim applied at 5 ppm was superior to 10 ppm malathion against several species of stored-product insects.[65] Although its effectiveness against *S. oryzae* was excellent for 12 months, it had decreased toxicity to *T. confusum*, *T. castaneum*, and *R. dominica* after 9 months on stored hard winter wheat.

Gardona [2 chloro-1-(2,4,5-trichlorophenyl) vinyl dimethyl phosphate], an organophosphate insecticide similar to malathion, is effective as a residual for some species of stored-product insects, especially *T. confusum* and *T. castaneum*.[66] *S. granarius* mortalities of 95 to 100% were obtained following a 20-day exposure to wheat with a residue of 8 ppm,[67] and gardona at 15 ppm reduced weevil reproduction significantly. Gardona at 10 ppm generally was not as effective as 10 ppm malathion in protecting wheat at 10, 12, and 13.5% moisture from damage by several species of stored-product insects.[68] Increases in moisture content also reduced the efficacy of gardona. Gardona persisted on corn and wheat for 8 months in farm storage;[69] 90% of it was residue in the bran.

Lindane (gamma isomer of benzene hexachloride — BHC) contains at least 99% of the gamma isomer of benzene hexachloride and is one of the most toxic isomers to insects and mites. Since 1945, formulations of BHC and of its gamma isomer have been tested widely, recommended, and used in various countries as a stored-product protectant with general success. Direct application to stored grains, residual applications in storage areas, and both discontinuous and continuous vaporization have been employed. However, its use is becoming more limited because of its long residual life, its ability to migrate to untreated areas, and its increased evidence of losing effectiveness as a result of insect resistance.[70-73]

Dichlorvos (2,2-dichlorovinyl dimethyl phosphate) satisfies the need for an insecticide with a relatively high initial toxicity to insects and short-lived residues for use as a grain protectant. This applies especially where grain fumigants result in excessive residues, poor insect control because of nonuniform distribution, or inadequate concentrations resulting from loosely constructed storage areas. Dichlorvos possesses some of these required characteristics to serve as a likely candidate.[74] In addition, many laboratory and field studies over the past several years have documented its biological activity to protect different grains against stored-product insects.[75-81]

Harvested grain, especially in tropical countries, is often stored in jute bags for extended periods. Conventional fumigation or residual insecticide applications are usually not feasible with small lots of produce stored in many places over wide areas. However, dichlorvos, dissolved in carbon tetrachloride and applied into the sacked grain with a motorized knapsack sprayer, produced high control of the infesting insects.[82]

Experimental Protectants

There is, and will continue to be, a need for evaluating new experimental chemicals that have potential for use as direct contact, residual, or vapor toxicants to combat stored-grain insects. Speirs and Lang[85] reported the relative effectiveness of 68 candidate insecticides. The effectiveness of 48 insecticides was compared against five species of stored-grain insects by Strong and Sbur.[86] Strong followed with laboratory evaluations of several organophosphorous insecticides on *Attagenus alfierii*[87] and the flour beetles *T. confusum* and *T. castaneum*.[88] McDonald and Speirs[89] reported there new chemicals with greater toxicity than malathion to larvae of *Plodia interpunctella*.

Capric acid occurs in triglycerides within many foods, including milk. When mixed with stored foods and dehydrated foodstuff, it has effectively controlled *T. confusum* at concentrations considered safe for mammals.[90] Lyophilized citrus oils from lemon, grapefruit, lime, kumquat, and tangerine were also highly toxic to *Callosobruchus maculatus*[91] and moderately toxic to the rice weevil, *S. oryzae*.[92] Sorbic acid, butylated hydroxyanisole, and butylated hydroxytoluene acted as ovicides to *T. confusum* when applied at 2 and 5% as additives in their diet.[93]

The control of dermestids looks promising using sex pheromones to attract adults to areas treated with chemosterilants, insecticides, or disease agents.[94] The juvenile hormone analogue (methyl-3,7,11-trimethyl-7,11-dichlor-2-dodecenoate) was effective in controlling *Stegobium paniceum*.[95] Higher dosages were required for *R. dominica*. It has been considered unlikely that insects resistant to pheromones would develop. However, Dyte[96] reported a strain of *T. castaneum* that was resistant to many conventional insecticides as being resistant to the synthetic juvenile hormone identified as a cis/trans mixture of methyl-10,11-expoxy-7-ethyl-3,11-dimethyl-2,6-tridecadienoate.

FUMIGANTS

A fumigant is a chemical which exists as a gas, at ambient temperatures and pressures, or produces a gas from solids or liquids. As gases they diffuse through air, permeate products, and enter the respiratory systems of insects, a mode of entry of major importance. Compared to residual insecticides, fumigants have no lingering effectiveness. As soon as they diffuse away from the target area, immediate insect reinfestation may follow.

One reason for the wide acceptance of fumigants for control of stored-grain insects is the multiple application methods available. In addition, modifications of conventional methods are usually feasible as may be required under different climatic conditions and storage characteristics. The bioactivity of a fumigant is influenced by the method of application. The commodity or the space treated is usually enclosed by structures composed of metal, concrete, wood, fiberglass, or plastic with or without facilities for recirculating the gas to aid in achieving homogeneous gas concentrations.

An ideal fumigant should have the following characteristics:

1. Low cost per effective fumigation
2. High acute toxicity to all developmental stages of the target insects, but free from undue hazard to man
3. High volatility and good penetration power, but not excessively sorbed by grain
4. Adequate warning properties and easily detected
5. Noncorrosive, nonflammable, and nonexplosive under practical conditions with good storage life
6. Nonreactive with the commodity so as not to produce residual odors
7. Aerate readily and leave no harmful residues
8. Noninjurious to seed germination nor lower commercial grain grade
9. Nondamaging to milling qualities of grain or other processing properties
10. Readily available and simple to apply

The safe and effective use of fumigants depends, in part, on knowing their primary physical, chemical, and biological properties. Whitney[97] listed 17 fumigants and their properties, the group comprising about 99% of the stored-product insect fumigants currently used in the U.S. Pertinent properties of the major fumigants are provided (Table 1).

Concentration × Time

Most fumigations have been conducted on the basis of fumigant dosage followed by an estimate of the required exposure period. Such guidelines may be sufficient, but they do not always take into consideration the multiple factors that alter the required fumigant concentration. Thus fumigation success can be realized more often using analytical techniques to monitor fumigant concentration. With this information, sublethal concentrations in specific areas could be supplemented with additional amounts of fumigant. Additional dosage or an extension of the exposure time to provide a suitable concentration × time factor (C × T) product may provide control of the insects without exposing the product to excessive dosages resulting from follow-up fumigations.

Time, temperature, and dosage relationships for carbon disulfide, sulfuryl fluoride, acrylonitrile, methyl bromide, ethylene dibromide, carbon tetrachloride, methyl chloroform, ethylene dichloride, and chloropicrin were determined for *T. confusum*.[98] Other values and applications of C × T products have been investigated.[99-102] Because of the multitude of variables necessary in determining C × T products for stored-product insects, Thompson[103] developed dosage recommendations for methyl bromide, carbon tetrachloride mixed with ethylene dibromide, aluminum phosphide, and others. His C × T products were expressed as the weight of fumigant to be used per unit weight and volume of commodity.

Table 1
FUMIGANTS FOR GRAIN AND CEREAL PRODUCTS

Name and synonym	Chemical formula	Molecular weight	Sp. gravity as vapor (air = 1)	Explosive limits (% vol in air)	General comments
Acrylonitrile (vinyl cyanide)	CH_2CHCN	53.06	1.83	3—17	Spot fumigant
Chloropicrin (trichloronitromethane)	CCl_3NO_2	164.39	5.7	Nonflammable	Intense odor; used as a warning agent
Dichlorvos	$CCl_2CHOPO(OCH_3)_2$	221.00	7.6	Nonflammable	Space fumigant effective against wide variety of exposed stored-product insects
Hydrogen cyanide (hydrocyanic acid)	HCN	27.03	0.93	6—41	Space fumigant or recirculated through grain
Methyl bromide (bromoethane)	CH_3Br	94.95	3.27	Nonflammable	Space of commodity fumigant
Phosphine	PH_3	34.00	1.2	Spontaneously flammable	Highly toxic with excellent penetration properties

The thermal conductivity method of measuring methyl bromide in air was developed for practical use by Phillips and Bulger.[104] Models of the original unit have been improved, thus providing a means to study the distribution, penetration, sorption, and C × T patterns within bulk lots of shelled corn, wheat, grain sorghum, barley, rice, seeds, and related commodities. Various sources of information on the design, calibration, operation, and utilization of the thermal conductivity gas analyzer are available.[105-108]

Monro[109] explains the value, design, and operation of interference refractometers for measuring fumigant concentrations. These instruments are relatively simple to use, and the gas concentration values obtained are not affected by voltage variations, a potential problem with most electrical units. In essence, they consist of glass tubes containing specific chemicals that change color upon exposure to specific fumigants. Air plus the fumigant is drawn through the tubes using either a hand-operated pump or bellows. The length of color change within the tubes is proportional to the gas concentration, assuming the instrument has been properly calibrated. Heseltine and Royce[110] describe the usefulness of color indicators that react proportionately to the gas concentration and exposure period.

Halogenated hydrocarbon fumigants, such as methyl bromide, can be measured by use of halide leak detectors or halide lamps. Such lamps are generally available from refrigeration supply dealers and are sufficient for obtaining field estimates of gas concentrations. They work on the principle that a flame in contact with copper will be green to blue if an organic halide gas is present in the atmosphere surrounding the flame. As the organic halide gas concentration increases, the flame changes from greenish-blue to blue. Fuels used for the flame include kerosene, wood alcohol, acetylene, and propane. They should not be used in any area where there is a fire hazard. Halide meters are also available commercially and are designed primarily for measuring halogenated hydrocarbons in air at concentrations up to 500 ppm with 10% accuracy.

Precise spectrophotometric and gas-chromatographic methods of measuring minute gas concentrations of many fumigants have also been developed for acrylonitrile,[111] carbon disulfide,[112,113] carbon tetrachloride,[114,115] chloropicrin,[116,117] dichlorvos,[118,119] ethylene dibromide, ethylene dichloride, hydrogen cyanide and methyl bromide,[120] and phosphine.[121-123] Various chemical methods for many of the fumigants are also useful.[124,125]

Fumigation Under Tarpaulin

One common method of fumigating bagged grain or packaged cereal products is to enclose the commodity under gas-proof sheets that will retain the fumigant at concentrations and for periods lethal to stored-product insects. Polyethylene, vinyl films, and synthetic rubbers or materials coated with one or more of these products are effective in retaining methyl bromide.[126] Polyethylene or polyvinyl chloride sheets, at least 12.7 mm thick, are suitable for indoor fumigations. Nylon or cotton fabrics laminated with neoprene, polyvinyl chloride, or butyl rubber should be used outside.

A primary advantage of tarpaulin fumigations is the possibility of treating the commodity in place. Sealing at the sheet/floor juncture is accomplished by laying tubular sacks containing sand (sand snakes) or lengths of chain on the edge of the fumigation sheet. Allowing the tarpaulins to remain in place following fumigation protects the product against reinfestation.

Fumigations with methyl bromide[127] and phosphine[128] under gas-proof sheets are explained. Bowen[129] provides a formula for calculating the correct fumigant dosage. Methyl bromide and phosphine work best for fumigating large stacks because they penetrate quickly and diffuse rapidly. Liquid fumigants are effective for small stacks. Empty gas or oil drums may also be used to fumigate very small quantities of a commodity. Recommended dosages are suggested in Table 2. The fumigant can be introduced by pouring on a liquid fumigant or by inserting a packet of solid-type fumigant on the surface of the commodity. The chamber could be sealed with tarpaulin. A box composed of masonite panelwood or its equivalent

Table 2
DOSAGE TABLE FOR FUMIGANTS USED IN SMALLER CHAMBERS

mm of liquid/100 ft³

	1/16 lb/ 1000 ft³	1/4 lb/ 1000 ft³	1/2 lb/ 1000 ft³	3/4 lb/ 1000 ft³	1 lb/ 1000 ft³	2 lb/ 1000 ft³	3 lb/ 1000 ft³	4 lb/ 1000 ft³
Chloropicrin	1.7	6.9	13.7	20.6	27.4	54.9	82.3	109.7
Ethylene chlorobromide	1.7	6.7	13.4	20.1	26.8	53.6	80.5	107.3
Ethylene oxide at 7°C	3.2	12.7	25.5	38.2	51.0	102.0	153.0	204.0
Hydrocyanic acid	4.1	16.5	32.9	49.4	65.9	131.8	197.6	263.5
Methyl bromide at 0°C	1.6	6.5	13.0	19.5	26.1	52.2	78.3	14.4
Propylene oxide	3.4	13.6	27.3	40.9	54.5	109.1	163.6	218.2

painted with an asphalt-aluminum emulsion can serve as a fumigation chamber. Various tarpaulin application procedures and precautions are explained in Monro's text.[130]

Fumigation Safety Procedures

The probability that a fumigant will harm nontarget organisms, including man, when used properly is often referred to as its "toxicity hazard." However, fumigants with relatively low toxicity hazard potential may result in serious injuries if not used properly, especially if the fumigation procedure is complicated. Many commonly used grain fumigants have detectable odors at concentrations safe to humans for a single exposure. However, people become accustomed to the odor quickly and may fail to recognize harmful concentrations. The use of remote techniques for application, such as the recirculation of methyl bromide or the use of fixed spray nozzles for applying liquid grain fumigants, is of value in limiting contact of the applicator with the fumigant.

Most fumigants can produce skin blisters if contaminated clothing is not removed immediately. Even gas contact to the skin may be harmful. Suitable protection for the applicator's eyes is obtained with full-face gas masks. Any person using fumigants, including those responsible for persons handling them, must be well acquainted with the hazards involved.

The following safety list, adapted from a fumigation checklist prepared by Kansas State University, Manhattan, includes most of the important considerations that should receive attention before, during, and following a fumigation.

Preliminary Planning and Preparation

1. Understand the facility and commodity to be fumigated including the design of the structure, adjacent as well as connecting structures, both above and below ground; number of persons or animals expected to be at or near the area to be fumigated; the type of commodity, including its history and condition; emergency shut-off stations for electricity, water, and gas; and telephone with pertinent emergency numbers for fire or police departments, hospitals, poison control centers, and physicians.
2. Select and obtain a suitable fumigant.
3. Understand label directions.
4. Notify local medical, fire, and police authorities and other security personnel as to the fumigants to be used, proposed date and time of use, type of respiratory protection required, and the fire hazard rating.
5. Have alternate application or protective equipment and replacement parts available.
6. Inform all persons, directly or indirectly involved, of the fumigation hazards to life and property and the required safety measures and emergency procedures.
7. Display appropriate warning signs for posting treated areas and provide for the necessary security.
8. Have necessary first aid equipment and antidotes available.
9. Apply the fumigant from outside the structure.
10. Develop plans to ventilate the fumigated area.

Personnel and Protective Equipment

1. Assign two persons to each fumigation, especially in circumstances where entry into the fumigated area is essential. A two-way radio communication system is helpful.
2. Store all protective equipment so as to insure maximum life of the device and be readily accessible to employees.

3. Canister type gas masks are ineffective when sufficient concentrations of oxygen are not available, do not prevent sorption through the skin, will not remove sufficient toxic gases if the concentration is beyond the capability of the canister, require different types of canisters for different toxic gases, and should be mutilated after any single use period, especially if the gas mask is used by more than one person.

4. All fumigators and their personnel should be instructed in first aid and other emergency procedures including personal decontamination and antidotes.

5. Personnel handling fumigants should be cautioned to report all indications of illness or physical discomfort regardless of their apparent minor nature. These may include but not be restricted to any or all of the following symptoms: dizziness, nausea, headaches, and lack of coordination.

Application Procedures and Fumigation Period

1. All applications should be made in accordance with the fumigant manufacturers' labeled recommendations.

2. Consider prevailing winds and other pertinent weather factors in regard to fumigation effectiveness and safety.

3. Apply fumigants from outside the exposed areas where appropriate.

4. Provide watchmen, when required.

Postapplication Procedures

1. Before reentry, use a suitable gas detector to determine fumigant concentrations. Do not depend on odors. Various methods to monitor field fumigant concentrations are published.[131]

2. Provide security to prevent unauthorized entry.

3. Turn on ventilating or aerating fans where appropriate.

4. Double check for gas concentrations in areas which are expected to aerate slowly.

5. Remove warning signs when the gas concentration is within safe limits for human exposure.

6. Return unused chemicals in proper and clearly labeled containers to storage area. Dispose of empty containers as per local recommendations.

Fumigant Characteristics

Methyl Bromide (CH_3Br)

This fumigant became popular during the 1930s to disinfest plants, vegetables, and some fruits; however, it has also been used extensively for stored products, mills, ships, and railway cars. Methyl bromide will penetrate most products at concentrations lethal to the insect pests under normal atmospheric pressure. It is highly toxic to most stages of stored-product insects, although the pupal stage may be an exception.[132,133] Even the age of the pupal stage can alter its efficiency,[134,135] as young pupae appear most resistant.

Concentrations of methyl bromide in various grains at different moisture contents were determined by Whitney and Walkden.[136] Bioassay indicated adult *S. oryzae* to be the most susceptible, followed by immature *S. oryzae* and adult *T. confusum*. Higher concentrations for short periods were more effective than lower concentrations for longer periods. Mostafa et al.[137] found that the eggs of four species of stored-product insects became more susceptible to methyl bromide with age.

Methyl bromide and ethylene dibromide were included in fumigation studies conducted by Lindgren et al.[138] and Krohne and Lindgren.[139] Although both were effective separately, Kazmier and Fuller[140] reported that mixtures of the two were more effective than methyl bromide against all life stages of *T. confusum*. The effectiveness of methyl bromide was reported limited as a fumigant in silo bins because of unsatisfactory distribution of the gas.[141] However, penetration was effective when methyl bromide was applied with carbon dioxide.[142]

Methyl bromide is marketed in steel cylinders varying from 5 to 1800 lb (2 to 816 kg) and in cans containing $1^1/_2$ lb (0.675 kg). As methyl bromide is dispensed from its container, some evaporative cooling occurs. One of several forms of heat exchange should be supplied. Passing the methyl bromide through a coil of copper pipe, immersed in a container of hot water, may suffice. Supplemental heat is required as the liquid methyl bromide should not contact such commodities as grain or cereal products. The release of methyl bromide at relatively low temperatures may result in its uneven distribution and heterogeneous bioactivity.

Methyl bromide is nonflammable and nonexplosive. It has no corrosive action in its pure form on most metals except aluminum. However, it reacts with many plastic and organic materials. Monro[143] provides a list of the materials that may be damaged by exposure. Polyethylene and neoprene seem to be the least affected. Rubber is strongly attached. It still may be used in conjunction with many synthetic rubber and plastic sheets that serve as tarpaulins for fumigations. A review of the properties and usage of methyl bromide as a fumigant was published by Thompson.[144]

Methyl bromide is highly toxic to humans, being readily absorbed both through the lungs and through the skin. After exposure, the skin may become red and itch, followed by the development of blisters. The symptoms of poisoning from inhalation of methyl bromide vary considerably both in their degree and in their time appearance 24 hr after exposure. The most common early symptoms are fatigue and complaints about blurred vision. This may be followed by temporary blindness, conjunctivitis, and swelling of the eyelids. Exposed individuals often complain about ringing in the ears, dizziness, confusion, and fainting spells. Exposure to high concentrations may cause sudden death from respiratory paralysis and heart failure.

Phosphine (Hydrogen Phosphide, PH_3)

Using phosphine generated from aluminum phosphide as a fumigant for bulk grain was established in Germany about 1937. Originally, the method was to insert packets of aluminum phosphide into grain. In the 1950s a new formulation became available that consisted of aluminum phosphide and aluminum carbonate in tablets. One formulation on the market to produce phosphine consists of a compressed mixture of aluminum phosphide, ammonium carbonate, and paraffin. Phosphine, carbon dioxide, and ammonia are produced when this formulation comes in contact with moisture. The carbon dioxide and ammonia help to combat the flammability of phosphine, with the ammonia also serving as a warning gas. Aluminum hydroxide remains as a residue. Decomposition of the tablet takes about 36 hr at 25°C if sufficient moisture is available. About one third of each tablet, by weight, is diffused as phosphine. Phosphine smells like garlic, but this odor cannot be used for warning purposes since it disappears under some fumigation conditions.[145] The developmental history of phosphine for use on raw and processed agricultural products was reviewed by Hazleton.[146] Magnesium phosphide was approved for marketing by the Environmental Protection Agency in 1979.

The effectiveness of phosphine is benefited by its low molecular weight and low boiling point, characteristics that promote its rapid diffusion and penetration into grain,[147-149] through cereal products in mills,[150] in railway cars,[151] into commodities under tarpaulins,[152] or through bagged flour in plywood overpacks.[153]

Phosphine can be applied to grain by prorating the tablets, or whatever formulation is selected, on grain as it flows into storage or by injecting it into stored grain using a special metal probe. The procedure calls for pushing the probe to the bottom of the grain mass and then applying the formulation via the center of the tube as it is withdrawn. This also can be accomplished using ordinary metal piping.

For warehouses or in railway cars, the formulation can be dispersed onto sheets of paper or in envelopes to allow phosphine generation, but providing a recovery of inert ingredients if necessary. Phosphine has no adverse effects on food materials and, under normal conditions, will not affect the germination of seeds.

Phosphine is one of the most toxic fumigants to stored-product insects.[154] Although highly toxic to many insects, it is markedly less so to certain stages of some species.[155] A (C × T) product of 10 mg hr/ℓ of phosphine is sufficient to control adult *Sitophilus* spp., but young pupae of these species require a (C × T) product of 300 mg hr/ℓ. Ozer[156] conducted a detailed study of the toxicity of phosphine to both *S. granarius* and *S. oryzae*. His results show small effects on mortality with increases in concentration, a finding which is borne out by the difficulty experienced in some field trials.

Phosphine, tested against 13 species of stored-product beetles by Hole et al.,[157] was least effective against *Sitophilus* spp. except for diapausing larvae of *Trogoderma granarium*. Long exposures (16 days) at 15°C or below were required to be effective. Mixtures of methyl bromide and phosphine provide enhanced toxicity to *S. granarius* and *Tribolium confusum*, allowing possibilities for expanded uses.[158] Other relatively new uses of phosphine include the fumigation of sacked milled rice,[159] in open-ended bin spouts,[160] aboard ships in transit,[161] and of cereal grain and processed products when generated from Detia Ex-B®.[162]

Phosphine is also highly toxic to humans. Its inhalation causes restlessness that may be followed by tremors, fatigue, and sleepiness. The victim becomes nauseated and often suffers from vomiting and diarrhea, thirst, headache, dizziness, ringing in the ears, fainting spells, and a burning sensation in the chest. Eventually the victim may become sleepy, pass into rigor, and develop convulsions before death. Continued exposure to sublethal concentrations may lead to chronic poisoning characterized by bronchitis, gastrointestinal disturbances, destruction of the teeth, and disturbances of speech, vision, and motor functions.

The hazards of using phosphine as a fumigant are low because of its slow release following exposure of the solid formulation to moisture. Wearing an approved gas mask while dispersing this formulation is not required for the safety of the applicator. Having the gas mask readily available is recommended. All other fumigant applications, other than by automatic remote application systems, require the applicator to be wearing the proper gas mask at all times.

Chloropicrin (CCl₃NO₂)

Chloropicrin (CCl_3NO_2)

Chloropicrin is highly toxic to insects, is a powerful tear gas, and is highly toxic to man. The established threshold limit is 0.1 ppm. Concentrations as low as 1 ppm produce an intense smarting of the eyes. Continued exposure may cause serious lung injury.

Chloropicrin-treated commodities are often unpleasant to handle as they may be extremely irritating from even low quantities diffusing from the treated material. It is also corrosive to metals.

Its primary use other than acting as a "warning gas" is as a soil fumigant, but it also is registered for use on harvested grain.

Dichlorvos (O,O-dimethyl-2,2-dichlorovinyl phosphate)

Dichlorvos (O,O-dimethyl-2,2-dichlorovinyl phosphate)

Strips of polyvinyl chloride (PVC) impregnated with dichlorvos are recommended to control adult *Plodia interpunctella* in the overhead space of grain bins using one strip per 1000 ft³ of space. Other than conventional fumigants, it is the only insecticide that is registered for use on stored soybeans. Adult *Anagasta kuhniella* were eliminated and 75 to 80% of the larvae were killed in empty metal bins using these PVC strips.[163] Warehouse infestations of *Ephestia elutella*[164] and *C. cautella*[165] were also controlled using dichlorvos in this strip formulation. Excellent control of adult *P. interpunctella* was obtained during 24-hr exposures to the vapors from a dosage of one strip per 28.3 m³ of space over shelled corn.[166] Malathion-resistant *P. interpunctella* were no exception.[167] *Tenebriodes mauritanicus* appear relatively tolerant as they were not controlled even following extended exposure to dichlorvos.[168]

Potential Fumigants

There is a constant need to develop and evaluate new chemicals as potential fumigants for controlling stored-product insects. Of special interest is the possibility of identifying new fumigants that are specific for the target insects and less hazardous to the applicator or the consumer than current formulations. Many of the candidate chemicals that have been researched and reported have never been marketed. A few deserve reevaluation.

Soles and Harein[169] reported the relatively high fumigant toxicity of *N*-(alpha-methylacetonitrile)-morpholine and acetate of dimethyl 2,2-dichloro-1-hydroxyvinylphosphonate to four species of stored-product insects. The candidate fumigants 1,2,3-tribromopropene, ethylenimine, and crotyl bromide (86% 1-bromo-2-butene and 14% 3-bromo-1-butene) were also more toxic at LD_{95} than carbon tetrachloride to stored-product insects.[170] Cooper and Gillenwater[171] reported that five of the six candidate fumigants included in their testing program were more effective than methyl bromide against *Tribolium confusum, Laemophleous serricorne*, and *A. megatoma*. The level of effectiveness by all six indicated that further evaluation as potential grain and space fumigants was warranted.

Aeration

The primary purpose of aeration is to force air through bulk-stored grain to equalize grain temperatures and moisture content. However, fumigants can also be introduced and diffused satisfactorily using the same aeration system. In fact, such aeration systems, whether permanent or temporary, provide a practical method of applying certain fumigants to stored grain. The fumigant is usually more uniformly distributed and the dosage requirements somewhat less for aeration than for other methods with the same fumigant.

Upright storage facilities can be fumigated using either a single-pass or a recirculation aeration technique. The aeration fans are operated for one complete air change for the single-pass method. This can usually be obtained in 10 min at 1/20 ft³/min/bu or 20 min at 1/40 ft³/min/bu. Single-pass applications, combined with gravity penetration, have been used in flat storage facilities, although the recirculation method is frequently the most effective. Storey[172] compared gravity, penetration, single pass, and closed recirculation with methyl bromide in flat storages. The fumigant is moved through the grain several times using the recirculation method. There are aeration systems that can recirculate the fumigant by moving it down one bin and up through another. The recirculation method has been adapted for use in many different types of storage facilities.[173] However, Air flow resistance, which varies depending on the type of grain, its moisture, and dockage content, must be taken into consideration.[174] Most of the liquid fumigants are dispersed satisfactorily by aeration, although methyl bromide is the most widely used fumigant under such circumstances. Storey[175] studied the distribution of methyl bromide and a combination of carbon disulfide and carbon tetrachloride through grain sorghum and wheat in silo-type elevator tanks using aeration systems. The type of grain, fumigant dosage, and air flow rate affected the fumigation efficiency. Phosphine should not be used with aeration because of its degradation at reduced pressures.

VARIABLES THAT INFLUENCE THE EFFICIENCY OF FUMIGATION

Primary factors that decrease or increase the distribution and biological activity of fumigants into grain are temperature, moisture, fumigant formulation and dosage, storage structure, dockage, and characteristics of the insect population and the grain kernels. The size, shape, and structure of the kernels determine the capacity of the grain to compact in storage and thus retard the movement of gases through the grain mass. Also, the kernels influence the kind, rate, and extent of sorption of the gases and their release into the atmosphere.

Temperature and Moisture

It is well-known that the temperature and moisture of grain are two of the major factors producing the greatest variations in the efficiency of a fumigant. When grain is fumigated, the gas comes into equilibrium with the temperature of the grain in a short time. An increase in temperature results in greater molecular activity of the gas to facilitate the diffusion and penetration of the fumigant with decreasing sorption. The opposite effect occurs as temperatures decrease. Sublethal gas concentrations may result if the grain temperature is excessively high or low. When the temperature of the grain is 46°C or higher, the fumigant vaporizes rapidly and may escape before lethal gas concentrations can be obtained in all areas. However, most stored-grain insects cannot survive in grain at this relatively high temperature, thus reducing the need for fumigating. Stored grain with a temperature of 10°C or below need not be fumigated, as most insect pests are relatively inactive at this temperature. Unfortunately, some stored-grain insect pests are able to survive low temperatures that prohibit the practical use of fumigants,[176] and some develop concentrated populations to create a favorable environment with a cool grain mass.[177] In general, fumigants with a relatively low boiling point can be used more effectively at reduced temperatures.

As mentioned earlier, temperatures will not be constant throughout a grain mass. Dry grain heating can occur in grain with a moisture content of 15% or below as the result of insects. These "hot spots" may reach 40°C. This increased temperature accelerates insect activity, and the hot spots may spread throughout the grain mass. Such temperatures usually will not damage the grain, but secondary effects resulting from translocation of water to the surface of the grain mass may be costly. Wet grain heating is due to the metabolism of microorganisms, especially storage molds, and occurs in grains of more than 15% moisture. The kernels become discolored, develop a musty odor, and deteriorate in nutritive value. Fumigants are useful to reduce heating produced by insects, but are not effective against microorganisms.

Temperatures within bulk stored grains change slowly. The only exceptions are the layers of kernels next to the bin wall and at or near the surface. The lag between the wall and center temperatures is excessive. Even with a sharp and significant change in the outside temperatures, changes within the bin are small. For example, beginning in September in the U.S., the temperature in the center of binned bulk grain remains higher than that near the walls until late March. The reverse of this process occurs throughout the spring and summer. Moisture migration, or movement, within the bin is a direct response to temperature variation. Temperature gradients create convection currents causing cool air to move downward near the outside bin walls, then across toward the center where the air is warmed and rises toward the surface. The higher temperature at the center enhances the moisture loss from the grain to the warmer air, and the moisture is carried by convection currents to the cooler area near the surface. Moisture is lost from the air by condensation and is absorbed by the surface grain. Wet grain heating may result. The relationship of storage temperature and grain moisture content to insect heating, reduction in germination, and damp grain heating, established by Burges and Burrell,[178] has been used as a guideline for safe storage.

Increasing the moisture content of grain generally increases the sorptive capacity of the kernels, with a resulting loss of fumigant concentration. In general, a 25% increase in fumigant dosage is, or should be, allowed to compensate for absorption and penetration when the grain moisture content exceeds 14.5%. However, sufficient moisture is mandatory for fumigation with phosphine. With low amounts of moisture available, the rate of phosphine generation may be too little and too late to obtain adequate insect mortality.

Methods of Application

Some fumigants may be applied directly to the grain stream as the bin is being filled. There is an advantage to this method of application since it disperses the fumigant more evenly throughout the grain mass. Gas such as methyl bromide is convenient to apply if the storage is equipped with an aeration system to distribute the gas.

Phosphine and other solid-type formulations are especially well adapted for application to grain as it is being moved, although it can also be probed into the grain after being binned.

Two standard methods of applying fumigants are gravity penetration and forced distribution. With gravity penetration, the fumigant is usually distributed over the surface grain or it is probed into the grain. The fumigant can also be introduced either continuously or intermittently. Sometimes, an extra amount is added for the first and last lots of grain where the greatest concentrations of insects usually occur and where the fumigant is most easily lost.

Another means of fumigating deep bins is the layering method. This is accomplished by placing a certain amount of grain in a bin, applying a prorated amount of the total dosage of fumigant to the surface of the grain, and introducing an additional amount of grain at each desired level until the bin is filled.

A guide for various fumigation application rates for bulk grain is covered in Table 3.

Dockage

Dockage is an important factor as it reduces fumigation efficiency resulting from decreased penetration, increased sorption, and channeling of vapors through the grain mass. Mortalities of *S. sasakii* decreased significantly when fumigating wheat of 5% or more moisture.[180] In addition, dockage generally favors insect survival and development as it provides some protection for the insects from the toxic concentrations of the fumigants.

Storage Facilities

Grain storage structures and materials vary widely in the U.S., not to mention in the tropics of developing countries. In North America, the construction material generally includes metal, concrete, and wood. All types have to be sealed if successful fumigation is expected. Gas-proof tarpaulins over storage structures have been used in the U.S. to ensure a lethal gas concentration for *T. granarium*.

The most gas-tight storages are constructed of metal, as enclosures fashioned from sheet metal welded at the seams are the most effective. Such tanks may have been constructed originally for storing oil and later converted into grain storages. Other storages in this general group are usually circular and constructed from corrugated metal strips that are bolted together at the joints. The gas-tight quality of these circular metal bins is improved if the joints are caulked when the bins are assembled. Rectangular buildings constructed of corrugated and flat metal and bolted together at the seams fall in this same category.

Storage structures fashioned from concrete are popular. It is important that the walls be finished so that they are smooth and absent of pockets, grooves, or any other irregularities. Such irregularities provide places where grain may remain static to provide food for insects. Cracks and crevices in the walls or foundations also serve as hiding places for insects. Larvae of the dermestid beetles are particularly attracted to such areas, increasing the problem of fumigation since the fumigant vapors may not penetrate into the crevices in sufficient concentration to kill the insects. The concrete walls may also become porous in time and permit the escape of excessive amounts of fumigant vapors.

Grain stored in wooden facilities is difficult to fumigate because such structures are porous and permit an excessive loss of fumigant. As a result of this leakage, fumigant dosage recommendations for wooden bins may be twice the amount recommended for metal or concrete bins. Wood storage structures also provide cracks into which grain may lodge and

Table 3
FUMIGANTS AND DOSAGE RECOMMENDED FOR BULK FUMIGATION OF GRAIN IN UPRIGHT AND FLAT STORAGES

Fumigant	Dosage per m³	per 1000 bu	Minimum exposure (days)	Remarks
Upright Storage				
Direct mixing				
Calcium cyanide	154.5 g	12.0 lb	7	May stain white maize and polished rice
Chloropicrin	25.7 g	2.0 lb	1	should be removed by aeration after 24 hr
Aluminum phosphide	2.5 tablets	90 tablets	3	5 days at 10—15°C
	9 pellets	300 pellets		4 days at 16—20°C
				3 days at 21°C or above
Recirculation				
Methyl bromide	25.7 g	2.0 lb	1	Should be removed by aeration after 24 hr
Hydrocyanic acid	38.6g	3.0 lb	1	Grain should be thoroughly aerated before removal
Mixture ETO:CO₂ (10:90)	386.1 g	30.0 lb	1	Do not use on seed
Flat Storage				
Application by probe				
Aluminum phosphide	2.5 tablets	90 tablets	3	5 Days at 10—15°C
				4 Days at 15—20°C
				3 Days at 21°C or above
Surface application (gravity distribution)				
Methyl bromide	32.2 g	2.5 lb	1	Applied under gas-tight sheet in South Africa; must be removed by aeration after 24 hr
Recirculation				
Methyl bromide	38.6 g	3.0 lb	1	Should be removed by aeration after 24 hr
Hydrocyanic acid	38.6 g	3.0 lb	1	Should be thoroughly aerated before grain is moved
Mixture chloropicrin:methyl chloride (85:15)	38.6 g	3.0 lb	1	Should be removed by aeration after 24 hr

where insects may find refuge and multiply. Hall[181] summarized the different types of storage facilities in developing countries.

Frequency of Application and Exposure Period

The number and timing of fumigant applications depends chiefly on the need for fumigation as indicated by the presence of living insects. The effectiveness of previous treatments, the opportunities for reinfestation, and the rate and type of reinfestation also should influence the frequency of fumigations. The desired length of exposure and dosage depends on many factors. Exposure time and dosage are inversely related so that shorter exposures require higher dosages and vice versa.

Insect Species and Stage

Increased fumigant doses usually are required for massive insect populations. Such populations are often associated with relatively high-moisture grain. Large insect populations also create greater quantities of dust, frass, and damaged kernels. Insect colonies that have been established for extended periods are usually living in settled, well-compacted grain. Likewise, they will be found in pockets with considerable accumulations of their own by-products around them.

Some insect species are more resistant to a given fumigant than other species, and certain stages (or ages of a stage) of a given species are more resistant than other stages. It should be noted, however, that the different species do not behave alike as to the stage of greatest resistance to susceptibility. Information on this point can be determined only by experience.

Resistance or susceptibility of an insect may result from such physiological activities as its respiratory rate, its ability to close the openings to its respiratory system, and the quality and quantity of food consumed prior to fumigation.

There are several definitions for resistance and tolerance. Resistance occurs when an insect population demonstrates consistently greater survival following similar exposures to an insecticide. Tolerance is usually restricted to exposures where the insecticide has penetrated into the tissues of the insect but fails to produce death.

Only eight species of stored-grain insects were known to have developed resistance to insecticides by 1965.[182,183] The pesticides include malathion, lindane, pyrethrins, and methyl bromide, some of the most popular materials for chemical control of stored-grain insects. At least 13 species had developed pesticide resistance by 1970.[184] Resistance has not developed as quickly among stored-grain insects compared to public health and agricultural insect pests. Innately tolerant stages of an insect may survive an exposure to an insecticide, reproduce, and increase the selection of more tolerant progeny, thus providing the potential for the development of further resistant species.

Monro et al.[185] reported an induced tolerance of *S. granarius* to methyl bromide by using the survivors of an insect population following exposures as parents for the succeeding generation. This tolerance was retained for 23 generations after cessation of exposure. An increase of tolerance to other fumigants, not related to methyl bromide, was also noted. Other laboratory population selections of stored-grain insects have also developed resistance to phosphine.[186]

Resistance to specific fumigants and cross resistance to other fumigants is a definite possibility under field conditions as well. If the mechanisms of developing resistance are understood and this information is considered when establishing control programs, procedures and conditions that could support the development of economically important resistant insect populations may be avoided.

Lindgren and Vincent[187] were only able to show small differences in the tolerances of *T. confusum* and *T. castaneum* to the fumigants ethylene dibromide, methyl bromide, and hydrogen cyanide. Two years later, they reported the relative susceptibility of the same insect cultures to malathion and pyrethrins as determined by the topical application method described by March and Metcalf.[188]

A pyrethrin-resistant strain of *S. granarius* increased from $18 \times$ resistant than a standard strain, as reported in 1960,[189] to 52 in 1963.[190] Positive cross tolerance of *S. granarius* was also noted to dieldrin, carbaryl, lindane, malathion, allethrin, synergized pyrethrins, and synergized allethrins. These tolerances were interpreted in relation to various environmental stresses. *S. granarius* also developed resistance to Bagon® and fenthion.[191]

Although there are many resistant possibilities in addition to those noted above, malathion-resistant stored-grain insects are of greatest concern because it is a widely accepted treatment — widely accepted because of an endless collection of data regarding its toxicity to the target insects, relatively low mammalian toxicity, low cost, and acceptable residual characteristics under many conditions of application. This concern for malathion by the grain and cereal industry reaches levels of even greater importance as it is realized that possible substitutes are difficult and costly to obtain, especially with the impact of increased legal restrictions.

Malathion-resistant *T. castaneum* are widely dispersed throughout international trade, and both malathion- and lindane-resistant strains are being spread in ships to additional ports and new countries.[192] A malathion cross-resistant strain of *T. castaneum* was later reported[193] to be cross resistant to malaoxen, diethyl malathion, phenthoate, and acethion. Cross resistance of *T. castaneum* to many insecticides is also listed by Champ and Champbell-Brown[194] and Bhatia et al.[195] Even a cross resistance from several conventional insecticides to synthetic juvenile hormones has been reported for *T. castaneum*.[196S]

INFESTATION IN TRANSIT

A major step in efforts to provide food is to have adequate facilities and management technology to protect it during distribution. This requirement applies to grain from the point of harvest, through multiple grain handling and cereal product processing channels, to wholesalers and retailers and eventually to the ultimate consumer.

In terms of volume within developed countries, virtually all grain moves through distribution channels in railway boxcars, trucks, and ships. Although an increasing number of special-purpose railway boxcars, trucks, and ships are being used by industry, most of the grain is transported in vehicles built for multipurpose use. This includes grain destined for humans or animals plus an endless variety of nonfood commodities. Most of the vehicles used today were not designed to deter insect infestations.

The common railway boxcar is designed to carry anything from soup to nuts. A wooden liner is usually provided to protect packaged materials from breakage and moisture. Even the roof may be lined with plywood to protect the product from condensation. End linings are generally installed vertically from floor to ceiling, providing four to six vertical compartments as the liners are bolted against the corrugated steel ends of the railway boxcars.

Commodities, such as grain or cereal products, soon accumulate behind the wall and end or ceiling liners, especially if the product is blown into the railway boxcar. In addition, the doorpost and floors of these vehicles soon become cracked, broken, or gouged, providing areas for accumulations of products and a home for stored-grain insects.

A combination of air blowing, followed by adequate vacuum cleaning, appears to work best to remove the residual commodity from railway boxcars. They should *not* be cleaned with water. The wet and moldy product that is not removed provides an optimum environment for many species of insects and microorganisms.

There is little chance of killing all insects residing behind the wall liners in railway boxcars, especially if they also have the protection provided by accumulations of grain and various cereal products. However, until railway boxcars are designed to eliminate such harborage sites, it would be advantagous to inhibit the migration of stored-grain insects to the commodities in transit. Dichlorvos dispensed as an aerosol may provide such protection, as Schesser[197] obtained 100% control of *T. confusum* exposed to dichlorvos in railway boxcars treated with 6 oz of 6.5% dichlorvos per 141.5 m^3 of space.

It is not an uncommon practice to use barges to backhaul nonfood cargoes following a shipment of grain in bulk. A major concern, in addition to subsequent insect infestation, is improper cleaning before reloading with food or feed. The practice of cleaning barges with raw untreated river water and not allowing adequate drying time also invites mold growth and mycotoxin development.

Pyrethrins synergized with piperonyl butoxide, malathion, methyoxychlor, or dichlorvos, if applied correctly, are useful as protective sprays when applied to the walls and floor of empty railway boxcars, ships, and trucks. Having residual characteristics, they help to reduce the migration and subsequent infestation of insects emerging from sites in the structure of the vehicle.

Malathion is one of the most effective residual insecticides to combat insect infestations of stored grain or cereal products in railway boxcars. Schesser[198-201] tested different insecticide and fumigant formulations using various application techniques in an attempt to develop an improved method. One formulation consisted of a mixture of dichlorvos and malathion, the dichlorvos to produce lethal vapors and the malathion to serve as a contact toxicant. Schesser[202] found a formulation containing equal amounts of dichlorvos and malathion, to produce a total concentration of 2.5%, effective in disinfesting empty railway boxcars infested with *T. castaneum.*

The type of surface receiving the insecticidal treatment makes considerable difference in the resulting protection provided for the grain against possible insect infestation. Concrete and other alkaline surfaces often reduce the toxicity of insecticides. This was noted first by Burkholder[203] in the treatment of malathion on concrete and latex-painted surfaces to control dermestids. Later Burkholder and Dicke found that painting or water proofing various surfaces decreased the dosage required for adequate insecticidal activity.[204] Incorporating 0.5% sodium carboxymethyl cellulose in malathion spray formulation also improved the persistence of its residues on an alkaline, cement substrate.[205] Similar results were noted using talc or calcium carbonate.

Sun[206] pointed out multiple factors that influence the toxicity of fumigants to insects, including temperature, moisture, diet, stage and age of development, and insect habits. These and many other variables are diagramatically expressed in Figure 1.

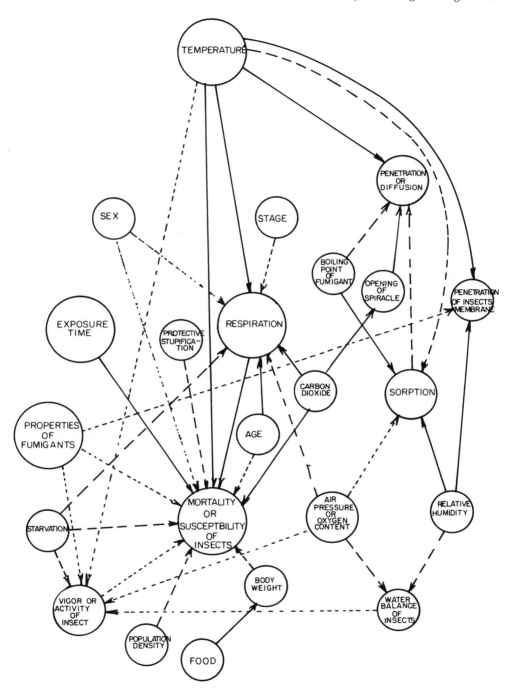

FIGURE 1.   Relation of variables and their counter and parallel effects on the susceptibility of insects to fumigants.

REFERENCES

1. **Hall, D. W.,** Handling and Storage of Food Grains in Tropical and Sub-Tropical Areas, FAO Agric. Dev. Paper 90, Food and Agriculture Organization, Rome, 1970.
2. **Joffe, A.,** Moisture migration in horizontally stored bulk maize: the influence of grain-infesting insects under South African conditions, *S. Afr. J. Agric. Sci.,* 1, 175, 1958.
3. **Freeman, J. A.,** Infestation of grain in international trade — a review of problems and control methods, *J. Sci. Food Agric.,* 11, 623, 1957.
4. **Sinha, R. N. and Utida, S.,** Climatic areas potentially vulnerable to stored product insects in Japan, *Appl. Entomol. Zool.,* 2, 124, 1967.
5. **Sinha, R. N.,** Insects associated with stored products in Canada, *Can. Insect Pest Rev.,* Suppl. 2, 1, 1965.
6. **Linsley, E. G.,** Natural sources, habits and reservoirs of insects associated with stored food products, *Hilgardia,* 16, 187, 1944.
7. **Woodroffe, G. E.,** Natural sources of domestic insects: a fundamental approach for the stored products entomologist, *Sanitarian,* p. 1, 1961.
8. **Cutler, J. R. and Hosie, G.,** Bird nests as sources of infestation of *Ptinus tectus* Boleldieu (Coleoptera, Ptinidae) and the distribution of this insect in stacks of bagged flour, *J. Stored Prod. Res.,* 2, 27, 1966.
9. **Strong, R. G.,** Distribution and relative abundance of stored-product insects in California: a method of obtaining sample populations, *J. Econ. Entomol.,* 63, 591, 1970.
10. **LaHue, D. W. and Kadoum, A.,** Residual effectiveness of emulsion and encapsulated formulations of malathion and fenitrothion against four stored grain beetles, *J. Econ. Entomol.,* 72, 234, 1979.
11. **Mensah, G. W. K., Watters, F. L., and Webster, G. R. B.,** Translocation of malathion, bromophos, and iodofenphos into stored grain from treated structural surfaces, *J. Econ. Entomol.,* 72, 385, 1979.
12. **LaHue, D. W.,** Evaluation of Malathion, Synergized Pyrethrum and Diatomaceous Earth on Shelled Corn as Protectants Against Insects in Small Bins, USDA Marketing Res. Rep. 768, U.S. Department of Agriculture, Washington, D.C., 1966.
13. **White, G. D., Berndt, W. L., Schesser, J. H., and Fifield, C. C.,** Evaluation of Four Inert Dusts for the Protection of Stored Wheat in Kansas from Insect Attack, USDA Agric. Res. Serv. 51-8, U.S. Department of Agriculture, Washington, D.C., 1966.
14. **LaHue, D. W. and Fifield, C. C.,** Evaluation of Four Insect Dusts on Wheat as Protectants against Insects in Small Bins, USDA Marketing Res. Rep. 780, U.S. Department of Agriculture, Washington, D.C., 1967.
15. **McHaughey, W. H.,** Diatomaceous earth for confused flour beetle and rice weevil control in rough, brown and milled rice, *J. Econ Entomol.,* 65, 1427, 1971.
16. **Venugopal, J. S. and Majumder, S. K.,** Active Mineral in Insecticidal Clays, Abstr. Symp. Pesticides, Mysore, India, 1964.
17. **Majumder, S. K. and Venugopal, J. S.,** Pesticidal minerals, in *Pesticides,* Acad. Pest Control Sci., Mysore, India, 1968, 190.
18. **Bano, A. and Majumder, S. K.,** Tricalcium phosphate as an insecticide, in *Pesticides, Acad. Pest* Control Sci., Manipal Power Press, Mysore, India, 1968, 177.
19. **Ashman, F.,** An assessment of the value of dilute dust insecticide for the protection of stored maize in Kenya, *J. Appl. Ecol.,* 3, 169, 1966.
20. **Richards, G.,** personal communication, 1973.
21. **Ebeling, W.,** Use of mineral dusts for protection against insect pests with special reference to cereal grains, in *Grain Sanitation,* Majumder, S. K. and Venugopal, J. S., Eds., Acad. Pest Control Sci., Mysore, India, 1969, 103.
22. **LaHue, D. W.,** Evaluation of Malathion, Synergized Pyrethrum and Diatomaceous Earth on Shelled Corn as Protectants against Insects in Small Bins, USDA Marketing Res. Rep. 768, U.S. Department of Agriculture, Washington, D.C., 1966.
23. **LaHue, D. W.,** Evaluation of Malathion, Synergized Pyrethrum and a Diatomaceous Earth as Protectants Against Insects in Sorghum Grain in Small Bins, USDA Marketing Res. Rep. 781, U.S. Department of Agriculture, Washington, D.C., 1967.
24. **LaHue, D. W.,** Evaluation of Malathion, Diazinon, a Silica Aerogel and a Diatomaceous Earth as Protectants on Wheat Against Lesser Grain Borer Attack in Small Bins, USDA Marketing Res. Rep. 860, U.S. Department of Agriculture, Washington, D.C., 1970.
25. **LaHue, D. W.,** Grain protectants for seed corn, *J. Econ. Entomol.,* 69, 652, 1976.
26. **Minett, W. and Williams, P.,** Influence of malathion distribution on the protection of wheat grain against insect infestation, *J. Stored Prod. Res.,* 7, 233, 1971.
27. **Miles, J. W., Guerrant, G. O., Goette, M. B., and Churchill, F. C.,** Studies on the Chemistry, Methods of Analysis and Storage Stability of Malathion Formulations, Tech. Rep Series 475, Expert Committee on Insecticides, Geneva, 1971.

28. **Parkin, E. A., Scott, E. I. C., and Forester, R.,** The resistance of field strains of beetles, *Pest Infest. Res. (Slough, England),* 1961-1962, 34.

29. **Hayward, L. A. W.,** Ground Nuts, Rep. W. Afr. Stored Prod. Res. Unit, 1961-1962, 12.

30. **Mansah, G. W. K. and Watters, F. L.,** Comparison of four organophosphorous insecticides on stored wheat for control of susceptible and malathion-resistant strains of the red flour beetle, *J. Econ. Entomol.,* 72, 456, 1979.

31. **LaHue, D. W.,** Control of malathion-resistant Indian-meal moths *Plodia interpunctella* with dichlorvos resin strips, *Proc. N. Central Branch Entomol. Soc. Am.,* 24, 117, 1969.

32. **Bang, U. H. and Floyd, E. G.,** Effectiveness of malathion in protecting stored polished rice from damage by several species of stored grain insects, *J. Econ. Entomol.,* 55, 188, 1962.

33. **Parkin, E. A.,** The protection of stored seeds from insects and rodents, *Proc. Int. Seed Test Assoc.,* 28, 893, 1963.

34. **Moore, S. and Decker, G. C.,** Control of the Angoumois grain moth (*Sitotroga cerealella*) in stored earcorn with malathion in Illinois, 1959-1960, *J. Econ. Entomol.,* 54, 479, 1961.

35. **Lindgren, D. L., Krohne, H. E., and Vincent, L. E.,** Malathion and chlorthion for control of insects infesting stored grain, *J. Econ. Entomol.,* 47, 705, 1954.

36. **Kadoum, A. M. and LaHue, D. W.,** Degradation of malathion on wheat and corn of various moisture contents, *J. Econ. Entomol.,* 72, 228, 1979.

37. **Watters, F. L.,** Effects of grain moisture content on residual toxicity and repellency of malathion, *J. Econ. Entomol.,* 52, 131, 1959.

38. **LaHue, D. W.,** Evaluation of Several Formulations of Malathion as a Protectant of Grain Sorghum Against Insects in Small Bins, USDA Marketing Res. Rep. 828, U.S. Department of Agriculture, Washington, D.C., 1969.

39. **LaHue, D. W.,** Laboratory Evaluation of Dichlorvos as a Short-Term Protectant for Wheat, Shelled Corn, and Grain Sorghum Against Stored Grain Insects, USDA Agric. Res. Serv. 51-37, U.S. Department of Agriculture, Washington, D.C., 1970.

40. **King, D. R., Morrison, E. O., and Sundman, J. A.,** Bioassay of chemical protectants and surface treatments for the control of insects in stored sorghum grain, *J. Econ. Entomol.,* 55, 506, 1962.

41. **Green, A. A. and Tyler, P. S.,** A field comparison of malathion, dichlorvos and fenitrothion for the control of *Oryzaephilus surinamensis* (L.) (Coleoptera, Silvanidae) infesting stored barley, *J. Stored Prod. Res.,* 1, 273, 1966.

42. **Strong, R. G., Sbur, D. E., and Partida, G. J.,** The toxicity and residual effectiveness of malathion and diazinon used for protection of stored wheat, *J. Econ. Entomol.,* 60, 500, 1967.

43. **Lemon, R. W.,** Laboratory evaluation of malathion, bromophos and fenitrothion for use against beetles infesting stored products, *J. Stored Prod. Res.,* 2, 197, 1967.

44. **Tyler, P. S. and Green, A. A.,** The effectiveness of fenitrothion and malathion as grain protectants under severe practical conditions, *J. Stored Prod. Res.,* 4, 119, 1968.

45. **Mallis, A.,** *Handbook of Pest Control,* McNair-Dorland, New York, 1969.

46. **Lloyd, C. J. and Hewlett, P. S.,** Relative susceptibility to pyrethrum in oil of Coleoptera and Lepidoptera infesting stored products, *Bull. Entomol. Res.,* 49, 177, 1958.

47. **Brooks, I. C., Haus, J., Blumenthal, R. R., and Davis, B. S., Jr.,** SBP-1382 — a new synthetic pyrethroid, *Soap Chem. Spec.,* March-April 1969.

48. **Davies, M. S., Chadwick, P. R., Holborn, J. M., Stewart, D. C., and Wickhem, J. C.,** Effectiveness of the (+)-trans-chrysanthemie acid ester of (±)-allenthrolone (bio-allethrin) against four insect species, *Pest. Sci.,* 1, 225, 1970.

49. **LaHue, D. W.,** Evaluation of Malathion, Synergized Pyrethrum and Diatomaceous Earth on Shelled Corn as Protectants Against Insects in Small Bins, USDA Marketing Res. Rep. 768, U.S. Department of Agriculture, Washington, D.C., 1966.

50. **Brooke, J. P.,** Protection of grain in storage, *World Crops,* 13, 27, 1961.

51. **Strong, R. G., Sbur, D. E., and Partida, G. J.,** The toxicity and residual effectiveness of malathion and diazinon used for protection of stored wheat, *J. Econ. Entomol.,* 60, 500, 1967.

52. **LaHue, D. W.,** Evaluation of Malathion, Diazinon, a Silica Aerogel and a Diatomaceous Earth as Protectants on Wheat Against Lesser Grain Borer Attack in Small Bins, USDA Marketing Res. Rep. 860, U.S. Department of Agriculture, Washington, D.C., 1970.

53. **Telford, H. S., Zwick, R. W., Sikorowski, P., and Weller, M.,** Laboratory evaluation of diazinon as a wheat protectant, *J. Econ. Entomol.,* 57, 272, 1964.

54. **Roan, C. C. and Srivastava, B. P.,** Dissipation of diazinon residues in wheat, *J. Econ. Entomol.,* 58, 996, 1965.

55. **Lemon, R. W.,** Laboratory evaluation of some organophosphorus insecticides against *Tribolium confusm* Duv. and *Tribolium castaneum* (Hbst.) (Coleoptera, Tenebrionidae), *J. Stored Prod. Res.,* 1, 247, 1966.

56. **Lemon, R. W.,** Laboratory evaluation of malathion, bromophos and fenitrothion for use against beetles infesting stored products, *J. Stored Prod. Res.,* 2, 197, 1967.

57. **Green, A. A. and Tyler, P. S.,** A field comparison of malathion, dichlorvos and fenitrothion for the control of *Oryzaephilus surinamensis* (L.) (Coleoptera, Silvanidae) infesting stored barley, *J. Stored Prod. Res.,* 1, 273, 1966.

58. **Kane, J. and Green, A. A.,** The protection of bagged grain from insect infestation using fenitrothion, *J. Stored Prod. Res.,* 4, 59, 1968.

59. **Tyler, P. S. and Green, A. A.,** The effectiveness of fenitrothion and malathion as grain protectants under severe practical conditions, *J. Stored Prod. Res.,* 4, 119, 1968.

60. **Harein, P. K.,** Effect of ronnel upon the adult rice weevil, *Sitophilus oryze, J. Econ. Entomol.,* 53, 372, 1960.

61. **Harein, P. K. and Gillenwater, H. B.,** Exploratory tests with bromodan as a protectant for wheat against stored-product insects, *J. Econ. Entomol.,* 59, 413, 1966.

62. **Green, A. A., Tyler, P. S., Kane, J., and Rowlands, D. G.,** An assessment of bromophos for the protection of wheat and barley, *J. Stored Prod. Res.,* 6, 217, 1970.

63. **McDonald, L. L. and Gillenwater, H. B.,** Relative toxicity of Bay 77488 and Dursban against stored-product insects, *J. Econ. Entomol.,* 60, 1195, 1967.

64. **Strong, R. G.,** Relative susceptibility of five stored-product moths to some organophosphorous insecticides, *J. Econ. Entomol.,* 62, 1036, 1969.

65. **LaHue, D. W. and Dicke, E. B.,** Phoxim as an insect protectant for stored grains, *J. Econ. Entomol.,* 64, 1530, 1971.

66. **Strong, R. G.,** Relative susceptibility of confused and red flour beetles to twelve organophosphorous insecticides with notes on adequacy of the test method, *J. Econ. Entomol.,* 63, 258, 1970.

67. **Harein, P. K. and Rao, H. R. G.,** Dichlorvos and gardona as protectants for stored wheat against granary weevil infestation in laboratory studies, *J. Econ. Entomol.,* p. 1402, 1972.

68. **LaHue, D. W.,** Gardona as a protectant against insects in stored wheat, *J. Econ. Entomol.,* 66, 485, 1973.

69. **Hall, R. C., Ballee, D. L., Bennett, G. W., and Fahay, J. E.,** Persistence and distribution of gardona and dichlorvos in grain and grain products, *J. Econ. Entomol.,* 66, 315, 1973.

70. **Champ. B. R. and Cribb, J. N.,** Lindane resistance in *Sitophilus oryzae* (L.) and *Sitophilus zeamaiz* Motsch. (Coleoptera, Curculionidae) in Queensland, *J. Stored Prod. Res.,* 1, 9, 1965.

71. **Dyte, C. E. and Blackman, D. G.,** The spread of insecticide resistance in *Tribolium castaneum* (Herbst) (Coleoptera, Tenebrionidae), *J. Stored Prod. Res.,* 6, 255, 1970.

72. **Parkin, E. A.,** The onset of insecticide resistance among field strains of stored-product insects, *12th Int. Congr. Entomol.,* p. 657, 1965.

73. **Parkin, E. A.,** The onset of insecticide resistance among field populations of stored-product insects, *J. Stored Prod. Res.,* 1, 1, 1965.

74. **Mattson, A. M., Spillane, J. T., and Pearce, G. W.,** Dimethyl 2,2-dichlorovinyl phosphate (DDVP), an organic phosphorous compound highly toxic to insects, *J. Agric. Food Chem.,* 3, 319, 1955.

75. **Strong, R. G. and Sbur, D. E.,** Evaluation of insecticides as protectants against pests of stored grain seeds, *J. Econ. Entomol.,* 54, 235, 1961.

76. **Strong, R. G. and Sbur, D. E.,** Influence of grain moisture and storage temperature on the effectiveness of five insecticides as grain protectants, *J. Econ. Entomol.,* 57, 44, 1964.

77. **Strong, R. G. and Sbur, D. E.,** Protective sprays against internal infestations of grain beetles in wheat, *J. Econ. Entomol.,* 57, 544, 1964.

78. **Jay, E. G., Gillenwater, H. B., and Harein, P. K.,** The toxicity of several dichlorvos (DDVP) and naled formulations to the adult confused flour beetle, *J. Econ. Entomol.,* 57, 415, 1964.

79. **Kirkpatrick, R. L., Harein, P. K., and Cooper, C. V.,** Laboratory tests with dichlorvos as a wheat protectant against rice weevils, *J. Econ. Entomol.,* 61, 356, 1968.

80. **LaHue, D. W.,** Laboratory Evaluation of Dichlorvos as a Short-Term Protectant for Wheat, Shelled Corn, and Grain Sorghum Against Stored Grain Insects, USDA Agric. Res. Serv. 51-37, U.S. Department of Agriculture, Wasington, D.C., 1970.

81. **Harein, P. K. and Rao, H. R. G.,** Dichlorvos and gardona as protectants for stored wheat against granary weevil infestations in laboratory studies, *J. Econ. Entomol.,* p. 1402, 1972.

82. **Green, A. A. and Wilkin, D. R.,** The control of insects in bagged grain by the injection of dichlorvos, *J. Stored Prod. Res.,* 5, 11, 1969.

83. **Cogburn, R. R.,** Pirimiphos-methyl as a protectant for stored rough rice: small bin tests, *J. Econ. Entomol.,* 69, 369, 1976.

84. **Quinlan, J. K., White, G. D., Wilson, J. L., Davidson, L. I., and Hendricks, L. H.,** Effectiveness of chlorpyrifos-methyl and malathion as protectants for high moisture stored wheat, *J. Econ. Entomol.,* 72, 90, 1979.

85. **Speirs, R. D. and Lang, J. H.,** Contact, Residue and Vapor Toxicity of New Insecticides to Stored-Product Insects. II, USDA Marketing Res. Rep., U.S. Department of Agriculture, Washington, D.C., 1970.

86. **Strong, R. G. and Sbur, D. E.,** Evaluation of insecticides for control of stored-product insects, *J. Econ. Entomol.,* 61, 1034, 1968.

87. **Strong, R. G.,** Relative susceptibility of *Attagenus alfierii* and *A. megatoma* larvae to several organo-phosphorous insecticides, *J. Econ. Entomol.,* 63, 286, 1970.

88. **Strong, R. G.,** Relative susceptibility of confused and red flour beetles to twelve organophosphorous insecticides with notes on adequacy of the test method, *J. Econ. Entomol.,* 63, 258, 1970.

89. **McDonald, L. L. and Speirs, R. D.,** Toxicity of five new insecticides to stored-product insects, *J. Econ. Entomol.,* 65, 529, 1972.

90. **House, H. L. and Grahm, A. R.,** Capric acid blended into foodstuff for control of an insect pest, *Tribolium confusum* (Coleoptera, Tenebrionidae), *Can. Entomol.,* 99, 994, 1967.

91. **Su, H. C. F., Speirs, R. D., and Mahany, P. G.,** Citrus oils as protectants of black-eyed peas against cowpea weevil: laboratory evaluation, *J. Econ. Entomol.,* 65, 1433, 1972.

92. **Su, H. C. F., Speirs, R. D., and Mahany, P. G.,** Toxicity of citrus oils to several stored-product insects: laboratory evaluation, *J. Econ. Entomol.,* 65, 1438, 1972.

93. **Baker, J. E. and Mabie, J. M.,** *Tribolium confusum*: food additives as ovicides, *J. Econ. Entomol.,* 66, 765, 1973.

94. **Burkholder, W. E. and Dicke, R. J.,** The toxicity of malathion and fenthion to dermestid larvae as influenced by various surfaces, *J. Econ. Entomol.,* 59, 253, 1966.

95. **Bhatnagar-Thomas, P. L.,** Control of insect pests of stored grains using a juvenile hormone analogue, *J. Econ. Entomol.,* 66, 277, 1973.

96. **Dyte, C. E.,** Resistance to synthetic juvenile hormone in a strain of the flour beetle, *Tribolium castaneum, Nature (London),* 238, 48, 1972.

97. **Whitney, W. K.,** Fumigation hazards as related to the physical, chemical and biological properties of fumigants, *Pest Control,* 7, 16, 1962.

98. **Kenaga, E. E.,** Time, temperature, and dosage relationships of several insecticidal fumigants, *J. Econ. Entomol.,* 54, 537, 1961.

99. **Whitney, W. K. and Walkden, H. H.,** Concentrations of Methyl Bromide Lethal to Insects in Grain, USDA Agric. Marketing Serv. Rep. 511, U.S. Department of Agriculture, Wasington, D.C., 1961.

100. **Harein, P. K. and Krause, G. F.,** Dosage-time relationships between 80:20 (CCl_4:CS_2) and the adult rice weevil, *Sitophilus oryzae, J. Econ. Entomol.,* 54, 521, 1964.

101. **Estes, P. M.,** The effects of time and temperature on methyl bromide fumigation of adults of *Sitophilus granarius* and *Tribolium confusum, J. Econ. Entomol.,* 58, 611, 1965.

102. **Howe, R. W. and Hole, B. D.,** Predicting the dosage of fumigant needed to eradicate insect pests from stored products, *J. Appl. Ecol.,* 4, 337, 1967.

103. **Thompson, R. H.,** Specifications Recommended by the United Kingdom Ministry of Agriculture, Fisheries and Food for the Fumigation of Cereals and Other Foodstuffs Against Pests of Stored Products, Public OEPP, Ser. D, 15, 1970, 9.

104. **Phillips, G. L. and Bulger, J. W.,** Analysis of Methyl Bromide by Measurements of Thermal Conductivity, USDA Bur. Entomol. Plant Quarantine No. E851, U.S. Department of Agriculture, Washington, D.C., 1953.

105. **Phillips, G. L.,** Experiments on Distributing Methyl Bromide in Bulk Grains with Aeration Systems, USDA Agric. Marketing Serv. 150, U.S. Department of Agriculture, Washington, D.C., 1957.

106. **Kenaga, E. E.,** Calibration of thermal conductivity units for use with commodity fumigants, *Down Earth,* 14, 6, 1958.

107. **Heseltine, H. K.,** The use of thermal conductivity meters in fumigation research and control, *Pest Infest. Res. Bull.,* p. 2, 1961.

108. **Koucherova, S. G. and Lisitsyn, F. T.,** The gas analyzer and its use in quarantine fumigation, *Zaschita Rastenii,* 7, 52, 1962.

109. **Monro, H. A. U.,** Manual of Fumigation for Insect Control, FAO Manual 79, Food and Agriculture Organization, Rome, 1969.

110. **Heseltine, H. K. and Royce, A.,** A concentration-time product indicator for fumigations, *Pest Technol.,* 2, 88, 1960.

111. **Brieger, H., Rieders, F., and Hodes, W. A.,** Acrylonitrile: spectrophotometric determination, acute toxicity and mechanism of action, *Arch. Ind. Hyg.,* 6, 128, 1952.

112. **Berck, B.,** Determination of fumigant gases by gas chromatography, *J. Agric. Food Chem.,* 13, 373, 1965.

113. **Bielorai, R. and Alumot, E.,** Determination of residues of a fumigant mixture in cereal grain by electron capture gas chromatography, *J. Agric. Food Chem.,* 14, 622, 1966.

114. **Berck, B.,** Sorption of ethylene dibromide, ethylene dichloride and carbon tetrachloride by cereal products, *J. Agric. Food Chem.,* 13, 248, 1965.

115. **Bielorai, R. and Alumot, E.,** Determination of residues of a fumigant mixture in cereal grain by electron capture gas chromatography, *J. Agric. Food Chem.,* 14, 622, 1966.

116. **Kanazawa, J.,** Determination of chloropicrin in fumigants by gas-liquid chromatography, *Agric. Biol. Chem.,* 27, 159, 1963.
117. **Berck, B.,** Determination of fumigant gases by gas chromatography, *J. Agric. Food Chem.,* 13, 373, 1965.
118. **Webley, D. J. and McKone, C. E.,** The estimation of Dichlorvos Vapour, Misc. Rep. 424, Tropical Pest Institute, Arusha Tanganyika, 1963.
119. **Heuser, S. G. and Scudamore, K. A.,** A rapid method for sampling dichlorvos vapour in air, *Chem. Ind.,* p. 2093, 1966.
120. **Berck, B.,** Determination of fumigant gases by gas chromatography, *J. Agric. Food Chem.,* 13, 373, 1965.
121. **Dumas, T.,** Determination of phosphine in air by gas chromatography, *J. Agric. Food Chem.,* 12, 257, 1964.
122. **Dumas, T.,** Microdetermination of phosphine in air by gas chromatography, *J. Agric. Food Chem.,* 17, 1165, 1969.
123. **Chakrabarti, B. and Wainman, H. E.,** Determination by gas chromatography of phosphine used in fumigation, *Chem. Ind. Technol.,* p. 300, 1972.
124. **Monro, H. A. U.,** Manual of Fumigation for Insect Control, FAO Manual 79, Food and Agriculture Organization, Rome, 1969.
125. **Bond, E. J.,** Chemical control of stored grain insects, in *Grain Storage — Part of a System,* Sinha, R. N. and Muir, W. E., Eds., Avi Publishing, Westport, Conn., 1973, 137.
126. **Phillips, G. L. and Nelson, H. D.,** Permeability of methyl bromide of plastic films and plastic and rubber-coated fabrics, *J. Econ. Entomol.,* 50, 452, 1957.
127. **Brown, W. B.,** Fumigation with Methyl Bromide under Gas-Proof Sheets, Pest Infest. Res. Bull. 1, 2nd ed., D.S.I.R., London, 1959.
128. **Brown, W. B., Heseltine, H. K., Wainman, H. E., Taylor, R. W., Wheatley, P. E., and Harris, A. H.,** Fumigation of Stacks of Bagged Produce with Phosphine, Pest Infestation Res. Rep., 1968-1969, 50.
129. **Bowen, M. F.,** Efficient utilization of paulins in fumigation and as covers for the protection of storage, *J. Econ. Entomol.,* 54, 270, 1961.
130. **Monro, H. A. U.,** Manual of Fumigation for Insect Control, FAO Manual 79, Food and Agriculture Organization, Rome, 1969.
131. **Monro, H. A. U.,** Manual of Fumigation for Insect Control, FAO Manual 79, Food and Agriculture Organization, Rome, 1969.
132. **Howe, R. W. and Hole, B. D.,** The susceptibility of the developmental stages of *Sitophilus granarius* (L.) (Coleoptera, Curculionidae) to methyl bromide, *J. Stored Prod. Res.,* 2, 13, 1966.
133. **Monro, H. A. U., Cunningham, C. R., and King, J. E.,** Hydrogen cyanide and methyl bromide as fumigants for insect control in empty cargo ships, *Sci. Agric. (Canada),* 32, 241, 1952.
134. **Godden, E. and Howe, R. W.,** The susceptibility of the developmental stages of *Tribolium castaneum* to methyl bromide, *Tribolium Inf. Bull.,* 8, 76, 1965.
135. **Bennett, R. G.,** The influence of age and concentration of fumigant on the susceptibility of pupae of *Tribolium castaneum* (Herbst) (Coleoptera, Tenebrionidae) to methyl bromide, *J. Stored Prod. Res.,* 5, 119, 1969.
136. **Whitney, W. K. and Walkden, H. H.,** Concentrations of Methyl Bromide Lethal to Insects in Grain, USDA Agric. Marketing Serv. Rep. 511, U.S. Department of Agriculture, Washington, D.C., 1961.
137. **Mostafa, S. A. S., Kamel, A. H., El-NaHal, A. K. M., and El-Borollosy, F. M.,** Toxicity of carbon bisulfide and methyl bromide to the eggs of four stored-product insects, *J. Stored Prod. Res.,* 8, 193, 1972.
138. **Lindgren, D. L., Krohne, H. E., and Vincent, L. E.,** Malathion and chlorthion for control of insects infesting stored grain, *J. Econ. Entomol.,* 47, 705, 1954.
139. **Krohne, H. E. and Lindgren, D. L.,** Susceptibility of life stages of *Sitophilus oryzae* to various fumigants, *J. Econ. Entomol.,* 51, 157, 1958.
140. **Kazmier, H. E. and Fuller, R. G.,** Ethylene dibromide:methyl bromide mixtures as fumigants against the confused flour beetle, *J. Econ. Entomol.,* 52, 1081, 1959.
141. **Brown, W. B. and Heseltine, H. K.,** Fumigation of grain in silo bins, *Milling,* 112, 229, 1949.
142. **Calderon, M. and Carmi, Y.,** Fumigation trials with a mixture of methyl bromide and carbon dioxide in vertical bins, *J. Stored Prod. Res.,* 8, 315, 1973.
143. **Monro, H. A. U.,** Manual of Fumigation for Insect Control, FAO Manual 79, Food and Agriculture Organization, Rome, 1969.
144. **Thompson, R. H.,** A review of the properties and usage of methyl bromide as a fumigant, *J. Stored Prod. Res.,* 1, 353, 1966.
145. **Bond, E. J. and Dumas, T.,** Loss of warning odor from phosphine, *J. Stored Prod. Res.,* 3, 389, 1967.
146. **Hazleton, L. W.,** The arrival of phostoxin, *Pest Control,* 36, 26, 1968.
147. **Heseltine, H. K. and Thompson, R. H.,** The use of aluminum phosphide tablets for the fumigation of grain, *Milling,* 129, 676, 1957.

148. **McGregor, H. E.,** Evaluation of phosphine gas as a fumigant for shelled yellow corn stored in concrete silo-type storage, *Northwestern Miller*, 265, 38, 1961.

149. **Rout, G. and Mohanty, R. N.,** Studies on hydrogen phosphide against the rice weevil, *J. Econ. Entomol.*, 60, 276, 1967.

150. **Cogburn, R. R.,** Fumigation of bucket elevators with phosphine gas to control rice weevil and red flour beetle adults, *J. Econ. Entomol.*, 60, 1485, 1967.

151. **Schesser, J. H.,** Phosphine fumigation of processed cereal products in rail cars, *Am. Miller Process.*, p. 8, January 1967.

152. **Cogburn, R. R. and Tilton, E. W.,** Studies of phosphine as a fumigant for stacked rice under gas-tight tarpaulins, *J. Econ. Entomol.*, 56, 706, 1963.

153. **Gillenwater, H. B.,** personal communication, 1973.

154. **Bond, E. J. and Monro, H. A. U.,** The toxicity of various fumigants to the cadelle *Tenebriodes mauritanicus, J. Econ. Entomol.*, 54, 451, 1961.

155. **Lindgren, D. L., Vincent, L. E., and Strong, R. G.,** Studies on hydrogen phosphide as a fumigant, *J. Econ. Entomol.*, 51, 900, 1958.

156. **Ozer, M.,** Phostoxin in degisik doz, muddet ve isida Calandra granaria L. Ve *Calandra Oryzae* L. nin biyolojik safhalarina karsi tokosik etkisi, *Koruma*, 2, 19, 1961.

157. **Hole, B. D., Bell, C. H., Mills, K. A., and Goodship, G.,** The toxicity of phosphine to all developmental stages of thirteen species of stored product beetles, *J. Stored Prod. Res.*, 12, 235, 1976.

158. **Bond, E. J.,** Toxicity of mixtures of methyl bromide and phosphine to insects, *J. Econ. Entomol.*, 71, 341, 1978.

159. **Cogburn, R. R.,** Detia Ex-B® for phosphine fumigation in sacked milled rice, *J. Econ. Entomol.*, 67, 436, 1974,.

160. **Bond, E. J., Sellen, R. A., and Dumas, T.,** Control of insects with phosphine in open-ended bin spouts, *J. Econ. Entomol.*, 70, 22, 1976.

161. **Leesch, J. G., Redlinger, L. M., and Gillenwater, H. B., Davis, R., and Zehner, J. M.,** An in-transit shipboard fumigation of corn, *J. Econ. Entomol.*, 71, 928, 1978.

162. **Schesser, J. H.,** Fumigation of cereal grains and processed products in transport vehicles with phosphine from Detia Ex-B, *J. Econ. Entomol.*, 70, 199, 1976.

163. **Conway, J.,** The control of *Anagasta kuhniella* (Zeller) (Lepidoptera, Phycitidae) in metal bins using dichlorvos slow release PVC strips, *J. Stored Prod. Res.*, 1, 381, 1966.

164. **Green, A. A., Kane, J., Heuser, S. G., and Scudamore, K. A.,** Control of *Ephestia elutella* (Hb.) (Lepidoptera, Phycitidae) using dichlorvos in oil, *J. Stored Prod. Res.*, 4, 69, 1968.

165. **McFarlane, J. A.,** Treatment of large grain stored in Kenya with dichlorvos slow-release strips for control of *Cadra cautella*, *J. Econ. Entomol.*, 63, 288, 1970.

166. **LaHue, D. W.,** Controlling the Indian-Meal Moth in Shelled Corn with Dichlorvos PVC Resin Strips, USDA ARS 51-52, U.S. Department of Agriculture, Washington, D.C., 1971.

167. **LaHue, D. W.,** Control of malathion-resistant Indian-meal moths *Plodia interpunctella* (with dichlorvos resin strips), *Proc. N. Central Branch Entomol. Soc. Am.*, 24, 117, 1969.

168. **Bond, E. J., Monro, H. A. U., Dumas, T., Benazet, J., and Turtle, E. E.,** Control of insects in empty cargo ships with dichlorvos, *J. Stored Prod. Res.*, 8, 11, 1972.

169. **Soles, R. L. and Harein, P. K.,** The fumigant toxicity of two new chemicals to stored-product insects, *J. Econ. Entomol.*, 55, 1014, 1962.

170. **Harein, P. K. and Soles, R. L.,** Fumigant toxicity of 1,2,3-tribromo-propene, ethylenimine, crotyl bromide and carbon tetrachloride to stored-product insects, *J. Econ. Entomol.*, 57, 369, 1964.

171. **Cooper, C. V. and Gillenwater, H. B.,** Preliminary evaluation of six candidate fumigants against stored-product insects, *J. Ga. Entomol. Soc.*, 7, 250, 1972.

172. **Storey, C. L.,** Comparative Study of Methods of Distributing Methyl Bromide in Flat Storages of Wheat: Gravity-Penetration, Single Pass and Closed Recirculation, USDA Marketing Res. Rep. 794, U.S. Department of Agriculture, Washington, D.C., 1967.

173. **Kline, G. L. and Converse, H. H.,** Operating Grain Aeration Systems in the Hard Winter Wheat Area, USDA Marketing Res. Rep. 480, U.S. Department of Agriculture, Washington, D.C., 1961.

174. **Shedd, C. K.,** Resistance of grains and seeds to air flow, *Agric. Eng.*, 34, 616, 1953.

175. **Storey, C. L.,** Distribution of Grain Fumigants in Silo-Type Elevator Tanks by Aeration Systems, USDA Marketing Res. Rep. 915, U.S. Department of Agriculture, Washington, D.C., 1971.

176. **Cunnington, A. M.,** Physical limits for complete development of the grain mite *Acarus siro* (L.) (Acarina, Acaridae), in relation to its world distribution, *J. Appl. Ecol.*, 2, 295, 1965.

177. **Monro, H. A. U.,** Manual of Fumigation for Insect Control, FAO Manual 79, Food and Agriculture Organization, Rome, 1969.

178. **Burges, H. D. and Burrell, N. J.,** Cooling bulk grain in the British climate to control storage insects and to improve keeping quality, *J. Sci. Food Agric.*, 15, 32, 1964.

179. **Harein, P. K. and Krause, G. F.,** Differential sorption of fumigant formulations by wheat and relative toxicity to adult rice weevil, *Sitophilus sasakii, J. Econ. Entomol.,* 54, 261, 1961.

180. **Harein, P. K.,** Effect of dockage on the efficiency of 80:20 (carbon tetrachloride:carbon disulfide by volume) as a fumigant for adult rice weevil, *Sitophilus sasakii* (Tak.), in wheat, *J. Kans. Entomol. Soc.,* 34, 195, 1961.

181. **Hall, D. W.,** Food storage in developing countries, *Trop. Sci.,* 11, 298, 1969.

182. **Parkin, E. A.,** The onset of insecticide resistance among field strains of stored-product insects, 12th Int. Congr. Entomol., 165, 657.

183. **Parkin, E. A.,** The onset of insecticide resistance among field populations of stored product insects, *J. Stored Prod. Res.,* 1, 1, 1965.

184. **Dyte, C. E.,** Insecticide resistance in stored-product insects with special reference to *Tribolium castaneum, Trop. Stored Prod. Inf.,* 20, 13, 1970.

185. **Monro, H. A. U., Musgrave, A. J., and Upitis, E.,** Induced tolerance of stored-product beetles to methyl bromide, *Ann. Appl. Biol.,* 49, 373, 1961.

186. **Monro, H. A. U., Upitis, E., and Bond, E. J.,** Resistance of a laboratory strain of *Sitophilus granarius* to phosphine, *J. Stored Prod. Res.,* 8, 199, 1972.

187. **Lindgren, D. L. and Vincent, L. E.,** The susceptibility of laboratory reared and field collected cultures of *Tribolium confusum* and *T. castaneum* to ethylene dibromide, hydrocyanic acid and methyl bromide, *J. Econ. Entomol.,* 58, 551, 1965.

188. **March, R. B. and Metcalf, R. L.,** Laboratory and field studies of DDT resistant house flies in southern California, *Calif. Dep. Agric. Bull.,* 38, 93, 1949.

189. **Parkin, E. A. and Lloyd, C. J.,** Selection of a pyrethrum-resistant strain of the grain weevil, *Calandra granaria* L., *J. Sci. Food Agric.,* 11, 471, 1960.

190. **Lloyd, C. J. and Parkin, E. A.,** Further studies on a pyrethrum-resistant strain of the granary weevil, *Sitophilus granarius* (L), *J. Sci. Food Agric.,* 9, 655, 1963.

191. **Kumar, V. and Morrison, F. E.,** Carbamate and phosphate resistance in adult granary weevils, *J. Econ. Entomol.,* 60, 1430, 1967.

192. **Dyte, C. E. and Blackman, D. G.,** The spread of insecticide resistance in *Tribolium castaneum* (Herbst) (Coleoptera, Tenebrionidae), *J. Stored. Prod. Res.,* 6, 255, 1970.

193. **Dyte, C. E. and Blackman, D. G.,** Laboratory evaluation of organophosphorous insecticides against susceptible and malathion-resistant strains of *Tribolium castaneum* (Herbst) (Coleoptera, Tenebrionidae), *J. Stored Prod. Res.,* 8, 103, 1972.

194. **Champ, B. R. and Champbell-Brown, M. J.,** Insecticide resistance in Australian *Tribiolium castaneum* (Herbst) (Coleoptera, Tenebrionidae). II. Malathion resistance in Eastern Australia, *J. Stored Prod. Res.,* 6, 11, 1970.

195. **Bhatia, S. K., Yadav, T. D., and Mookharjee, P. B.,** Malathion resistance in *Tribolium castaneum* (Herbst) in India, *J. Stored Prod. Res.,* 7, 227, 1971.

196. **Dyte, C. E.,** Resistance to synthetic juvenile hormone in a strain of the flour beetle, *Tribolium castaneum, Nature (London),* 238, 48, 1972.

197. **Schesser, H. H.,** Boxcar research with dichlorvos aerosol, *Proc. North Central Branch Entomol. Soc. Am.,* 27, 56, 1972.

198. **Schesser, J. H.,** A comparison of two fumigant mixtures for disinfesting empty railway freight cars, *Northwest Miller,* 274, 10, 1967.

199. **Schesser, J. H.,** A dichlorvos-malathion mixture for insect control in empty railcars, *Am. Miller Process,* 95, 7, 1967.

200. **Schesser, J. H.,** Phosphine fumigation of processed cereal products in railcars, *Am. Miller Process,* p. 8, January 1967.

201. **Schesser, J. H.,** Boxcar research with dichlorvos aerosol, *Proc. North Central Branch Entomol. Soc. Am.,* 27, 56, 1972.

202. **Schesser, J. H.,** A dichlorvos-malathion mixture for insect control in empty railcars, *Am. Miller Process.,* 95, 7, 1967.

203. **Burkholder, W. E.,** The toxicity of malathion to dermestids as influenced by various surfaces, *Proc. North Central Branch Entomol. Soc. Am.,* 16, 100, 1961.

204. **Burkholder, W. E. and Dicke, R. J.,** The toxicity of malathion and fenthion to dermestid larvae as influenced by various surfaces, *J. Econ. Entomol.,* 59, 253, 1966.

205. **Tyler, P. S. and Rowlands, D. G.,** Sodium carboxymethyl cellulose as a stabilizer for malation formulations, *J. Stored Prod. Res.,* 3, 109, 1967.

206. **Sun, U. P.,** An Analysis of Some Important Factors Affecting the Results of Fumigation Tests on Insects, Tech. Bull. 177, Agric. Exp. Stn., University of Minnesota, St. Paul, 1947.

CONTROL OF FUNGI

Clyde M. Christensen

The main factors that determine whether grains, seeds, and other agricultural products in storage will be invaded and damaged by fungi are moisture content, temperature, and time. Effective control involves the maintenance of conditions that will prevent damaging invasion by the fungi concerned. Most of the work in this field has been done with grains and seeds and their products, and the material presented here applies especially to these.

MOISTURE CONTENT

The moisture contents of common grains and seeds in equilibrium with different relative humidities are listed in Table 1. The moisture contents are on a wet weight basis, since that is the basis used in trade and commerce. The data given are approximations only, since the equilibrium moisture content of a given sample of grain at a given relative humidity will vary by as much as $\pm 1.0\%$ or more, depending on the conditions to which the sample has been exposed previously, and on other and still unknown factors. Storage fungi cannot grow at moisture contents in equilibrium with relative humidities of about 65% and below.

In practice, it is almost impossible to know precisely the moisture content throughout a given bulk of grains or seeds. A representative sample may tell the *average* moisture content of the bulk from which it was taken, but it does not tell the *range* in moisture content within the bulk. The moisture content of a 250-g sample is commonly used as a measure of the moisture content throughout a bulk of 1000 metric tons or more, so that the sample is only 1/4,000,000th of the bulk that it represents. Also, the meters or methods presently used on farms and in the trade to measure moisture contents in stored grains and their products are not always as accurate as they are thought to be. If different temperatures prevail in different portions of a bulk, moisture may be transferred from the warmer to the cooler portions. The higher the moisture content of the product, the greater the temperature differential, and the longer the time, the greater the moisture transfer. In shelled corn or maize, with a moisture content between 14.5 and 16.0% (the range encountered in nearly all corn in commerce) a difference of only 0.5% can mean the difference between safe storage and spoilage. The only way to know the range of moisture content within a given bulk at a given time is to remove samples from different portions of the bulk and determine the moisture content of each one separately with a meter or method known to give accurate results. One of the functions of forced aeration of stored grains is to maintain a uniform temperature throughout the bulk and thus reduce moisture transfer.

TEMPERATURE

The minimum, optimum, and maximum temperatures for growth of most fungi that invade and damage stored grains and seeds and their products are, respectively, 0 to 5, 30 to 35, and 50 to 55°C. At the lower limits of moisture that permit invasion by storage fungi — moisture contents in equilibrium with relative humidities of 70 to 80% — low temperatures are very effective in limiting the growth of and damage caused by storage fungi. A combination of 15.3 to 15.5% moisture and 30°C, for example, in stored corn will result in considerable "damage" (brown germs, which constitute a grading factor) within 6 months, whereas with the same moisture content but stored at 20°C, it will remain sound and free of damage for more than a year. However, at moisture contents in equilibrium with relative humidities between 85 and 95%, and at temperature between 5 and 10°C, a slight difference

Table 1
EQUILIBRIUM MOISTURE CONTENTS OF COMMON GRAINS, SEEDS, AND FEED INGREDIENTS AT RELATIVE HUMIDITIES OF 65 TO 95 + %, AND THE FUNGI LIKELY TO BE ENCOUNTERED AT EACH MOISTURE CONTENT

| Relative humidity (%) | Moisture content (% wet wt of materials)[a] | | | Fungi |
	Starchy cereal seeds,[b] alfalfa pellets	Soybeans and soybean meal	Peanut meal, copra	
65—70	12.5—13.5	12.0—12.5	5.5—6.0	*Aspergillus halophilicus*
70—75	14.5—15.0	14.0—14.5	6.0—6.5	*A. restrictus, A. glaucus, Sporendonema*
75—80	15.0—15.5	14.5—15.0	7.0—7.5	*A. candidus, A. ochraceus, A. versicolor,* plus the above
80—85	18.0—18.5	17.0—17.5	8.5—9.5	*A. flavus,* a few species of *Penicillium,* plus the above
85—90	19.0—20.0	18.5—19.5	10.0—12.0	Several species of *Penicillium,* plus the above
95—100	22.0—24	20.0—22.0	15.0—16.0	All advanced decay fungi, yeasts, and bacteria

[a]   The figures are close approximations, but some variation can be expected in practice.
[b]   Wheat, barley, oats, rye, rice, maize, sorghum.

in either moisture content or temperature can make a great difference in the rate of growth and damage caused by storage fungi.

At the lower limits of moisture content that permit them to grow — moisture contents in equilibrium with relative humidities of 68 to 75% — the storage molds may grow so slowly that they do not cause any detectable rise in temperature. In the 1950s, there were many cases of relatively extensive development of "sick" wheat (dark brown to black germs) in commercial bins; most of these were not accompanied by heating. They were caused by invasion of the germs of the kernels, over a period of months, by *Aspergillus restrictus,* a fungus adapted to environments of high osmotic pressure and which grows too slowly to produce any detectable rise in temperature in the material in which it is growing. Usually, however, invasion by storage fungi sufficient to cause significant commercial damage to quality is accompanied by heating. For this reason, periodic monitoring of temperatures throughout the grain mass or bulk is essential in maintenance of quality.

TIME

Obviously, a combination of moisture content and temperature that permits no, or only slow, growth of storage fungi will make for a much longer safe storage life than a combination that permits or promotes more rapid growth of the fungi responsible for damage.

EVALUATING CONDITION AND STORABILITY

In long-time storage of grains and seeds and their products, temperature monitoring and periodic sampling and testing to determine moisture content, damage, and number, kinds, and degree of invasion by storage fungi enable those in charge of the materials to detect incipient damage long before it becomes of commercial significance and to avoid losses.

Appendix

PREFERRED UNITS FOR EXPRESSING PHYSICAL QUANTITIES (AND THE CONVERSION FACTORS)

ASAE Engineering Practice EP285.6

USE OF SI (METRIC) UNITS*

SECTION 1 — PURPOSE AND SCOPE

1.1   This Engineering Practice is intended as a guide for uniformity incorporating the international System of Units (SI). It is intended for use in implementing ASAE policy, "Use of SI Units in ASAE Standards, Engineering Practices, and Data." This Engineering Practice includes a list of preferred units and conversion factors.

SECTION 2 — SI UNITS OF MEASURE

2.1   SI consists of seven base units, two supplementary units, a series of derived units consistent with the base and supplementary units. There is also a series of approved prefixes for the formation of multiples and submultiples of the various units. A number of derived units are listed in paragraph 2.1.3 including those with special names. Additional derived units without special names are formed as needed from base units or other derived units, or both.

2.1.1   Base and supplementary units. For definitions refer to International Organization for Standardization ISO 1000, SI Units and Recommendations for the Use of Their Multiples and of Certain Other Units.

Base Units
    Meter (m) — unit of length
    Second (s) — unit of time
    Kilogram (kg) — unit of mass
    Kelvin (K) — unit of thermodynamic temperature
    Ampere (A) — unit of electric current
    Candela (cd) — luminous intensity
    Mole (mol) — the amount of a substance
Supplementary units
    Radian (rad) — plane angle
    Steradian (sr) — solid angle
2.1.2   SI unit prefixes

Multiples and submultiples	Prefix	SI Symbol
10^{18}	exa	E
10^{15}	peta	P

---

*   Prepared under the general direction of ASAE Committee on Standards (T-1); reviewed and approved by ASAE division standardizing committees: Power and Machinery Division Technical Committee (PM-03), Structures and Environment Division Technical Committee (SE-03), Electric Power and Processing Division Technical Committee (EPP-03), Soil and Water Division Standards Committee (SW-03), and Education and Research Division Steering Committee; adopted by ASAE December 1964; reconfirmed for one year, December 1969, December 1970, December 1971, December 1972; revised by the Metric Policy Subcommittee December 1973; revised March 1976; revised and reclassified as an Engineering Practice, April 1977; revised April 1979, revised editorially December 1979; revised September 1980, February 1982; revised editorially January 1985.

10^{12}	tera	T
10^9	giga	G
10^6	mega	M
10^3	kilo	k
10^2	hecto	h
10^1	deka	da
10^{-1}	deci	d
10^{-2}	centi	c
10^{-3}	milli	m
10^{-6}	micro	μ
10^{-9}	nano	n
10^{-12}	pico	p
10^{-15}	femto	f
10^{-18}	atto	a

2.1.3   Derived units are combinations of based units or other derived units as needed to describe physical properties, for example, acceleration. Some derived units are given special names; others are expressed in the appropriate combination of SI units. Some currently defined derived units are tabulated in Table 1.

SECTION 3 — RULES FOR SI USAGE

3.1   General. The established SI units (base, supplementary, derived, and combinations thereof with appropriate multiple or submultiple prefixes) should be used as indicated in this section.

3.2   Application of prefixes. The prefixes given in paragraph 2.1.2 should be used to indicate orders of magnitude, thus eliminating insignificant digits and decimals, and providing a convenient substitute for writing powers of 10 as generally preferred in computation. For example:

12 300 m or 12.3 $\times$ 10^3 m becomes 12.3 km, and

0.0123 mA or 12.3 $\times$ 10^{16} A becomes 12.3 μA

It is preferable to apply prefixes to the numerator of compound units, except when using kilogram (kg) in the denominator, since it is a base unit of SI and should be used in preference to the gram. For example:

Use 200 J/kg, not 2 dJ/g

With SI units higher order such as m^2 or m^3, the prefix is also raised to the same order. For example:

mm^2 is $(10^{-3}$ m$)^2$ or 10^{-6} m^2

3.3   Selection of prefix. When expressing a quantity by a numerical value and a unit, a prefix should be chosen so that the numerical value preferably lies between 0.1 and 1000, except where certain multiples and submultiples have been agreed for particular use. The same unit, multiple, or submultiple should be used in tables even though the series may exceed the preferred range of 0.1 to 1000. Double prefixes and hyphenated prefixes should not be used. For example:

use GW (gigawatt) and kMW

3.4   Capitalization. Symbols for SI units are only capitalized when the unit is derived from a proper name; for example, N for Isaac Newton (except liter, L). Unabbreviated units are not capitalized; for example kelvin and newton. Numerical prefixes given in paragraph 2.1.2 and their symbols are not capitalized; except for the symbols M (mega), G (giga), T (tera), P (peta), and E (exa).

3.5   Plurals. Unabbreviated SI units form their plurals in the usual manner. SI symbols are

Table 1
DERIVED UNITS

Quantity	Unit	SI Symbol	Formula
Acceleration	Meter per second squared	—	m/s^2
Activity (of a radioactive source)	Disintegration per second	—	(disintegration)/s
Angular acceleration	Radian per second squared	—	rad/s^2
Angular velocity	Radian per second	—	rad/s
Area	Square meter	—	m^2
Density	Kilogram per cubic meter	—	kg/m^3
Electrical capacitance	Farad	F	$A \cdot s/V$
Electrical conductance	Siemens	S	A/V
Electrical field strength	Volt per meter	—	V/m
Electrical inductance	Henry	H	$V \cdot s/A$
Electrical potential difference	Volt	V	W/A
Electrical resistance	Ohm	Ω	V/A
Electromotive force	Volt	V	W/A
Energy	Joule	J	$N \cdot m$
Entropy	Joule per kelvin	—	J/K
Force	Newton	N	$kg \cdot m/s^2$
Frequency	Hertz	Hz	(cycle)/s
Illuminance	Lux	lx	lm/m^2
Luminance	Candela per square meter	—	cd/m^2
Luminous flux	Lumen	lm	$cd \cdot sr$
Magnetic field strength	Ampere per meter	—	A/m
Magnetic flux	Weber	Wb	$V \cdot s$
Magnetic flux density	Tesla	T	Wb/m^2
Magnetomotive force	Ampere	A	—
Power	Watt	W	J/s
Pressure	Pascal	Pa	N/m^2
Quantity of electricity	Coulomb	C	$A \cdot s$
Quantity of heat	Joule	J	$N \cdot m$
Radiant intensity	Watt per steradian	—	W/sr
Specific heat	Joule per kilogram-kelvin	—	$J/kg \cdot K$
Stress	Pascal	Pa	N/m^2
Thermal conductivity	Watt per meter-kelvin	—	$W/m \cdot K$
Velocity	Meter per second	—	m/s
Viscosity, dynamic	Pascal-second	—	$Pa \cdot s$
Viscosity, kinematic	Square meter per second	—	m^2/s
Voltage	Volt	V	W/A
Volume	Cubic meter	—	m^3
Wavenumber	Reciprocal meter	—	(wave)/m
Work	Joule	J	$N \cdot m$

always written in singular form. For example:

50 newtons or 50 N

25 millimeters or 25 mm

3.6 Punctuation. Whenever a numerical value is less than one, a zero should precede the decimal point. Periods are not used after any SI unit symbol, except at the end of a sentence. English speaking countries use a dot for the decimal point, others use a comma. Use spaces instead of commas for grouping numbers into threes (thousands). For example:

6 357 831.376 88

not 6,357,831.367,88

3.7 Derived units. The product of two or more units in symbolic form is preferably indicated by a dot midway in relation to unit symbol height. The dot may be dispensed with when there is no risk of confusion with another unit symbol. For example:

Use N · m or N m, but not mN

A solidus (oblique stroke, /) a horizontal line, or negative powers may be used to express a derived unit formed from two others by division. For example:

$$\text{m/s}, \frac{\text{m}}{\text{s}}, \text{ or } \text{m} \cdot \text{s}^{-1}$$

Only one solidus should be used in a combination of units unless parentheses are used to avoid ambiguity.

3.8   Representation of SI units in systems with limited character sets. For computer printers and other systems which do not have the characters available to print SI units correctly, the methods shown in ISO 2955, Information Processing — Representation of SI and Other Units for Use in Symbols with Limited Character Sets, is recommended.

SECTION 4 — NON-SI UNITS

4.1   Certain units outside the SI are recognized by ISO because of their practical importance in specialized fields. These include units for temperature, time, and angle. Also included are names for some multiples of units such as "liter" (L)* for volume, "hectare" (ha) for land measure and "metric tone" (t) for mass.

4.2   Temperature. The SI base unit for thermodynamic temperature is kelvin (K). Because of the wide usage of the degree Celsius, particularly in engineering and nonscientific areas, the Celsius scale (formerly called the centigrade scale) may be used when expressing temperature. The Celsius scale is related directly to the kelvin scale as follows:

   one degree Celsius (1°C) equals one kelvin (1 K), exactly

A Celsius temperature (t) is related to a kelvin temperature (T), as follows:

   $t = T - 273.15$

4.3   Time. The SI unit for time is the second. This unit is preferred and should be used when technical calculations are involved. In other cases use of the minute (min), hour (h), day (d), etc. is permissible.

4.4   Angles. The SI unit for plane angle is the radian. The use of arc degrees (°) and its decimal or minute ('), second (") submultiples is permissible when the radian is not a convenient unit. Solid angles should be expressed in steradians.

SECTION 5 — PREFERRED UNITS AND CONVERSION FACTORS

5.1   Preferred units for expressing physical quantities commonly encountered in agricultural engineering work are listed in Table 2. These are presented as an aid to selecting proper units for given applications and to promote consistency when interpretation of the general rules of SI may not produce consistent results. Factors for conversion from old units to SI units are included in Table 2.

SECTION 6 — CONVERSION TECHNIQUES

6.1   Conversion of quantities between systems of units involves careful determination of the number of significant digits to be retained. To convert "1 quart of oil" to "0.946 352 9 liter of oil" is, of course, unrealistic because the intended accuracy of the value does not warrant expressing the conversion in this fashion.

---

*   The International symbol for liter is either the lowercase "l" or the uppercase "L". ASAE recommends the use of uppercase "L" to avoid confusion with the numeral "1".

Table 2
PREFERRED UNITS FOR EXPRESSING PHYSICAL QUANTITIES

Quantity	Application	From: old units	To: SI units	Multiply by:
Acceleration, angular	General	rad/s²	rad/s²	
Acceleration, linear	Vehicle	(mile/h)/s	(km/h)/s	1.609 344ᵃ
	General (includes acceleration of gravity)ᵇ	ft/s²	m/s²	0.304 8ᵃ
Angle, plane	Rotational calculations	r (revolution)	r (revolution)	
		rad	rad	
	Geometric and general	° (deg)	°	
		' (min)	° (decimalized)	1/60ᵃ
		' (min)	'	
		" (sec)	° (decimalized)	1/3600ᵃ
		" (sec)	"	
Angle, solid	Illumination calculations	sr	sr	
Area	Cargo platforms, roof and floor area, frontal areas, fabrics, general	in.²	m²	0.000 645 16ᵃ
		ft²	m²	0.092 903 04ᵃ
	Pipe, conduit	in.²	mm²	645.16ᵃ
		in.²	cm²	6.451 6ᵃ
	Small areas, orifices, cross section area of structural shapes	ft²	m²	0.092 903 04ᵃ
		in.²	mm²	645.16ᵃ
	Brake and clutch contact area, glass, radiators, feed opening	in.²	cm²	6.451 6ᵃ
	Land, pond, lake, reservoir, open water channel (small)	ft²	m²	0.092 903 04ᵃ
	(large)	acre	ha	0.404 687 3(d)
	(very large)	mile²	km²	2.589 998
Area per time	Field operations	acre/h	ha/h	0.404 687 3
	Auger sweeps, silo unloader	ft²/s	m²/s	0.092 903 04ᵃ
Bending moment	(See moment of force)			
Capacitance, electric	Capacitors	μF	μF	
Capacity, electric	Battery rating	A·h	A·h	
Capacity, heat	General	Btu/°F	kJ/Kᶜ	1.899 101
Capacity, heat, specific	General	Btu/(lb·°F)	kJ/(kg·K)ᶜ	4.186 8ᵃ

Table 2 (continued)
PREFERRED UNITS FOR EXPRESSING PHYSICAL QUANTITIES

Quantity	Application	From: old units	To: SI units	Multiply by:
Capacity, volume	(See volume)			
Coefficient of heat transfer	General	Btu/(h·ft²·°F)	W/(m²·K)ᶜ	5.678 263
Coefficient of linear expansion	Shrink fit, general	°F⁻¹, (1/°F)	K⁻¹, (1/K)ᶜ	1.8ᵃ
Conductance, electric	General	mho	S	1ᵃ
Conductance, thermal	(See coefficient of heat transfer)			
Conductivity, thermal	Material property	mho/ft	S/m	3.280 840
Conductivity, electric	General	Btu·ft/(h·ft²·°F)	W/(m·K)ᶜ	1.730 735
Consumption, fuel	Off highway vehicles (see also efficiency, fuel)	gal/h	L/h	3.785 412
Consumption, oil	Vehicle performance testing	qt/(1000 miles)	L/(1000 km)	0.588 036 4
Consumption, specific, oil	Engine testing	lb/(hp·h)	g/(kW·h)	608.277 4
		lb/(hp·h)	g/MJ	168.965 9
Current, electric	General	A	A	
Density, current	General	A/in.²	kA/m²	1.550 003
		A/ft²	A/m²	10.763 91
Density, magnetic flux	General	Kilogauss	T	0.1ᵃ
Density, (mass)	Solid, general; agricultural products, soil, building materials	lb/yd³	kg/m³	0.593 276 3
		lb/in.³	kg/m³	27 679.90
		lb/ft³	kg/m³	16.018 46
	Liquid	lb/gal	kg/L	0.119 826 4
	Gas	lb/ft³	kg/m³	16.018 46
	Solution concentration	—	g/m³, mg/L	—
Density of heat flow rate	Irradiance, general	Btu/(h·ft²)	W/m²	3.154 591ᵈ
Consumption, fuel	(See flow, volume)			
Consumption, specific fuel	(See efficiency, fuel)			
Drag	(See force)			
Economy, fuel	(See efficiency, fuel)			

Quantity	To convert from	to	Multiply by
Efficiency, fuel			
Highway vehicles			
Economy	mile/gal	km/L	0.415 143 7
Consumption	—	L/(100 km)	c
Specific fuel consumption	lb/(hp·h)	g/MJ	168.965 9
Off-highway vehicles			
Economy	hp·h/gal	kW·h/L	0.196 993 1
Specific fuel consumption	lb/(hp·h)	g/MJ	168.965 9
Specific fuel consumption	lb/(hp·h)	kg/(kW·h)[f]	0.608 277 4
Energy, work, enthalpy, quantity of heat			
Impact strength	ft·lbf	J	1.355 818
Heat	Btu	kJ	1.055 056
	kcal	kJ	4.186 8[a]
Energy, usage, electrical	kW·h	kW·h	
	kW·h	MJ	3.6
Mechanical, hydraulic, general	ft·lbf	J	1.355 818
	ftpdl	J	0.042 140 11
	hp·h	MJ	2.684 520
	hp·h	kW·h	0.745 699 9
Energy per area Solar radiation	Btu/ft²	MJ/m²	0.011 356 528
Energy, specific General	cal/g[g]	J/g	4.186 8[a]
	Btu/lb	kJ/kg	2.326[a]
Enthalpy (See energy)			
Entropy (See capacity, heat)			
Entropy, specific (See capacity, heat, specific)			
Floor loading (See mass per area)			
Flow, heat, rate (See power)			
Flow, mass, rate			
Gas, liquid	lb/min	kg/min	0.453 592 4
	lb/s	kg/s	0.453 592 4
Dust flow	g/min	g/min	
Machine work capacity, harvesting, materials handling	ton (short)/h	t/h, Mg/h[a]	0.907 184 7
Flow, volume			
Air, gas, general	ft³/s	m³/s	0.028 316 85
	ft³/min	m³/min	1.699 011
Liquid flow, general	gal/s (gps)	L/s	3.785 412
	gal/s (gps)	m³/s	0.003 785 412
	gal/min (gpm)	L/min	3.785 412
Seal and packing leakage, sprayer flow	oz/s	mL/s	29.573 53
	oz/min	mL/min	29.573 53

Table 2 (continued)
PREFERRED UNITS FOR EXPRESSING PHYSICAL QUANTITIES

Quantity	Application	From: old units	To: SI units	Multiply by:
	Fuel consumption	gal/h	L/h	3.785 412
	Pump capacity, coolant flow, oil flow	gal/min (gpm)	L/min	3.785 412
	Irrigation sprinkler, small pipe flow	gal/min (gpm)	L/s	0.063 090 20
	River and channel flow	ft³/s	m³/s	0.028 316 85
Flux, luminous	Light bulbs	lm	lm	
Flux, magnetic	Coil rating	maxwell	Wb	0.000 000 01 [a]
Force, thrust, drag	Pedal, spring, belt, hand lever, general	lbf	N	4.448 222
		ozf	N	0.278 013 9
		pdl	N	0.138 255 0
		kgf	N	9.806 650
		dyne	N	0.000 01 [a]
	Drawbar, breakout, rim pull, winch line pull,[b] general	lbf	kN	0.004 448 222
Force per length	Beam loading	lbf/ft	N/m	14.593 90
	Spring rate	lbf/in.	N/mm	0.175 126 8
Frequency	System, sound and electrical	Mc/s	MHz	1[a]
		kc/s	kHz	1[a]
		Hz, c/s	Hz	1[a]
	Mechanical events, rotational	r/s (rps)	s⁻¹, r/s	1[a]
	Engine, power-take-off shaft, gear speed	r/min (rpm)	min⁻¹, r/min	1[a]
	Rotational dynamics	r/min (rpm)	min⁻¹, r/min	1[a]
		rad/s	rad/s	1[a]
Hardness	Conventional hardness numbers, BHN, R, etc. not affected by change to SI			
Heat	(See energy)			
Heat capacity	(See capacity, heat)			
Heat capacity, specific	(See capacity, heat, specific)			
Heat flow rate	(See power)			
Heat flow — density of	(See density of heat flow)			
Heat, specific	General	cal/g·°C	kJ/kg·K	4.186 8 [a]
		Btu/lb·°F	kJ/kg·K	4.186 8 [a]

Quantity	Application	From	To	Factor
Heat transfer coefficient	(See coefficient of heat transfer)			
Illuminance, illumination	General	fc	lx	10.763 91
Impact strength	(See strength, impact)			
Impedance, mechanical	Damping coefficient	lbf·s/ft	N·s/m	14.593 90
Inductance, electric	Filters and chokes, permeance	H	H	
Intensity, luminous	Light bulbs	candlepower	cd	1[a]
Intensity, radiant	General	W/sr	W/sr	
Leakage	(See flow, volume)			
Length	Land distances, maps, odometers	mile	km	1.609 344[a,h]
	Field size, turning circle, braking distance, cargo platforms, rolling circumference, water depth, land leveling (cut and fill)	rod	m	5.029 210[h]
		yd	m	0.914 4
		ft	m	0.304 8[a]
	Row spacing	in.	cm	2.54[a]
	Engineering drawings, product specifications, vehicle dimensions, width of cut, shipping dimensions, digging depth, cross section of lumber, radius of gyration, deflection	in.	mm	25.4[a]
	Precipitation, liquid, daily and seasonal, field drainage (runoff), evaporation and irrigation depth	in.	mm	25.4[a]
	Precipitation, snow depth	in.	cm	2.54[a]
	Coating thickness, filter particle size	mil	μm	25.4[a]
		μin.	μm	0.025 4[a]
		micron	μm	1[a]
	Surface texture			
	Roughness, average	μin.	μm	0.025 4[a]
	Roughness sampling length, waviness height and spacing	in.	mm	25.4[a]
	Radiation wavelengths, optical measurements (interference)	μin.	nm	25.4[a]
Length per time	Precipitation, liquid per hour	in./h	mm/h	25.4[a]
	Precipitation, snow depth per hour	in.h	cm/h	2.54[a]
Load	(See mass)			
Luminance	Brightness	footlambert	cd/m²	3.426 259
Magnetization	Coil field strength	A/in.	A/m	39.370 08

Table 2 (continued)

PREFERRED UNITS FOR EXPRESSING PHYSICAL QUANTITIES

Quantity	Application	From: old units	To: SI units	Multiply by:
Mass	Vehicle mass, axle rating, rated load, tire load, lifting capacity,[i] tipping load, load, quantity of crop, counter mass, body mass general	ton (long)	t, Mg[j]	1.016 047
		ton (short)	t, Mg[j]	0.907 184 7
		lb	kg	0.453 592 4
	Small mass	slug	kg	14.593 90
		oz	g	28.349 52
Mass per area	Fabric, surface coatings	oz/yd²	g/m²	33.905 75
		lb/ft²	kg/m²	4.882 428
		oz/ft²	g/m²	305.151 7
	Floor loading	lb/ft²	kg/m²	4.882 428
	Application rate, fertilizer, pesticide	lb/acre	kg/ha	1.120 851
	Crop yield, soil erosion	ton (short)/ acre	t/ha[j]	2.241 702
Mass per length	General, structural members	lb/ft	kg/m	1.488 164
		lb/yd	kg/m	0.496 054 7
Mass per time	Machine work capacity, harvesting, materials handling	ton (short)/h	t/h, Mg/h[j]	0.907 184 7
Modulus of elasticity	General	lbf/in.²	MPa	0.006 894 757
Modulus of rigidity	(See modulus of elasticity)			
Modulus, section	General	in.³	mm³	16 387.06
		in.³	cm³	16.387 06
Modulus, bulk	System fluid compression	psi	kPa	6.894 757
Moment, bending	(See moment of force)			
Moment of area, second	General	in.⁴	mm⁴	416 231.4
		in.⁴	cm⁴	41.623 14
Moment of force, torque, bending moment	General, engine torque, fasteners, steering torque, gear torque, shaft torque	lbf·in.	N·m	0.112 984 8
		lbf·ft	N·m	1.355 818
		kgf·cm	N·m	0.098 066 5[a]
Moment of inertia, mass	Locks, light torque	ozf·in.	mN·m	7.061 552
Moment of mass	Flywheel, general	lb·ft²	kg·m²	0.042 140 11
Moment of momentum	Unbalance	oz·in.	g·m	0.720 077 8
	(See momentum, angular)			

Moment of section	(See moment of area, second)			
Momentum, linear	General	lb·ft/s	kg·m/s	0.138 255 0
Momentum, angular	Orsional vibration	lb·ft^2/s	kg·m^2/s	0.042 140 11
Permeability	Magnetic core properties	H/ft	H/m	3.280 840
Permeance	(See inductance)			
Potential, electric	General	V	V	
Power	General, light bulbs	W	W	
	Air conditioning, heating	Btu/min	W	17.584 17
		Btu/h	W	0.293 071 1
	Engine, alternator, drawbar, power take-off, hydraulic and pneumatic systems, heat rejection, heat exchanger capacity, water power, electrical power, body heat loss	hp (550 ft·lbf/s)	kW	0.745 699 9
Power per area	Solar radiation	Btu/ft^{2}h	W/m^2	3.154 591
Pressure	All pressures except very small	lbf/in.^{2} (psi)	kPa	6.894 757
		in.Hg (60°F)	kPa	3.376 85
		in.H_2O (60°F)	kPa	0.248 84
		mmHg (0°C)	kPa	0.133 322
		kgf/cm^2	kPa	98.066 5
		bar	kPa	100.0[a]
		lbf/ft^2	kPa	0.047 880 26
		atm (normal = 760 torr)	kPa	101.325[a]
	Very small pressures (high vacuum)	lbf/in.^{2} (psi)	Pa	6 894.757
Pressure, sound level	Acoustical measurement — when weighting is specified show weighting level in parenthesis following the symbol, for example dB(A)	dB	dB	
Quantity of electricity	General	C	C	
Radiant intensity	(See intensity, radiant)			
Resistance, electric	General	Ω	Ω	
Resistivity, electric	General	Ω·ft	Ω·m	0.304 8[a]
		Ω·ft	Ω·cm	30.48[a]
Sound pressure level	(See pressure, sound, level)			
Speed	(See velocity)			
Spring rate, linear	(See force per length)			
Spring rate, torsional	General	lbf·ft/deg	N·m/deg	1.355 818

Table 2 (continued)

PREFERRED UNITS FOR EXPRESSING PHYSICAL QUANTITIES

Quantity	Application	From: old units	To: SI units	Multiply by:
Strength, field, electric	General	V/ft	V/m	3.280 840
Strength, field, magnetic	General	Oersted	A/m	79, 577 47
Strength, impact	Materials testing	ft·lbf	J	1.355 818
Stress	General	lbf/in.^{2}	MPa	0.006 894 757
Surface tension	(See tension, surface)			
Temperature	General use	°F	°C	$t_C = (t_F - 32)/1.8$[a]
	Absolute temperature, thermodynamics, gas cycles	°R	K	$T_K = T_R/1.8$[a]
Temperature interval	General use	°F	K[c]	$1\ K = 1°C = 1.8°F$[a]
Tension, surface	General	lbf/in.	mN/m	175 126.8
		dyne/cm	mN/m	1[a]
Thermal diffusivity	Heat transfer	ft^2/h	m^2/h	0.092 903 04
Thrust	(See force)			
Time	General	s	s	
		h	h	
		min	min	
	Hydraulic cycle time	s	s	
	Hauling cycle time	min	min	
Torque	(See moment of force)			
Toughness, fracture	Metal properties	ksi·in.$^{0.5}$	MPa·m$^{0.5}$	1.098 843
Vacuum	(See pressure)			
Velocity, angular	(See velocity, rotational)			
Velocity, linear	Vehicle	mile/h	km/h	1.609 344[a]
	Fluid flow, conveyor speed, lift speed, air speed	ft/s	m/s	0.304 8[a]
	Cylinder actuator speed	in./s	mm/s	25.4[a]
	General	ft/s	m/s	0.304 8[a]
		ft/min	m/min	0.304 8[a]
		in./s	mm/s	25.4[a]

Quantity	Description	From	To	Factor
Velocity, rotational	(See frequency)			
Viscosity, dynamic	General liquids	Centipoise	mPa·s	1ᵃ
Viscosity, kinematic	General liquids	Centistokes	mm²/s	1ᵃ
Volume	Truck body, shipping or freight, bucket capacity, earth, gas, lumber, building, general	yd³	m³	0.764 554 9
		ft³	m³	0.028 316 85
	Combine harvester grain tank capacity	Bushel	L	35.239 07
	Automobile luggage capacity	ft³	L	28.316 85
	Gas pump displacement, air compressor, air reservoir, engine displacement			
	Large	in.³	L	0.016 387 06
	Small	in.³	cm³	16.387 06
	Liquid — fuel, lubricant, coolant, liquid wheel ballast	gal	L	3.785 412
		qt	L	0.946 352 9
		pt	L	0.473 176 5
		pt	L	0.473 176 5
	Small quantity liquid	oz	mL	29.573 53
	Irrigation, reservoir	acre·ft	m³	1 233.489ʰ
		acre·ft	dam³	1.233 489ʰ
	Grain bins	bushel (U.S.)	m³	0.035 239 07
Volume per area	Application rate, pesticide	gal/acre	L/ha	9.353 958
Volume per time	Fuel consumption (also see flow)	gal/h	L/h	3.785 412
Weight	May mean either mass or force — avoid use of weight			
Work	(See energy)			
Young's modulus	(See modulus of elasticity)			

Notes: 1. Quantities are arranged in alphabetical order by principal nouns. For example, surface tension is listed as tension, surface.

2. All possible applications are not listed, but others such as rates can be readily derived. For example, from the preferred units for energy and volume the units for heat energy per unit volume, kJ/m³, may not be derived.

3. Conversion factors are shown to seven significant digits, unless the precision with which the factor is known does not warrant seven digits.

ᵃ Indicates exact conversion factor.

ᵇ Standard acceleration of gravity is 9.806 650 m/s² exactly (Adopted by the General Conference on Weights and Measures).

ᶜ In these expressions, K indicates temperature intervals. Therefore K may be replaced with °C if desired without changing the value or affecting the conversion factor. kJ/(kg·K) = kJ/(kg·°C).

ᵈ Conversions of Btu are based on the International Table Btu.

ᵉ Convenient conversion: 235.215 ÷ (mile per gal) = L/(100 km).

Table 2 (continued)
PREFERRED UNITS FOR EXPRESSING PHYSICAL QUANTITIES

f   ASAE S209 and SAE J708, Agricultural Tractor Test Code, specify kg/(kW·h). It should be noted that there is a trend toward use of g/MJ as specified for highway vehicles.

g   Not to be confused with kcal/g. kcal often called calorie.

h   Official use in surveys and cartography involves the U.S. survey mile based on the U.S. survey foot, which is longer than the international foot by two parts per million. The factors used in this standard for acre, acre foot, rod are based on the U.S. survey foot. Factors for all other old length units are based on the international foot. (See ANSI/ASTM Standard E380-76, Metric Practice).

i   Lift capacity ratings for cranes, hoists, and related components such as ropes, cable chains, etc. should be rated in mass units. Those items such as winches, which can be used for pulling as well as lifting, shall be rated in both force and mass units for safety reasons.

j   The symbol t is used to designate metric ton. The unit metric ton (exactly 1 MG) is in wide use but should be limited to commercial description of vehicle mass, freight mass, and agricultural commodities. No prefix is permitted.

6.2   All conversions, to be logically established, must depend upon an intended precision of the original quantity — either implied by a specific tolerance, or by the nature of the quantity. The first step in conversion is to establish this precision.

6.3   The implied precision of a value should relate to the number of significant digits shown. The implied precision is plus or minus one half unit of the last significant digit in which the value is stated. This is true because it may be assumed to have been rounded from a greater number of digits, and one half of the last significant digit retained is the limit of error resulting from rounding. For example, the number 2.14 may have been rounded from any number between 2.135 and 2.145. Whether rounded or not, a quantity should always be expressed with this implication of precision in mind. For instance, 2.14 in. implies a precision of ±0.005 in., since the last significant digit is in units of 0.01 in.

6.4   Quantities should be expressed in digits which are intended to be significant. The dimension 1.1875 in. may be a very accurate one in which the digit in the fourth place is significant, or it may in some cases be an exact decimalization of a fractional dimension, 1 3/16 in., in which case the dimension is given with too many decimal places relative to its intended precision.

6.5   Quantities should not be expressed with significant zeros omitted. The dimension 2 in. may mean "about 2 in.," or it may, in fact, mean a very accurate expression which should be written 2.0000 in. In the latter case, while the added zeros are not significant in establishing the value, they are very significant in expressing the proper intended precision.

SECTION 7 — RULES FOR ROUNDING

7.1   Where feasible, the rounding of SI equivalents should be in reasonable, convenient, whole units.

7.2   Interchangeability of parts, functionally, physically, or both, is dependent upon the degree of round-off accuracy used in the conversion of the U.S. customary to SI value. American National Standards Institute ANSI/ASTM E380-76, Metric Practice, outlines methods to assure interchangeability.

7.3   Rounding numbers. When a number is to be rounded to fewer decimal places the procedure shall be as follows:

7.3.1   When the first digit discarded is less than 5, the last digit retained shall not be changed. For example, 3.463 25, if rounded to three decimal places, would be 3.463; if rounded to two decimal places, would be 3.46.

7.3.2   When the first digit discarded is greater than 5, or it is a 5 followed by at least one digit other than 0, the last figure retained shall be increased by one unit. For example, 8.376 52, if rounded to three decimal places, would be 8.377; if rounded to two decimal places, would be 8.38.

7.3.3   Round to closest even number when first digit discarded is 5, followed only by zeros.

7.3.4   Numbers are rounded directly to the nearest value having the desired number of decimal places. Rounding must not be done in successive steps to less places. For example:
     27.46 rounded to a whole number = 27. This is correct because the "0.46" is
     less than one half. 27.46 rounded to one decimal place is 27.5. This is correct
     value. But, if the 27.5 is in turn rounded to a whole number, this is successive
     rounding and the result, 28, is incorrect.

7.4   Inch-millimeter linear dimensioning conversion. 1 inch (in.) = 25.4 millimeters (mm) exactly. The term "exactly" has been used with all exact conversion factors. Conversion factors not so labeled have been rounded in accordance with these rounding procedures. To maintain intended precision during conversion without retaining an unnecessary number of digits, the millimeter equivalent shall be carried to one decimal place more than the inch value being converted and then rounded to the appropriate significant figure in the last decimal place.

CITED STANDARDS

ASAE S209, Agricultural Tractor Test Code

ANSI/ASTM E380-76, Metric Practice

ISO 1000, SI Units and Recommendations for the Use of Their Multiples and of Certain Other Units

ISO 2955, Information Processing — Representation of SI and Other Units for Use in Systems with Limited Character Sets

Index

INDEX

D